Investigation on Corrosion Behaviour of Metallic Materials

Investigation on Corrosion Behaviour of Metallic Materials

Guest Editors

Biserka Runje
Marin Kurtela

MDPI

Basel • Beijing • Wuhan • Barcelona • Belgrade • Novi Sad • Cluj • Manchester

Guest Editors

Biserka Runje
Faculty of Mechanical
Engineering and Naval
Architecture
University of Zagreb
Zagreb
Croatia

Marin Kurtela
Faculty of Mechanical
Engineering and Naval
Architecture
University of Zagreb
Zagreb
Croatia

Editorial Office
MDPI AG
Grosspeteranlage 5
4052 Basel, Switzerland

This is a reprint of the Special Issue, published open access by the journal *Coatings* (ISSN 2079-6412), freely accessible at: https://www.mdpi.com/journal/coatings/special_issues/Corrosion_Metallic.

For citation purposes, cite each article independently as indicated on the article page online and as indicated below:

Lastname, A.A.; Lastname, B.B. Article Title. *Journal Name* **Year**, *Volume Number*, Page Range.

ISBN 978-3-7258-7723-2 (Hbk)
ISBN 978-3-7258-7724-9 (PDF)
https://doi.org/10.3390/books978-3-7258-7724-9

Contents

About the Editors

Biserka Runje

Biserka Runje is a full professor at the University of Zagreb, Faculty of Mechanical Engineering and Naval Architecture. Her research and professional activities focus on quality management, quality assurance and control, dimensional metrology and nanometrology, as well as statistical modelling. She has authored or co-authored more than 200 scientific and professional papers published in international and national journals and conference proceedings, contributing significantly to the development and application of measurement science and quality engineering. Since 2025, she has been Deputy Editor-in-Chief of the international journal *Metalurgija*.

Marin Kurtela

Marin Kurtela was employed from 2011 to 2014 at the Department of Process Management at Brodarski Institut ltd., where as a research assistant and later as a team leader, he participated in the organization and successful implementation of more than 30 different projects, related to the design and performance of turbine control systems. From 2014 to 2016, he was employed in a private company, where as an independent designer of ship equipment and hydraulic systems, he participated in the organization and implementation of several dozen projects related to the design, supervision and construction of deck equipment, both for the domestic shipbuilding industry and for the international market of luxury yachts. Since 2016, he has been working at the Chair of Materials Protection at the Faculty of Mechanical Engineering and Naval Architecture of the University of Zagreb where he earned his PhD in 2021 and currently works as an assistant professor. He has authored and co-authored over 70 scientific and professional papers in the field of corrosion and materials protection, including more than 20 scientific papers published in CC and SCI-expanded journals. In 2020 he became a member of the technical committee TO 504, Corrosion and Protection of Metals and Alloys of the Croatian Standards Institute. His research focuses primarily on corrosion-resistant materials and failure analysis, corrosion inhibitors, and organic coatings.

coatings

Editorial

Investigation on Corrosion Behaviour of Metallic Materials

Biserka Runje [1] and Marin Kurtela [2,*]

1 Department of Measurement and Control, Faculty of Mechanical Engineering and Naval Architecture, University of Zagreb, 10000 Zagreb, Croatia; biserka.runje@fsb.unizg.hr
2 Department of Welded Structures, Faculty of Mechanical Engineering and Naval Architecture, University of Zagreb, 10000 Zagreb, Croatia
* Correspondence: marin.kurtela@fsb.unizg.hr; Tel.: +385-1-6168-343

The rapid industrial and technological progress of modern society has intensified the exploitation and conversion of natural resources into consumer and industrial goods. This continuous expansion of human activity exerts increasing pressure on the environment and often undermines ecological awareness and responsibility [1,2]. To achieve sustainable development, it is therefore essential to integrate environmental criteria and sustainability principles into all areas of human endeavour, including production, transportation, culture, and governance [3,4].

While many environmental challenges can be mitigated through technological innovation, a deeper transformation in societal values and understanding of the relationship between humanity and nature is equally necessary to establish a rational and harmonious coexistence [5,6]. This is especially true in industries where material degradation directly impacts safety, economy, and ecology. Corrosion of metallic materials remains a global concern, resulting in major economic losses, safety hazards, and environmental consequences [1,7]. The development of sustainable corrosion-protection systems has therefore become a key focus of research and industrial application [8].

Latest advances emphasise the use of green corrosion inhibitors (GCIs), i.e., environmentally friendly, biodegradable or naturally derived compounds that prevent corrosion through adsorption, film formation or complexation mechanisms [2,3,9]. Studies have shown that GCIs can effectively replace traditional toxic inhibitors while maintaining or enhancing performance [10,11]. In parallel, research on eco-sustainable coatings has gained attention, where inhibitors are incorporated into organic or hybrid matrices to enable controlled release and self-healing functionality [12–15]. The multifunctionality of GCIs enables their integration into next-generation protective systems such as nanostructured, self-healing, and biopolymer-based coatings [6,13].

In recent years, metal and metal oxide nanoparticles (MNPs), including Ag, Cu, ZnO, TiO_2, and Al have been extensively incorporated into composite coating systems to improve their anticorrosion, antimicrobial, and mechanical performance. The uniform dispersion of nanoparticles within the coating matrix significantly enhances barrier integrity, decreases corrosion current densities, inhibits microbial biofilm formation, and increases surface hardness and wear resistance [16–18]. Despite these advantages, several challenges persist, such as nanoparticle aggregation, insufficient interfacial adhesion, uncontrolled ion release, environmental and cytotoxicity concerns, and limitations in large-scale manufacturing [19,20]. Contemporary research, therefore, focuses on developing multifunctional coating architectures through green synthesis methods, surface functionalization strategies, and comprehensive long-term durability assessments [16,18,21]. Overall, MNP-based

 https://doi.org/10.3390/coatings15121403

composite coatings demonstrate substantial improvements in corrosion protection and mechanical stability, while ongoing investigations aim to optimise their long-term reliability, environmental safety, and economic feasibility [17,20,22]. This Special Issue on "Investigation on Corrosion Behaviour of Metallic Materials" compiles recent studies that address these emerging advances and challenges within this evolving research field.

The article by Žbulj et al. investigated the use of Lady's mantle flower extract (LMFE) as an eco-friendly corrosion inhibitor for carbon steel in CO_2-saturated brine. Electrochemical tests, potentiodynamic polarisation and electrochemical impedance spectroscopy showed that LMFE achieved over 90% inhibition efficiency at optimal concentrations of 4 g/L (static) and 5 g/L (flow). Surface analysis (SEM and FTIR) confirmed that LMFE adsorbs onto the steel, forming a protective film. The extract was found to be highly biodegradable ($BOD_5/COD = 0.96$) and only mildly toxic (19.34%), indicating its suitability as a green, biodegradable corrosion inhibitor for use in petroleum and similar industries [23].

In another article, Wang et al. investigated a mixed corrosion inhibitor composed of imidazoline (IM), sodium molybdate, and sodium dodecylbenzenesulfonate (SDBS) for protecting mild steel in a 3.5 wt.% NaCl solution. Using orthogonal experimental design and electrochemical methods (EIS and polarisation tests), the research found that the optimal mixture IM (100 mg/L), sodium molybdate (50 mg/L), and SDBS (100 mg/L), produced the best corrosion inhibition, achieving an impedance of 1.8 $k\Omega \cdot cm^2$. SEM and EDS analyses confirmed the formation of a dense, uniform protective film on the steel surface. Among the inhibitors, IM had the strongest influence on corrosion resistance, while SDBS enhanced film formation, and molybdate provided supplementary protection [24].

In their work, Yan et al. examined how varying cobalt (Co) content affects the structure and performance of NiTiAlCrCoxN films deposited on stainless steel using magnetron sputtering. All films had a face-centred cubic (FCC) structure with a (200) preferred orientation. Increasing Co content led to decreased hardness but increased elastic modulus. The film with a Co molar ratio of 1.4 showed the best cavitation erosion resistance, having the lowest mass loss (0.72 mg) and highest elastic modulus (253.22 GPa). This improvement was attributed to enhanced solid solution strengthening and better elasticity that reduced microjet impact during erosion [25].

Deng et al. proposed a novel anticorrosion method combining impressed current cathodic protection (ICCP) with graphene-based coatings for marine atmospheric environments. The graphene coating acts as an intermediate conductive layer, forming an electrolytic cell with the metal substrate that improves corrosion resistance, especially at coating defects where an electrolyte film forms. Experiments show the optimal protection voltage is 0.6 V and the best graphene coating thickness is 20 μm. This combined method provides better corrosion protection and lower corrosion rates than coatings alone, making it effective for long-term marine applications [26].

Stojanović et al. investigated the physical and chemical properties of high-temperature silicone-based polymer coatings, both solvent-borne and water-borne, applied to carbon steel with different surface roughness. The research evaluates corrosion protection, adhesion, hardness, thermal stability, and electrochemical behaviour under conditions simulating wood-burning stoves, including salt spray, humidity, and high-temperature cycling tests. Key findings include better adhesion on sandblasted surfaces, superior performance of inorganic zinc ethyl silicate coatings, generally higher barrier properties of solvent-based coatings compared to water-based ones, and excellent thermal stability of all coatings tested. The coatings maintained good appearance and adhesion after high-temperature exposure, making them suitable for high-temperature applications despite challenges in salty environments. Overall, the study concludes that the choice of surface preparation and coating type critically influences performance, and that silicone-based high-temperature coatings, especially inorganic zinc ethyl silicate, are suitable for indoor high-temperature applications such as wood-burning stoves. However, caution is advised in salty environments due to reduced durability [27].

In another article, Landek et al. demonstrated that DC plasma nitriding of AISI 316L stainless steel at 430 °C enhances surface hardness and alters surface roughness, with longer nitriding durations leading to increased hardness. However, extended nitriding also results in the formation of a brittle, hard layer that accelerates wear under micro-abrasion conditions. While nitriding slightly reduces corrosion resistance compared to untreated steel, the overall corrosion stability remains largely unaffected by the nitriding duration. Therefore, optimising nitriding time is essential to achieve a balanced improvement in wear resistance without significantly compromising corrosion performance [28].

In their work, Li et al. introduced a bionic surface structure model inspired by the pit-shaped microtextures found on the abdomen of the dung beetle, designed to enhance the wear resistance and extend the operational lifespan of drill pipes. Using finite element analysis (FEA), the stress distribution and wear behaviour of both the bionic pitted structure and a conventional smooth structure were simulated. The results demonstrated that the bionic design significantly optimised the stress distribution, leading to an 81.3% increase in service life compared with the ordinary structure. To experimentally validate the simulation, the surface of a 7075 aluminium alloy drill pipe was modified by laser cladding with nickel (Ni) powder. Wear tests were subsequently carried out to examine the wear mechanisms and surface damage characteristics of the cladded layer. Microstructural, compositional, and microhardness analyses revealed that the cladding layer predominantly consisted of the Al_3Ni_2 intermetallic phase, which imparted high surface hardness. A distinct transition zone was observed between the cladding layer and the substrate, consisting primarily of softer aluminium, thereby forming a gradient interface that improved the component's ability to withstand cyclic mechanical loading. Moreover, the bionic pit structure demonstrated an inherent capacity to trap and store wear debris, thereby mitigating abrasive interactions and reducing secondary wear. This self-adaptive feature contributed to an additional 70% improvement in service life, underscoring the synergistic effects of bionic surface engineering and laser cladding technology in enhancing the durability and performance of drill pipes under demanding operational conditions [29].

In another article, Stojanović et al. compared the properties of industrial topcoats dried under atmospheric conditions and using infrared (IR) radiation. Five different solvent-borne topcoats were tested, four based on polyurethane (PUR) and one based on polysiloxane (PSX). The results show that IR-dried topcoats generally have better properties compared to atmospherically dried ones. The drying time was significantly reduced from 5–8 h to just 10 min when using IR radiation. The adhesion of the topcoat to the metal substrate was also improved with IR drying. Mechanical properties like hardness and impact resistance were similar or slightly better for IR-dried topcoats. Electrochemical measurements showed higher initial corrosion resistance for IR-dried coatings, though some topcoats showed better long-term stability with atmospheric drying. Thermal stability was generally improved for IR-dried topcoats. The glass transition temperature (Tg) increased for PUR-based topcoats with IR drying, while it decreased for the PSX-based topcoat, indicating the sensitivity of PSX to IR radiation. FTIR analysis confirmed higher crosslinking for IR-dried topcoats. Overall, the results demonstrate the benefits of using IR radiation for curing industrial topcoats, though the specific polymer composition and presence of certain additives can affect the outcomes [30].

The article, by Samardžija et al., investigated the use of aluminium (Al), nickel (Ni), and silver (Ag) nanoparticles incorporated into epoxy coatings to improve their anticorrosion, migration, and antibacterial properties. The nanoparticles were characterised using SEM, EDS, and XRF analyses. Electrochemical impedance spectroscopy (EIS) was used to evaluate the anticorrosion performance of the nanocomposite coatings in a 3.5 wt.% NaCl solution. The migration of the nanoparticles from the epoxy coating into simulated wastewater was also studied. The antibacterial activity of the nanoparticles and nanocomposite coatings was tested against Gram-positive (*B. subtilis*) and Gram-negative (*P. aeruginosa*) bacteria. The results showed that the nanocomposite with 1% Ag nanoparticles exhibited the best corrosion resistance, followed by the 1% Ni and 1% Al nanocomposites. The Al and Ag nanoparticles demonstrated effective antibacterial activity, while the Ni nanoparticles showed limited antibacterial effects. The Al nanoparticles showed the highest migration from the epoxy coating into the wastewater, indicating their potential for antimicrobial activity. Overall, the 1% Al NP epoxy nanocomposite showed good anticorrosion and antibacterial properties, making it a promising candidate for applications in pipelines [31].

In their article, Samardžija et al. investigated the characterisation of epoxy resin and epoxy paint nanocomposites containing 1 wt.% aluminium nanoparticles (Al NPs) using Scanning Electrochemical Microscopy (SECM). Cross-sectional SECM spectra, along with SEM images and EDS elemental maps, were employed to analyse the distribution of aluminium and chlorine in the 1% Al NP epoxy resin coating and aluminium and oxygen in the 1% Al NP epoxy paint nanocomposite. The SECM analysis revealed notable variations in impedance among the samples: the epoxy resin exhibited a relatively homogeneous electrochemical response, whereas the epoxy paint displayed localised impedance peaks, attributed to the presence of additional components within the paint matrix. The analyses confirmed that the Al NPs were uniformly dispersed in both the epoxy resin and epoxy paint, without agglomeration or air bubbles, indicating excellent compatibility within the polymer matrices. Cross-sectional SECM spectra and EDS maps revealed the distribution of aluminium and chlorine in the epoxy resin coating and aluminium and oxygen in the epoxy paint coating. Electrochemical testing confirmed that the incorporation of Al NPs significantly enhanced the impedance and corrosion resistance of the coatings. This improvement is attributed to a cathodic passivation mechanism, in which oxides formed on the Al NP surfaces over time, effectively sealing pores and providing increased protection against aggressive environments. Among the tested formulations, nanocomposites containing 1.0 wt.% Al NPs exhibited the best combination of mechanical and anticorrosive performance. These findings indicate that the incorporation of Al nanoparticles enhances the corrosion resistance of the coatings, as reflected by the reduced coating capacitance (CPEcoat) values and improved electrochemical stability [32].

In their work, Yan et al. investigated the localised corrosion behaviour of Q125 casing steel in residual acid solutions during oil reservoir acidising, focusing on the effectiveness of Mannich base (Mb) inhibitors. It was found that higher inhibition efficiency occurs in fresh acid compared to residual acid, where localised corrosion and pitting are prevalent due to decreased inhibitor coverage. Analysis revealed that corrosion products like $FeCO_3$ hindered inhibitor adsorption, promoting corrosion, while electrochemical tests demonstrated that the Mb inhibitor significantly improved corrosion resistance in both fresh and residual acid solutions. The findings underscore the necessity of effective corrosion protection for Q125 casing steel in acidic environments to enhance oil recovery processes [33].

The reviewed literature by Yang [34] provided a comprehensive examination of the development, mechanisms, and performance of plasma electrolytic oxidation (PEO) coatings for enhancing the corrosion resistance of magnesium (Mg) alloys. PEO, also known as micro-arc oxidation, is identified as an efficient and environmentally benign surface treatment capable of producing dense, ceramic-like oxide layers that markedly improve protection in corrosive environments. These coatings typically exhibit a two-layered structure consisting of a porous outer region and a dense inner barrier layer, the latter serving as the primary defence against corrosive attack. Corrosion of PEO-coated Mg alloys generally initiates at inherent structural defects such as micropores, cracks, and discharge channels. These defects facilitate the ingress of chloride ions, which dissolve MgO and lead to the formation of $Mg(OH)_2$ and subsequently $MgCl_2$, driving localised corrosion and eventual failure of the coating. Although the chemical composition of PEO coatings varies with electrolyte type, the fundamental corrosion mechanisms are similar. Differences between silicate- and phosphate-based coatings arise mainly from variations in phase composition and discharge behaviour, which influence corrosion pathways and protective performance. Overall, the review highlights two major avenues for improving PEO coating performance: structural refinement to reduce defect density and phase composition engineering to form more stable, corrosion-resistant phases. Advances in these areas continue to push PEO technology toward more robust, scalable, and industrially viable protective coatings for Mg alloys.

Author Contributions: Conceptualization, B.R. and M.K.; validation, B.R. and M.K.; resources, B.R. and M.K.; writing—original draft preparation, B.R. and M.K.; writing—review and editing, B.R. and M.K. All authors have read and agreed to the published version of the manuscript.

Funding: This research received no external funding.

Conflicts of Interest: The authors declare no conflicts of interest.

References

1. Verma, C.; Ebenso, E.E.; Quraishi, M.A.; Hussain, C.M. Recent Developments in Sustainable Corrosion Inhibitors: Design, Performance and Industrial-Scale Applications. *Mater. Adv.* **2021**, *2*, 3806–3850. [CrossRef]
2. Ali Khan, M.A.; Irfan, O.M.; Djavanroodi, F.; Asad, M. Development of Sustainable Inhibitors for Corrosion Control. *Sustainability* **2022**, *14*, 9502. [CrossRef]
3. de Souza Morais, W.R.; da Silva, J.S.; Queiroz, N.M.P.; de Paiva e Silva Zanta, C.L.; Ribeiro, A.S.; Tonholo, J. Green Corrosion Inhibitors Based on Plant Extracts for Metals and Alloys in Corrosive Environments: A Technological and Scientific Prospection. *Appl. Sci.* **2023**, *13*, 7482. [CrossRef]
4. Nwigwe, U.S.; Nwoye, C.I. Green Corrosion Inhibitors for Steel and Other Metals in Basic Media: A mini-review. *Res. Eng. Struct. Mater.* **2023**, *9*, 775–789. [CrossRef]
5. Casanova, L.; Ceriani, F.; Messinese, E.; Paterlini, L.; Beretta, S.; Bolzoni, F.M.; Brenna, A.; Diamanti, M.V.; Ormellese, M.; Pedeferri, M. Recent Advances in the Use of Green Corrosion Inhibitors to Prevent Chloride-Induced Corrosion in Reinforced Concrete. *Materials* **2023**, *16*, 7462. [CrossRef]
6. Răut, D.I.G.; Matei, E.; Avramescu, S.M. Recent Development of Corrosion Inhibitors: Types, Mechanisms, Electrochemical Behaviour, Efficiency, and Environmental Impact. *Technologies* **2025**, *13*, 103. [CrossRef]
7. Lavanya, M.; Machado, A.A. Surfactants as Biodegradable Sustainable Inhibitors for Corrosion Control in Diverse Media and Conditions: A Comprehensive Review. *Sci. Total Environ.* **2024**, *908*, 168407. [CrossRef]
8. Bandeira, R.M.; Lima, F.P.; Nunes, M.S.; dos Santos, E.C.; Santos Júnior, J.R.; de Matos, J.M.E.; Feitosa, C.M.; Rai, M.; Bhattarai, S.; Mulmi, D.D. The Green Plant-Based Corrosion Inhibitors—A Sustainable Strategy for Corrosion Protection. *Surf. Sci. Technol.* **2025**, *3*, 19. [CrossRef]
9. Galleguillos Madrid, F.M.; Soliz, A.; Cáceres, L.; Bergendahl, M.; Leiva-Guajardo, S.; Portillo, C.; Olivares, O.; Toro, N.; Jimenez-Arevalo, V.; Páez, M. Green Corrosion Inhibitors for Metal and Alloys Protection in Contact with Aqueous Saline. *Materials* **2024**, *17*, 3996. [CrossRef]
10. Bender, R.; Féron, F.; Mills, D.; Ritter, S.; Bäßler, R.; Bettge, D.; De Graeve, I.; Dugstad, A.; Grassini, S.; Hack, T.; et al. Corrosion Challenges Towards a Sustainable Society. *Mater. Corros.* **2022**, *73*, 730–1751. [CrossRef]

11. Verma, C.; Chauhan, D.S.; Aslam, R.; Banerjee, P.; Aslam, J.; Quadri, T.W.; Zehra, S.; Verma, D.K.; Quraishi, M.A.; Dubey, S.; et al. Principles and Theories of Green Chemistry for Corrosion Science and Engineering: Design and Application. *Green Chem.* **2024**, *26*, 4270–4357. [CrossRef]
12. Hossain, N.; Asaduzzaman Chowdhury, M.; Kchaou, M. An Overview of Green Corrosion Inhibitors for Sustainable and Environment Friendly Industrial Development. *J. Adhes. Sci. Technol.* **2021**, *35*, 673–690. [CrossRef]
13. Baitule, P.; Manivannan, R. Corrosion Protection in Acidic Medium for Mild Steel Using Various Plants Extract as Green Corrosion Inhibitor—A Review. *Key Eng. Mater.* **2021**, *882*, 50–63. [CrossRef]
14. Aguilar-Ruiz, A.A.; Dévora-Isiordia, G.E.; Sánchez-Duarte, R.G.; Villegas-Peralta, Y.; Orozco-Carmona, V.M.; Álvarez-Sánchez, J. Chitosan-based sustainable coatings for corrosion inhibition of aluminum in seawater. *Coatings* **2023**, *13*, 1615. [CrossRef]
15. Kohl, M.; Alafid, F.; Bouška, M.; Krejčová, A.; Raycha, Y.; Kalendová, A.; Hrdina, R.; Burgert, L. New corrosion inhibitors based on perylene units in epoxy ester resin coatings. *Coatings* **2022**, *12*, 923. [CrossRef]
16. Liu, Y.; Meng, F.; Wang, F.; Liu, L. Dual-action epoxy coating with anti-corrosion and antibacterial properties based on well-dispersed ZnO/basalt composite. *Compos. Commun.* **2023**, *42*, 101674. [CrossRef]
17. Muresan, L.M. Nanocomposite Coatings for Anti-Corrosion Properties of Metallic Substrates. *Materials* **2023**, *16*, 5092. [CrossRef]
18. Mercan, D.-A.; Tudorache, D.-I.; Niculescu, A.-G.; Mogoantă, L.; Mogoşanu, G.D.; Bîrcă, A.C.; Vasile, B.Ș.; Hudiță, A.; Voinea, I.C.; Stan, M.S.; et al. Characterization of Nanoparticles in Antimicrobial Coatings. *Nanomaterials* **2025**, *15*, 637. [CrossRef]
19. Ebenezer, P.; Kumara, S.P.S.N.; Senevirathne, S.W.M.A.; Bray, L.J.; Wangchuk, P.; Mathew, A.; Yarlagadda, P.K. Advancements in Antimicrobial Surface Coatings Using Metal/Metaloxide Nanoparticles, Antibiotics, and Phytochemicals. *Nanomaterials* **2025**, *15*, 1023. [CrossRef]
20. Selim, M.S.; El-hoshoudy, A.N.; Zaki, E.G.; EL-Saeed, A.M.; Farag, A.A. Durable graphene-based alkyd nanocomposites for surface coating applications. *Environ. Sci. Pollut. Res.* **2024**, *31*, 43476–43491. [CrossRef]
21. Nguyen-Tri, P.; Nguyen, T.A.; Carriere, P.; Xuan, C.N. Nanocomposite Coatings: Preparation, Characterisation, Properties, and Applications. *Int. J. Corros.* **2018**, 4749501. [CrossRef]
22. Jiang, C.; Caod, Y.; Xiaoabc, G.; Zhuabc, R.; Lu, L. A review on the application of inorganic nanoparticles in chemical surface coatings on metallic substrates. *RSC Adv.* **2017**, *7*, 7531–7539. [CrossRef]
23. Žbulj, K.; Bilić, G.; Simon, K.; Hrnčavić, L. Lady's Mantle Flower as a Biodegradable Plant-Based Corrosion Inhibitor for CO_2 Carbon Steel Corrosion. *Coatings* **2024**, *14*, 671. [CrossRef]
24. Wang, X.; Fan, C.; Sun, L.; Shang, H.; Zhang, D.; Xu, N.; Wang, B.; Xu, J. Performance of a Composite Inhibitor on Mild Steel in NaCl Solution: Imidazoline, Sodium Molybdate, and Sodium Dodecylbenzenesulfonate. *Coatings* **2024**, *14*, 652. [CrossRef]
25. Yan, H.; Cheng, F.; Si, L.; Yang, Y.; Dou, Z.; Liu, F. Effect of Co Contents on Microstructure and Cavitation Erosion Resistance of NiTiAlCrCoxN Films. *Coatings* **2024**, *14*, 603. [CrossRef]
26. Deng, P.; Shangguan, J.; Hu, J.; Huang, H.; Zhou, L. Anticorrosion Method Combining Impressed Current Cathodic Protection and Coatings in Marine Atmospheric Environment. *Coatings* **2024**, *14*, 524. [CrossRef]
27. Stojanović, I.; Škrlec, B.; Kurtela, M.; Alar, V.; Odeljan, M. Physical and Chemical Properties of High-Temperature Silicone-Based Polymer Coatings Applied on Different Surface Roughnesses. *Coatings* **2023**, *13*, 2100. [CrossRef]
28. Landek, D.; Kurtela, M.; Stojanović, I.; Jačan, J.; Jakovljević, J. Corrosion and Micro-Abrasion Properties of an AISI 316L Austenitic Stainless Steel after Low-Temperature Plasma Nitriding. *Coatings* **2023**, *13*, 1854. [CrossRef]
29. Li, X.; Gao, K.; Zhao, Y.; Xie, X.; Lü, X.; Zhang, C.; Ai, H. Wear Resistance Study of Bionic Pitted Ni Cladding Layer on 7075 Aluminium Alloy Drill Pipe Surface. *Coatings* **2023**, *13*, 1768. [CrossRef]
30. Stojanović, I.; Logar, M.; Fatović, I.; Alar, V.; Ristevski, D.R. Experimental Study of Atmospherically and Infrared-Dried Industrial Topcoats. *Coatings* **2023**, *13*, 1343. [CrossRef]
31. Samardžija, M.; Stojanović, I.; Vuković Domanovac, M.; Alar, V. Epoxy Coating Modification with Metal Nanoparticles to Improve the Anticorrosion, Migration, and Antibacterial Properties. *Coatings* **2023**, *13*, 1201. [CrossRef]
32. Samardžija, M.; Kurtela, M.; Vuković Domanovac, M.; Alar, M. Anticorrosion and Antibacterial Properties of Al NP–Epoxy Nanocomposite Coating on Grey Cast Iron. *Coatings* **2023**, *13*, 898. [CrossRef]
33. Yan, W.; Ma, Z.; Wang, Z.; Zhang, J.; Li, K.; Wen, L.; Li, C.; Jiang, X.; Xu, Z. Localized Corrosion Mechanism of Q125 Casing Steel in Residual Acid Solution during Oil Reservoir Acidizing. *Coatings* **2023**, *13*, 710. [CrossRef]
34. Yang, C.; Chen, P.; Wu, W.; Sheng, L.; Zheng, Y.; Chu, P.K. A Review of Corrosion-Resistant PEO Coating on Mg Alloy. *Coatings* **2024**, *14*, 451. [CrossRef]

coatings

MDPI

Article

Localized Corrosion Mechanism of Q125 Casing Steel in Residual Acid Solution during Oil Reservoir Acidizing

Wei Yan [1,*], Zelin Ma [1], Zhi Wang [1], Jingru Zhang [1], Kepei Li [1], Lei Wen [2], Chen Li [3], Xiaohui Jiang [1] and Zhengxian Xu [1]

1 State Key Laboratory of Oil and Gas Resources and Detection, China University of Petroleum, Beijing 102249, China
2 National Center for Materials Service Safety, University of Science and Technology Beijing, Beijing 100083, China
3 College of Mechanical Engineering, Guangxi University, Nanning 530004, China
* Correspondence: yanwei@cup.edu.cn

Abstract: This paper aims to investigate the localized corrosion mechanism of Q125 casing steel in residual acid solution with Mannich base type inhibitors during oil reservoir acidizing process. The corrosion behavior of Q125 casing steel in fresh acid (20% HCl) and residual acid solution (pH 1.0 and pH 3.0 HCl) with and without 3-(4-chlorophenylimino)-1-(piperidine-1-ylmethyl) indolin-2-one (Mannich base type, Mb) inhibitor was studied by electrochemical test, weight loss, and surface analysis. The morphology and composition of corrosion products were analyzed by SEM/EDS and XPS; the local corrosion rate of casing steel with or without inhibitor was obtained by 3D profilometry. It was determined that the inhibitor had higher inhibition efficiency in fresh acid conditions than in residual acid conditions. Under the condition of residual acid, the decrease in inhibitor molecular coverage on the substrate surface promotes the occurrence of local corrosion. Pitting corrosion was detected in the residual acid solution containing Mannich base inhibitor, which may be related to the fact that $FeCO_3$ hinders the adsorption of inhibitor molecules on the substrate surface.

Keywords: Q125 casing steel; corrosion; mannish base inhibitor; acidizing fracturing; localized corrosion

1. Introduction

Q125 casing steel is a highly specialized type of steel that is widely used in the oil and gas extraction industry due to its exceptional strength, toughness, and corrosion resistance [1]. The harsh environmental conditions, such as high temperatures, high pressures, and corrosive substances, make it essential for the materials used in these applications to possess excellent corrosion resistance and high strength. However, despite its outstanding properties, Q125 casing steel can still suffer from severe corrosion damage, particularly in acidic environments [2]. Therefore, developing effective corrosion protection measures to improve the material's resistance to corrosion is crucial.

As the demand for oil and gas continues to increase, there has been a growing emphasis on efficient and sustainable extraction methods [3–5]. Acid fracturing technology is one such method that can enhance reservoir permeability and improve oil recovery rates [6,7]. However, the process of acidification in a tubing string can lead to corrosion of the metal materials due to the chemical reaction between the acid and the metal [8,9]. This is particularly true for high temperature acidification systems, where hydrochloric acid with a mass fraction of 20% or greater is commonly used as the base fluid. To address this issue, the simplest and most effective approach is to add corrosion inhibitors to the acid solution [4,10–12]. In this regard, the Mannich base type corrosion inhibitors are widely used in acid fracturing due to their high corrosion inhibition efficiency and low toxicity [13]. Therefore, understanding the corrosion resistance properties of Q125 casing steel and the use of corrosion inhibitors is crucial for the efficient and sustainable extraction of oil and gas.

Mannich base type corrosion inhibitor is mainly obtained by the condensation reaction of ketone, aldehyde, and amine containing active hydrogen. The synthesis methods of Mannich base corrosion inhibitors are diverse, the molecular structures of the obtained corrosion inhibitors are different, and the corrosion inhibition effects of the Mannich base corrosion inhibitor molecules with different structures are also different. Ahamad et al. [14] studied the inhibition efficiency of four piperidinylmethylindoline-2-one derivatives in 1 M HCl solution. It was determined that the different substituents of the aniline group would affect the inhibition efficiency. Furthermore, Ahamad et al. [14] reported that the lower the energy of the HOMO (highest occupied molecular orbital), LUMO (lowest unoccupied molecular orbital), and HOMO-LUMO gaps of the corrosion inhibitor, the higher the inhibition efficiency of the corrosion inhibitor. The results of Ayeni et al. [13] and Ferkous et al. [15] also support this view. Pan et al. [16] studied the effect of hydrophobic group carbon chain length on the corrosion inhibition efficiency of N80 steel in 15% HCl solution and determined that when the inhibitor content was less than 0.6%, the Mannich base inhibitor with short carbon chain had better corrosion inhibition performance. Guo et al. [17] synthesized a double Mannich base inhibitor using N-octylamine, acetophenone, and formaldehyde as the main raw material and reported that a double Mannich base inhibitor compared to a single Mannich base corrosion inhibitor had higher corrosion inhibition efficiency.

However, most of the above studies focused on the inhibition effect of inhibitors under fresh acid conditions. Fresh acid reacts with alkaline ore to obtain residual acid, and the pH value of residual acid is higher than that of fresh acid. However, the effect of residual acid and residual carbonate products on the corrosion inhibition efficiency of inhibitors is not clear. Therefore, investigation of the effects of Mb on the corrosion behaviors of Q125 casing steel in fresh and residual acid solutions can not only reveal the internal causes of string failure, but also provide certain guiding significance for the development of low permeability reservoirs and reservoir reconstruction projects.

2. Materials and Methods

2.1. Materials and Pretreatment

The chemical composition of Q125 grade casing steel used in this experiment is 0.28 C, 0.29 Si, 0.55 Mn, 0.98 Cr, 0.31 Mo, 0.03 Nb, 0.15 V, 0.006 P, 0.002 S, and Fe in balance (wt.%). The exposed area of the working electrode is 10 mm × 10 mm; the sample size of weight loss test is 30 mm × 10 mm × 3 mm. Before the experiment, all the samples were ground to 600 mesh by SiC sandpaper step by step. The oil was deoiled with acetone, then dehydrated by ethanol, and placed in a vacuum drying oven for 24 h.

2.2. Test Procedures

The experiment was carried out in a 2 L glass cell. Three-electrode system was used in the electrochemical test. A saturated calomel (Hg/Hg_2Cl_2, SCE) reference electrode was connected by a lugin capillary tube, and the counter electrode was a platinum electrode. The pH value of the solution was measured by a pH electrode (METTLER TOLEDO, FE20, Columbus, OH, USA). A potentiostat was used for the electrochemical test (GAMRY, Reference 3000, Warminster, PA, USA). The experiment was carried out in the solution of residual acid and fresh acid with or without 1 wt.% Mb at atmosphere pressure and 90 °C. The chemical structure of the Mb corrosion inhibitor (Provided by the Beijing Comers Oilfield Chemicals Technology Co., Ltd., Beijing, China: MB-CS125) is shown in Figure 1.

When the electrode was immersed in the solution for 4 h, the electrochemical test was carried out. The polarization range of the polarization resistance was relative to the open circuit potential (OCP) ± 10 mV, and the scan rate was 0.125 mV s^{-1}.The test range of potentiodynamic scanning was ±200 mV, and the scanning rate was 0.2 mV s^{-1}. The dynamic potential scan first performed cathodic polarization, and when the polarization returned to OCP after the end of the polarization, the anode polarization scan was performed.

EIS was carried out in the range of open circuit potential ±5 mV, and the frequency range was 10^4 Hz to 10^{-2} Hz.

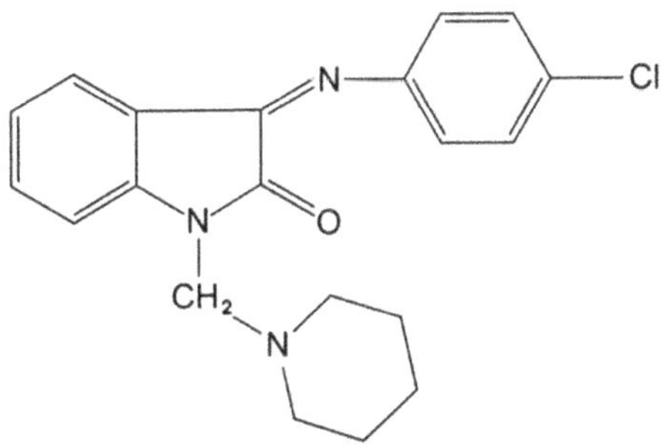

Figure 1. Chemical structure of Mb.

After the electrochemical test, the weight loss method specimens were immersed in the test solution for 4 h to study the influence of Mannich base type corrosion inhibitor on the corrosion behavior of Q125 grade casing steel under the conditions of fresh acid and residual acid. The corroded samples were cleaned with deionized water and dehydrated with ethanol. The corrosion products were removed from the corrosion samples by using Clark solution (50 g stannous chloride, 20 g antimony trioxide, and concentrated chlorogenic acid per 1000 mL) according to the standards of the American Society for Testing and Materials (ASTM) [18], and then dried weighed by an electronic balance. The corrosion rate was obtained by weight loss method according to Formula [19], and the inhibition efficiency is expressed in Formula (2):

$$V = \frac{W_0 - W}{S \times t} \quad (1)$$

$$Inhibition\ efficiency = \frac{V_0 - V_m}{V_0} \times 100\%, \quad (2)$$

where V is the corrosion rate, g m^{-2} h^{-1}; W_0 is the weight of the sample before corrosion, g; W is the mass of the sample after corrosion, g; S is the surface area of the sample, m^2; t is immersion time, h; V_0 is the corrosion rate without inhibitor, g m^{-2} h^{-1}; V_m is the corrosion rate with inhibitor, g m^{-2} h^{-1}.

The surface morphology was analyzed by Scanning Electron Microscope (SEM)/Energy Dispersive Spectrometer (EDS). The local corrosion was analyzed by three-Dimension (3D) topography. The components of corrosion products were analyzed by X-ray Photoelectron Spectroscopy (XPS), and the spectra were fitted by XPS peak software.

3. Results

3.1. Uniform Corrosion Rate

Figure 2 shows the weight loss results of casing steel in fresh and residual acid solutions with and without Mb inhibitor. The study determined that when there was no Mb inhibitor, the corrosion rate decreased as the hydrogen ion concentration in the solution decreased. However, even when the pH of the solution was 3.0, the casing steel still had a higher corrosion rate of 92.48 g m^{-2} h^{-1}. When Mb appeared, Mb could effectively inhibit the occurrence of uniform corrosion of casing steel. Although the corrosion inhibition efficiency decreased slightly as the hydrogen ion concentration of the solution decreased, it still had a higher corrosion inhibition effect, and the corrosion inhibition efficiency was 98.8% under the condition of pH 3.0.

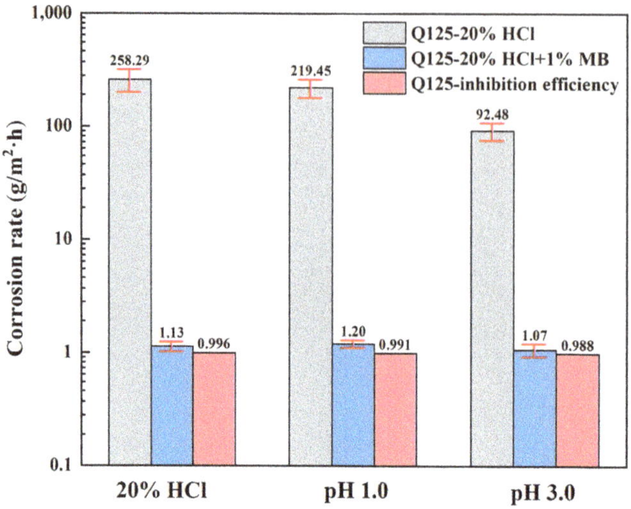

Figure 2. Uniform corrosion rate of casing steel immersed in fresh and residual acid solutions with or without Mb inhibitor for 4 h at 90 °C.

3.2. Electrochemical Measurements

The polarization curves of casing steel in fresh and residual acid solutions with and without Mb inhibitor for 4 h are shown in Figure 3. Under the condition of fresh acid, the appearance of Mb inhibited the anodic reaction and cathodic reaction of casing steel. The appearance of Mb caused the OCP to move up. Under the condition of residual acid, the OCP was significantly lower than that under the condition of fresh acid, and with the decrease in hydrogen ion content, the OCP decreased. When the corrosion inhibitor appeared, it could also inhibit the anode process and cathode reaction process, and the appearance of the corrosion inhibitor caused the OCP to move up, which was the same as that in the fresh acid condition, but the cathode reaction process in the residual acid condition was controlled by activation and diffusion. Compared with the condition of pH 3.0, the inhibition effect of Mb on the anodic reaction was more obvious at pH 1.0.

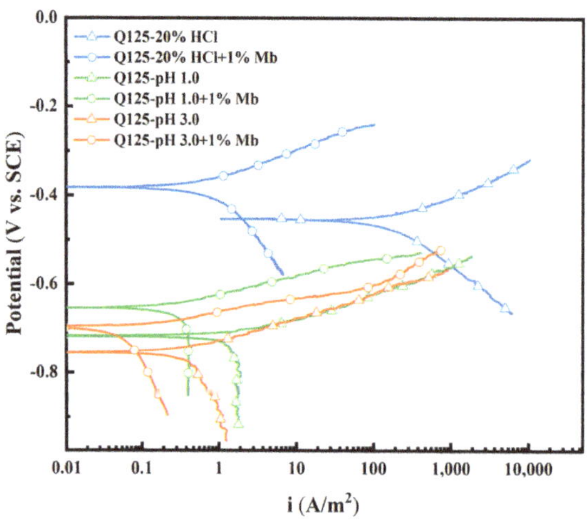

Figure 3. Polarization curves of casing steel immersed in fresh and residual acid solutions with or without Mb inhibitor for 4 h at 90 °C.

Table 1 presents the results of fitting electrochemical parameters obtained from potentiodynamic polarization curves recorded under various conditions depicted in Figure 3, where b_a is the anode Tafel slope, b_c is the cathode Tafel slope, r_p is the resistance of polarization, E_{corr} is the corrosion electrode potential, i_{corr} is the corrosion current density. Under residual acid conditions (pH 1.0 and pH 3.0), the cathode had a large Tafel slop. This phenomenon occurred because a large amount of $CaCO_3$ was added to the fresh acid (20% HCl) to neutralize the strong acid to obtain the residual acid solution. The addition of a large amount of $CaCO_3$ affected the chemical environment of the solution phase and thus the cathode reaction process. Moreover, the addition of Mb increased the Tafel slope of the anode and cathode compared with the fresh acid and residual acid without Mb. The trend of corrosion current density obtained by electrochemical test is consistent with the trend of corrosion current density obtained by weight loss method, showing significant attenuation characteristics. In addition, in terms of polarization resistance, it can also be seen that the addition of Mb inhibitor greatly increases the electrochemical corrosion resistance. Ferreira et al. [20] reported that when the displacement of E_{corr} was greater than 85 mV when the corrosion inhibitor appeared, the corrosion inhibitor belonged to the anodic or cathodic inhibition effect. When the displacement of the E_{corr} is less than 85 mV, the corrosion inhibitor is in a mixed inhibition effect. It could be seen from Table 1 that under the conditions of residual acid and fresh acid, the corrosion inhibitor has a mixed inhibitory effect.

Table 1. The fitting results of potentiodynamic polarization curve parameters under different conditions.

Condition	b_a mV/dec	b_c mV/dec	r_p	E_{corr}/mV	i_{corr}/mA cm^{-2}
20% HCl	61.55	95.24	8.2	−453.2	1.60
20% HCl + Mb	66.19	126.2	232.1	−381.2	0.081
pH 1.0	43.78	511	42.1	−717.2	0.42
pH 1.0 + Mb	47.02	548.7	463.3	−652.7	0.041
pH 3.0	51.56	235.5	140.1	−753.6	0.13
pH 3.0 + Mb	54.69	251.9	524.2	−694.7	0.037

Figure 4 shows the EIS results of casing steel under different conditions. When no Mb was added, the impedance spectrum exhibited a time constant, Figure 4b. When Mb was added, the impedance spectrum exhibited two (20% HCl + Mb and pH 1.0 + Mb, Figure 4c) or three (pH 3.0 + Mb, Figure 4d) time constants. The capacitive reactance arc in the high frequency region may be attributed to double layer capacitance. The capacitive arc in the middle- and low-frequency regions may be attributed to the corrosion product film or the adsorption film formed on the substrate surface by Mb. The inductive loop in the low-frequency region may be due to the adsorption process of Mb molecules. Under the same pH value, the diameter of the semicircular capacitor arc increases obviously when corrosion inhibitors appear, which indicates that Mb inhibited the progress of corrosion and reduced the corrosion rate.

Three kinds of equivalent circuit diagrams are shown in Figure 5 and the corresponding EIS fitting results are shown in Table 2 to demonstrate the obtained electromechanical parameters within different experimental conditions. Figure 5a was used to fit the EIS data of fresh acid and residual acid without Mb. Figure 5b was used to fit the EIS data of fresh acid and pH 3.0 with Mb (two time constants), and Figure 5c was used to fit the EIS data of pH 1.0 with Mb (three time constants). In the equivalent circuit diagram, Q_{dl} is the constant phase angle element representing double-layer capacitance; R_{ct} is charge transfer resistance; Q_f is the constant phase angle element representing the capacitance of the corrosion product film or the adsorption film formed by Mb. R_f is the resistance of the corrosion product film or the adsorption film formed by Mb; L and R_L represent inductance and inductive resistance, respectively; n_{dl} and n_f are the constant phase angle element exponents. With the increase in pH value, the impedance (R_{ct} + R_f) increased, indicating that the corrosion rate decreased with the increase in pH value without Mb. When Mb was added, compared with the same pH value without Mb, the impedance value at least doubled, indicating that Mb could significantly reduce the corrosion rate. When Mb was added to fresh acid, the proportion of impedance value increased most, indicating that the inhibitor had the highest inhibition efficiency under fresh acid conditions.

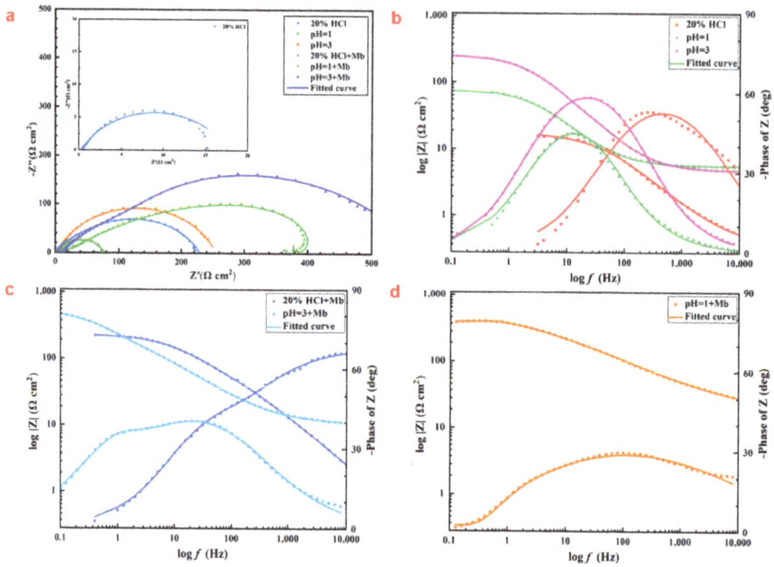

Figure 4. EIS of casing steel immersed in fresh and residual acid solutions with or without Mb inhibitor for 4 h at 90 °C. (**a**) Nyquist diagram; (**b–d**) Bode diagram.

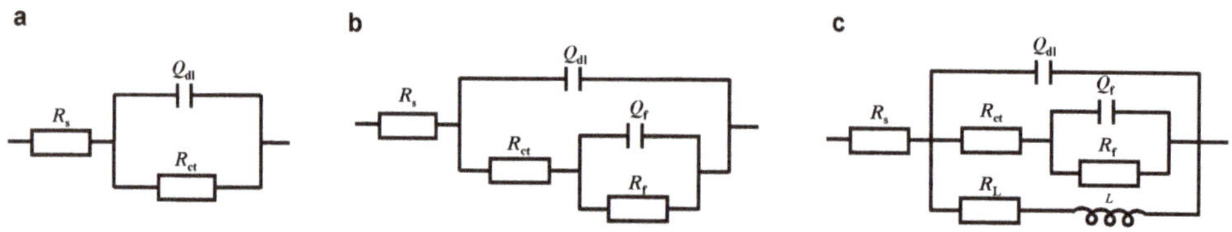

Figure 5. Fitting circuit of EIS results in Figure 4. Note: (**a**) 20% HCL, pH = 1, and pH = 3; (**b**) 20% HCL + Mb and pH = 3 + Mb; (**c**) pH = 1 + Mb.

Table 2. Fitting parameters of EIS of casing steel immersed in fresh and residual acid solution with or without Mb inhibitor for 4 h.

Condition	Q_{dl} (Ω^{-1} cm^{-2} s^{-ndl})	n_{dl}	R_{ct} (Ω cm^2)	Q_f (Ω^{-1} cm^{-2} s^{-ndl})	n_f	R_f (Ω cm^2)	L (H cm^{-2})	R_L (Ω cm^2)
20% HCl	9.67×10^{-5}	0.75	16.74					
20% HCl + Mb	9.34×10^{-4}	0.75	76.95	1.8×10^{-4}	0.74	151.7		
pH 1.0	1.57×10^{-3}	0.8	69					
pH 1.0 + Mb	3.93×10^{-7}	0.93	14.35	4.76×10^{-4}	0.48	504.7	1541	1188
pH 3.0	5.56×10^{-4}	0.8	250.9					
pH 3.0 + Mb	8.64×10^{-4}	0.61	330.2	6.89×10^{-4}	1	240.5		

3.3. Surface Morphology Analysis

Figure 6 shows the SEM results of casing steel in fresh and residual acid solutions with or without Mb inhibitor. In fresh acid solution, a loose corrosion product film was formed on the surface of the substrate. When the Mb inhibitor appeared, the surface of the substrate was relatively flat, and the trace of a polishing line could still be observed, which indicated that the presence of Mb inhibitor could effectively inhibit the corrosion behavior of HCl to casing steel, which was consistent with the weight loss method. At pH 1.0, a large number of streak-like corrosion products were formed on the surface of the substrate. Based on the EDS results in Table 3, it was determined that the main elements of the streak-like corrosion products were Fe, O, and C, which indicated that $FeCO_3$ might be formed on the substrate. When the Mb corrosion inhibitor appeared, a large number of spot-like corrosion products were detected on the substrate, and a small amount of streak-like corrosion products also appeared, but no traces of polishing lines were observed (Figure 6d). Under the condition of pH 3.0 without Mb (Figure 6e), stripe-like corrosion products also appeared on the substrate. Compared with the elemental composition of corrosion product film in areas A (Figure 6c) and B, it was determined that the content of C was higher under the condition of pH 3.0, which indicated that more $FeCO_3$ formed on the substrate with the increase in pH. In the solution containing a Mb inhibitor at pH 3.0, the corrosion products were covered on the surface of the substrate. It can be seen from Figure 6b,d,f that with the decrease in hydrogen ion concentration, more corrosion products are formed on the substrate.

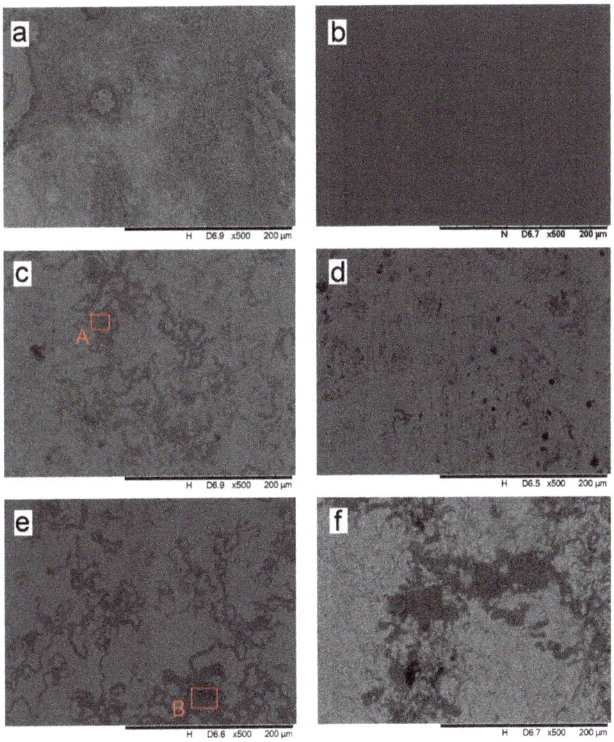

Figure 6. SEM of casing steel after being exposed to fresh and residual acid solutions with or without Mb inhibitor at 90 °C for 4 h: (**a**) 20% HCl; (**b**) 20% HCl + Mb; (**c**) pH 1.0; (**d**) pH 1.0 + Mb; (**e**) pH 3.0; (**f**) pH 3.0 + Mb.

Table 3. Elemental composition of the A and B regions (at.%).

Region	Fe	C	O	Cl
A	51.59	29.06	19.26	0.09
B	44.96	36.29	18.26	0.49

The elemental composition of the A (in Figure 6c) and B (in Figure 6d) regions are shown in Table 3.

3.4. 3D Profilometry

The 3D morphology results of the casing steel in the residual acid solution with or without Mb are shown in Figure 7. After removing the product film, the surface of the substrate was relatively flat in the residual acid without Mb. The corrosion type of casing steel was uniform corrosion under the condition of residual acid (Figure 7a,c). In the residual acid solution containing Mb at pH 1.0 (as shown in Figure 7b), pits were detected on the surface of the substrate, and the pitting rate reached 47.41 mm y^{-1}.

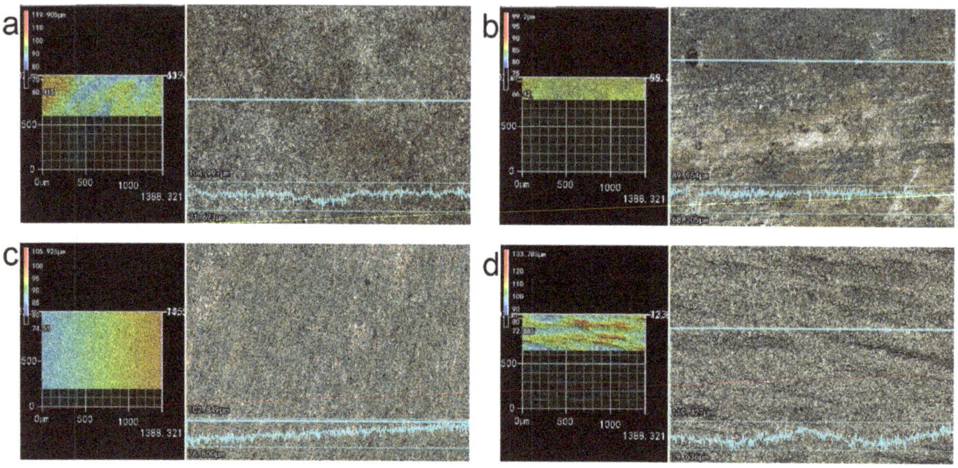

Figure 7. 3D profilometry of casing steel after exposed in residual acid solutions with or without Mb inhibitor for 4 h at 90 °C: (**a**) pH 1.0; (**b**) pH 1.0 + Mb; (**c**) pH 3.0; (**d**) pH 3.0 + Mb.

In the residual acid solution containing corrosion inhibitor at pH 3.0, although no pitting pits were observed on the substrate, the surface of the substrate was corrugated (Figure 7d), and the height difference between the peaks and troughs formed was close to 40 μm. The appearance of corrosion products might reduce the adsorption characteristics of corrosion inhibitors on the substrate so that the corrosion medium reacted with the substrate at the weak part of the adsorption film to form pits. The corrosion product morphology in Figure 6f combined with the 3D morphology in Figure 7d after removing the corrosion product film also supported this view.

3.5. XPS Analysis

The XPS results of C, O, Fe, and N elements of casing steel in hydrochloric acid solution with or without Mb at pH 3.0 are shown in Figure 8. The residual peak of C 1s at 284.6 eV was used to calibrate the binding energy of each substance. In the residual acid solution without Mb, the peak of C 1s, O 1s, and Fe $2p_{3/2}$ was at 288.0 eV, 531.4 eV, and 711.3 eV, respectively, indicating the appearance of $FeCO_3$ [21]. The peak of O 1s at 529.7 eV, the peak of Fe $p_{3/2}$ at 710.9 eV and 718.7eV, and the peak of Fe $2p_{1/2}$ at 724.5 eV corresponded to Fe_2O_3 [22]. When the Mb was added, the binding energy of C, O, and Fe elements was the same. In addition, the peak of N 1s was at 399.0 eV, which might be related to the residual Mb on the surface of the substrate or the corrosion products from the interaction between the Mb and the substrate.

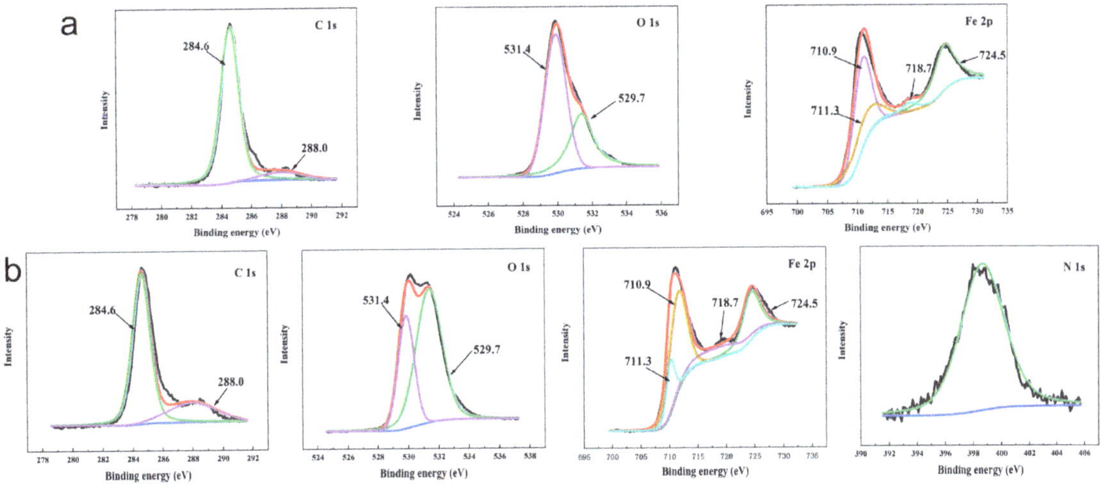

Figure 8. XPS results after immersion of casing steel in residual acid solution with or without Mb inhibitor at 90 °C for 4 h: (**a**) pH 3.0; (**b**) pH 3.0 + Mb.

4. Discussion

When no Mb was present, the morphology of corrosion products of casing steel in residual acid and the fresh acid solution was different (Figure 9a,b), which was mainly due to the addition of a large amount of $CaCO_3$ in the solution under the condition of residual acid. In the solution, CO_3^{2-} combines with Fe^{2+} to form $FeCO_3$ on the substrate, and the increase in solution pH value was beneficial to $FeCO_3$ film formation on the substrate, which was confirmed by the results of Figure 6c,e [23]. The increase in pH value of the solution led to the decrease in hydrogen concentration in the solution, which was the main reason for the significant difference in Tafel slope in the cathodic reaction process between residual acid and fresh acid. Moreover, the difference in corrosion product film protection between residual acid and fresh acid might affect the diffusion process of hydrogen ions, thus affecting the cathodic reaction process.

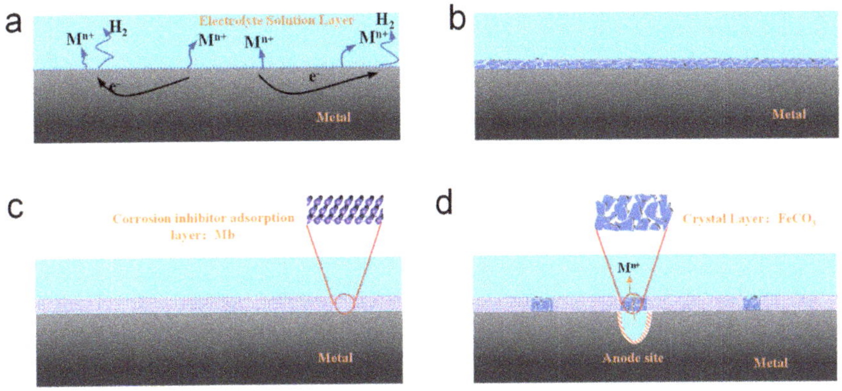

Figure 9. Corrosion/corrosion inhibition mechanism model diagram: (**a**) fresh acid; (**b**) residual acid; (**c**) fresh acid with Mb; (**d**) residual acid with Mb.

When Mb was added to fresh acid and residual acid solution, the decrease in corrosion rate occurred due to the adsorption of Mb molecules to the surface of the substrate to inhibit the corrosion of casing steel (Figure 9c). The adsorption of Mb on carbon steel surface belonged to mixed adsorption, which was also supported by current electrochemical results [14]. The sharing electrons were formed between p-electrons of the heterocyclic ring and vacant d orbitals of surface iron, and the adsorption of Mb molecules and mild steel belonged to chemical adsorption. In addition, the aromatic ring in the Mb molecule can also be used as an electron donating functional group to promote the chemisorption with the steel. Mb molecule might undergo protonation reaction in residual acid solution to obtain protonated Mb with a positive charge. The reaction is shown in Formula (3):

$$C_{20}H_{22}N_3OCl + xH^+ \rightarrow [C_{20}H_{22+x}N_3OCl]^{x+}. \quad (3)$$

Because the hydration degree of Cl^- was small, Cl^- was conducive to the formation of excessive negative charge near the substrate, which promoted the electrostatic adsorption between protonated Mb and steel, which belonged to physical adsorption.

Under the condition of residual acid, the appearance of Mb promoted the local corrosion of casing steel (Figure 9d). There was a certain corresponding relationship between the 3D morphology of corrosion product film removed in Figure 7b,d and the distribution of corrosion product film on the substrate surface in Figure 6d,e. It could be inferred that the $FeCO_3$ corrosion product hindered the adsorption of the Mb inhibitor on the substrate. The Q_{dl} value in Table 2 under the condition of residual acid containing Mb was smaller than that under the condition of fresh acid containing Mb, which supported this point. The results showed that the coverage rate of Mb molecule in the area where $FeCO_3$ corrosion product film exists was reduced, and the $FeCO_3$ corrosion product film had a less protective effect on the substrate than the adsorption film formed by Mb. The corrosion medium could react with the substrate through the $FeCO_3$ corrosion product film, thus promoting the occurrence of local corrosion.

5. Conclusions

In the present work, we obtain the following conclusions:

(1) Compared with the residual acid, the Mb inhibitor had higher inhibition efficiency under fresh acid conditions.

(2) Under fresh acid and residual acid conditions, Mb promoted the increase in Ecorr, and increased the Tafel slope of cathode and anode, resulting in an increase in reaction resistance. Therefore, Mb inhibitor showed mixed inhibition.

(3) Local corrosion of casing steel occurred in a residual acid solution containing Mb inhibitor. The occurrence of localized corrosion was related to the decrease in Mb inhibitor coverage on the substrate surface and the poor protection of $FeCO_3$ film.

(4) The $FeCO_3$ corrosion product hindered the adsorption of the Mb inhibitor on the substrate. The corrosion medium could react with the substrate through the $FeCO_3$ corrosion product film, thus promoting the occurrence of local corrosion.

Author Contributions: Data curation, W.Y. and Z.M.; Methodology, Z.W. and J.Z.; Project administration, L.W. and W.Y.; Resources, C.L. and K.L.; Writing—original draft, Z.M., X.J. and. Z.X.; Writing—review & editing, Z.M., K.L. and W.Y. All authors have read and agreed to the published version of the manuscript.

Funding: This research was founded by the Opening project fund of Materials Service Safety Assessment Facilities, MSAF-2021-102.

Institutional Review Board Statement: Not applicable

Informed Consent Statement: Not applicable.

Data Availability Statement: Data is contained within the article.

Conflicts of Interest: The authors declare no conflict of interest.

References

1. Lou, Q.; Du, W.; Han, X.; Liu, W.; Song, H. Study on the Mechanical Properties of 125S Casing for oil and gas wells. *J. Phys. Conf. Ser.* **2020**, *1605*, 012123. [CrossRef]
2. Li, Z.B.; Lu, L.; Bi, Z.Y.; Zhang, F.; Huang, X.J.; Liu, Y.M. Heat treatment process of Q125 grade SEW oil casing steel. *Cailiao Rechuli Xuebao/Trans. Mater. Heat Treat.* **2016**, *37*, 156–161.
3. Farrokhrouz, M.; Taheri, A.; Keshavarz, A. Numerical reactive flow transport simulation on core samples during acid fracturing in carbonaceous shale. *J. Nat. Gas Sci. Eng.* **2020**, *84*, 103615. [CrossRef]
4. Shen, C.; Alvarez, V.; Koenig, J.D.B.; Luo, J.-L. Gum Arabic as corrosion inhibitor in the oil industry: Experimental and theoretical studies. *Corros. Eng. Sci. Technol.* **2019**, *54*, 444–454. [CrossRef]
5. Zhang, K.; Chen, M.; Zhou, C.; Dai, Y.; Liu, F.; Li, J. Study of alternating acid fracturing treatment in carbonate formation based on true tri-axial experiment. *J. Pet. Sci. Eng.* **2020**, *192*, 107268. [CrossRef]
6. Zhang, R.; Hou, B.; Zhou, B.; Liu, Y.; Xiao, Y.; Zhang, K. Effect of acid fracturing on carbonate formation in southwest China based on experimental investigations. *J. Nat. Gas Sci. Eng.* **2020**, *73*, 103057. [CrossRef]
7. Sheng, M.; Khan, W.A.; Cheng, S.; Zhang, P.; Tian, S.; Xu, Q. Characteristics of micro-fracturing in shales induced by dilute acid. *J. Nat. Gas Sci. Eng.* **2021**, *88*, 103855. [CrossRef]
8. Kumar, H.; Yadav, V. Highly efficient and eco-friendly corrosion inhibitor for mild steel in 5 M HCl: An anti-pitting and anti-cracking agent. *Chem. Data Collect.* **2020**, *30*, 100552. [CrossRef]
9. Yadav, M.; Sharma, D.; Sarkar, T.K. Adsorption and corrosion inhibitive properties of synthesized hydrazine compounds on N80 steel/hydrochloric acid interface: Electrochemical and DFT studies. *J. Mol. Liq.* **2015**, *212*, 451–460. [CrossRef]
10. Akalezi, C.O.; Maduabuchi, A.C.; Enenebeaku, C.K.; Oguzie, E.E. Experimental and DFT evaluation of adsorption and inhibitive properties of *Moringa oliefera* extract on mild steel corrosion in acidic media. *Arab. J. Chem.* **2020**, *13*, 9270–9282. [CrossRef]
11. Lin, B.-L.; Shao, J.-J.; Xu, Y.-Y.; Lai, Y.-M.; Zhao, Z.-N. Adsorption and corrosion of renewable inhibitor of *Pomelo* peel extract for mild steel in phosphoric acid solution. *Arab. J. Chem.* **2021**, *14*, 103114. [CrossRef]
12. Zafari, S.; Sarabi, A.A.; Movassagh, B. A novel green corrosion inhibitor based on task-specific benzimidazolium ionic liquid for carbon steel in HCl. *Corros. Eng. Sci. Technol.* **2020**, *55*, 589–601. [CrossRef]
13. Ayeni, A.O.; Akinyele, O.F.; Hosten, E.C.; Fakola, E.G.; Olalere, J.T.; Egharevba, G.O.; Watkins, G.M. Synthesis, crystal structure, experimental and theoretical studies of corrosion inhibition of 2-((4-(2-hydroxy-4-methylbenzyl)piperazin-1-yl)methyl)-5-methylphenol—A Mannich base. *J. Mol. Struct.* **2020**, *1219*, 128539. [CrossRef]
14. Ahamad, I.; Prasad, R.; Quraishi, M.A. Adsorption and inhibitive properties of some new Mannich bases of Isatin derivatives on corrosion of mild steel in acidic media. *Corros. Sci.* **2010**, *52*, 1472–1481. [CrossRef]
15. Ferkous, H.; Djellali, S.; Sahraoui, R.; Benguerba, Y.; Behloul, H.; Çukurovali, A. Corrosion inhibition of mild steel by 2-(2-methoxybenzylidene) hydrazine-1-carbothioamide in hydrochloric acid solution: Experimental measurements and quantum chemical calculations. *J. Mol. Liq.* **2020**, *307*, 112957. [CrossRef]
16. Pan, Y.; Zhan, F.T.; Yang, Z. Effect of Carbon Chain Length of Hydrophobic Group of Mannich Bases on Their Corrosion Inhibition Performance. *Mater. Prot.* **2015**, *48*, 23–25+50. [CrossRef]
17. Guo, H.; Dong, X.; Yang, J.; Xu, X. Synthesis and performance evaluation of double Mannich base acidizing corrosion inhibitor. *Mod. Chem. Ind.* **2020**, *40*, 180–184. [CrossRef]
18. *ASTM G1-03(2017)e1*; Standard Practice for Preparing, Cleaning, and Evaluating Corrosion Test Specimens. ASTM Standard: Philadelphia, PA, USA, 2017; pp. 1–9. [CrossRef]
19. *ASTM G31-72(2004)*; Standard Practice for Laboratory Immersion Corrosion Testing of Metals. ASTM Standard: Philadelphia, PA, USA, 2004; pp. 1–8. [CrossRef]
20. Ferreira, E.S.; Giacomelli, C.; Giacomelli, F.C.; Spinelli, A. Evaluation of the inhibitor effect of l-ascorbic acid on the corrosion of mild steel. *Mater. Chem. Phys.* **2004**, *83*, 129–134. [CrossRef]
21. Tang, Y.; Guo, X.P.; Zhang, G.A. Corrosion behaviour of X65 carbon steel in supercritical-CO_2 containing H_2O and O_2 in carbon capture and storage (CCS) technology. *Corros. Sci.* **2017**, *118*, 118–128. [CrossRef]
22. Heuer, J.K.; Stubbins, J.F. An XPS characterization of $FeCO_3$ films from CO_2 corrosion. *Corros. Sci.* **1999**, *41*, 1231–1243. [CrossRef]
23. Guo, S.; Xu, L.; Zhang, L.; Chang, W.; Lu, M. Corrosion of alloy steels containing 2% chromium in CO_2 environments. *Corros. Sci.* **2012**, *63*, 246–258. [CrossRef]

coatings

MDPI

Article

Anticorrosion and Antibacterial Properties of Al NP–Epoxy Nanocomposite Coating on Grey Cast Iron

Marina Samardžija [1,*], Marin Kurtela [2], Marija Vuković Domanovac [3] and Vesna Alar [2]

1 Department of Chemistry, Faculty of Mining-Geology-Petroleum Engineering, University of Zagreb, 10110 Zagreb, Croatia
2 Department of Welded Structures, Faculty of Mechanical Engineering and Naval Architecture, University of Zagreb, 10000 Zagreb, Croatia; marin.kurtela@fsb.unizg.hr (M.K.); vesna.alar@fsb.unizg.hr (V.A.)
3 Department of Industrial Ecology, Faculty of Chemical Engineering and Technology, University of Zagreb, 10000 Zagreb, Croatia; mvukovic@fkit.unizg.hr
* Correspondence: marina.samardzija@rgn.unizg.hr; Tel.: +385-1-5535-912

Abstract: In this study, different concentrations of aluminium nanoparticles (Al NP) were incorporated into epoxy resin and epoxy paint. Here, we present a detailed systematic study of different methods of incorporating inorganic nanoparticles into epoxy coating. This work aims to obtain an epoxy coating with anticorrosion and antibacterial properties. The physical properties of coatings such as thickness, hardness, colour, and adhesion did not change with the addition of nanoparticles. According to the SEM and EDS analyses, the distribution effect of Al NPs in epoxy coating was better with ultrasonic homogenisation than with mechanical stirring. The EIS and SECM measurements were used to investigate corrosion resistance. The coating with 1.0 wt.% Al NP showed the best physical and chemical properties. SECM examination indicated that nanoparticles in epoxy resin increase the protection efficiency by 25.75% and in the epoxy paint by 40.89%. The results also showed the antibacterial activity of aluminium nanoparticles by inhibiting the growth of biofilm-forming bacteria such as *P. aeruginosa* and *B. subtilis*.

Keywords: electrochemical corrosion; microbiologically corrosion; Al nanoparticles; mechanisms

1. Introduction

Grey cast iron has been widely used in pipes serving as water mains [1] due to its good thermal conductivity, relatively low melting temperature, high damping capacity, and excellent castability [2]. Unfortunately, these properties deteriorate when metals interact with certain elements that recur within their environments, a process technically called corrosion [3]. Corrosion is the irreversible damage or destruction of a material and as such is a very expensive and dangerous phenomenon that causes serious problems in the world [4,5]. The National Association of Corrosion Engineers (NACE International) estimates global losses caused by corrosion at USD 2.5 trillion per year and the average annual cost of corrosion in the pipeline industry is estimated at USD 7 billion to monitor, replace, and maintain these assets [6,7]. The most conventional method for protecting metal structures against corrosion is to isolate the metal from corrosive agents [8]. Researchers are continuously seeking new innovative coatings with good corrosion resistance and high antibacterial activity [9]. Typically, municipal wastewater that flows through pipes is composed of multiple pathogens and non-pathogenic bacteria, organic/inorganic chemicals, suspended and dissolved compounds, and similar [10,11].

Microbiologically influenced corrosion (MIC) is a serious type of corrosion, as it accounts for approximately 20% of total economic losses [12]. The microbiological community that exists in wastewater is usually a combination of various types of bacteria, among which are pathogenic forms such as opportunistic pathogens (*Enterobacter cloacae*, *Enterococcus faecalis*, *Escherichia coli*, *Klebsiella pneumoniae*, *Proteus vulgaris* or *Pseudomonas aeruginosa*), obligate pathogens from *Salmonella* and *Shigella genera* and enteropathogenic strains of *Escherichia coli*. It can also contain viruses, protozoa, fungi, flatworms, roundworms, and similar [13,14]. The rod-shaped Gram-negative aerobic bacteria *P. aeruginosa* prefer moist environments such as urban runoffs and sewage effluents [15]. Furthermore, these bacteria have a high potential for developing great resistance to a wide range of currently available antimicrobial agents [16]. Likewise, the Gram-positive aerobic bacteria *B. subtilis* are widely present in the environment and show a capability to deal with heavy metals and dyes in polluted waters [17]. The presence of an energy source, carbon source, redox mediators, and water are important factors for bacterial colonization of any surface. Then, the metabolic activities of these colonized bacteria influence the electrochemical reactions initiating MIC. Secretion of extracellular polymeric substances (EPS) and biofilm formation on the material surfaces are crucial steps that enhance the probability of metal corrosion and MIC [18].

Recently, with the development of nanotechnology [19], the term "nanocomposite" has appeared with at least one dimension on the nanometre scale [20]. Nanoparticles have been widely used as resin fillers to block micropores and improve the corrosion resistance and the mechanical properties of the resins due to their great surface-to-volume proportion with respect to conventional macroscopic materials [4]. The purpose of the study by Manjumeena and co-authors [21] was to obtain a nanocomposite with dual properties. According to their study, silver nanoparticles show great antibacterial properties due to their ability to release Ag^{+} from nanoparticles which can damage the cell wall and cause oxidative stress. Talabi and co-authors [22] developed polymer nanocomposites with antibacterial properties. They showed that Cu^{2+} ions, which are released from copper nanoparticles, cause the denaturation of proteins in the bacterial membrane, by binding to DNA, which results in the cell not being able to replicate further. The surface-to-volume ratio of a nanoparticle is 35%–45% times higher as compared to large particles or atoms. This unique extrinsic property of the specific surface area of the nanoparticle is a contributory factor for its high value, and it also influences different intrinsic properties such as strong surface reactivity which is size dependent [23]. Due to all these properties of nanoparticles and the high viscosity of epoxy coating, it is hard to uniformly mix nano-size fillers into epoxy coating [15]. According to the available literature, there are different methods of incorporating nanoparticles into epoxy coating. In our last article, we employed strong mechanical stirring for the preparation of a nanocomposite of epoxy matrix and aluminium nanoparticles (Al NP) [24].

Pure aluminium is a metal that is soft, ductile, and corrosion-resistant [25]. Control over nano aluminium powder size has been crucial for changing properties. Slight changes in particle size can have a dramatic effect on surface area and therefore surface area-dependent properties such as rheology, powder mixing, dispersion, surface adsorption of condensed species, and bulk density [26]. Likewise, metal nanoparticles showed antibacterial behaviour against a wide series of bacteriological organisms [27]. Our interest is the development of a new anticorrosive and antibacterial nanocomposite coating to protect grey cast iron. The coating was fabricated with the addition of aluminium nanoparticles (Al NP) in a different ratio. In [24], the mechanical procedure for the preparation of the Al NP nanocomposite coating and its anticorrosive properties were demonstrated. In this paper, we investigated the influence of ultrasonic dispersion of Al NPs in an organic coating. Furthermore, in this study, we investigated the antibacterial properties of the metal powder of Al NPs. For all these purposes, we used electrochemical techniques (EIS and SECM), scanning electron microscopy (SEM), energy-dispersive X-ray spectroscopy (EDS) analyses, devices for establishing mechanical properties (Elcometer®456, PosiTector® SHD, Elcometer 510) and bacteria (*P. aeruginosa* and *B. subtilis*) for the determination of antibacterial properties.

2. Materials and Methods

2.1. Materials

In this study, grey cast iron was investigated. The elemental composition of grey cast iron is shown in Table 1.

Table 1. The chemical composition of the grey cast iron.

Element	C	Si	Mn	P	S	Fe
Percent (%)	2.5	1.5	1.05	0.5	0.07	balance

Grey cast iron plates with dimensions of 9.5 cm × 0.9 cm × 15 cm were used as substrates. The surface of the grey cast iron was cleaned with abrasive blasting and ethanol (70 wt.%). Two epoxy coatings were used in the work. The first coating was prepared using pure epoxy resin (Bisphenol A, West System, UK) and hardener (polyamine, West System, UK) in a ratio of 3.5 to 1. The second coating was prepared using epoxy paint (Hempel, Croatia) and hardener (polyamine, Hempel, Croatia) in a ratio of 4 to 1. Aluminium nanoparticles (Al NP) with an average particle size of 100 nm were provided by Guangzhou Hongwu Material Technology Co., Ltd., Guangzhou, China.

2.2. Preparation of the Epoxy Coatings

The obtained Al NPs were used to prepare nanocomposites of Al NP–epoxy coating. In this method, different concentrations of Al NPs (0.50, 0.75, and 1.0 wt.%) were added to the epoxy solution. The procedure for the epoxy coating modified by Al NP is illustrated in Figure 1.

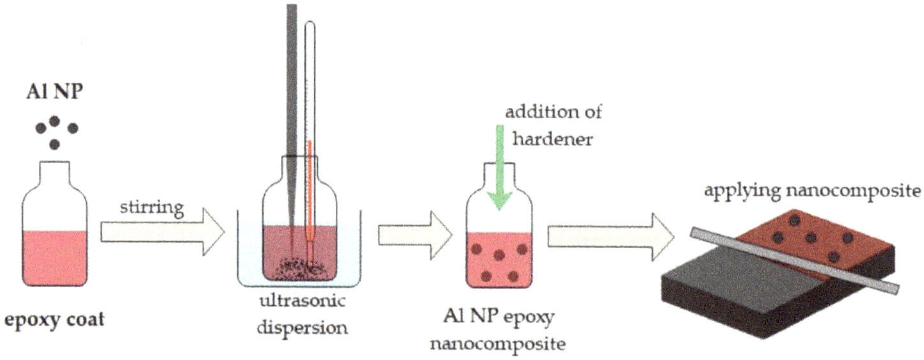

Figure 1. Preparation of the Al epoxy nanocomposite coating.

Al NPs were mixed with epoxy resin and epoxy paint (epoxy coating) using ultrasonic agitation for 20 min with a delay in the process due to the cooling of the nanocomposite. The hardener was added to the prepared samples and stirred until complete homogenisation. The obtained mixture was applied to the grey cast iron substrate using an applicator (150 μm). Then, the samples were dried under atmospheric conditions. After 24 h, another layer of nanocomposites was applied in the opposite direction (150 μm). The samples were left at room temperature (25 °C) for 7 days.

2.3. Coating Characterisation

The morphology of the Al NP samples was investigated by scanning electron microscope (SEM) (TESCAN Brno, Brno, Czech Republic). The size distribution and dispersion of nanoparticles, the homogeneity of the layer, and the occurrence of agglomeration were observed. The coated samples were also characterised using the energy-dispersive (EDS) detector.

Elcometer®456 (Elcometer Limited, Edge Lane, Manchester, UK) was used to assess the thickness of the nanocoating sample. Measurements were performed on ten different locations per sample. The change in the colour of the nanocomposite coating was determined using the RAL colour chart (RAL gGmbH, Siegburger, Germany). The hardness of the nanocomposite was tested according to ISO 868:2003 [28]. The testing was evaluated using PosiTector SHD Shore Hardness Durometer (DeFlesko Corporation, Ogdensburg, NY, USA). An adhesion test was performed to examine the effect of the coating strength of metal using an automatic Pull-Off Adhesion Tester (Elcometer 510, model T, Manchester, UK). The aluminium dollies (20 mm diameter) were glued on the Al NP nanocomposite using a two-part epoxy adhesive (Araldite resin and Araldite hardener). Complete curing of the adhesive was achieved by keeping the sample at 25 °C for 24 h.

The coating samples were exposed to the humidity chamber (Humidity Cabinet Model AB6), and the climatic chamber (Climatic chamber Kambic KK-190 CHLT, CiK Solutions GmbH, Karlsruhe, Germany). The humidity test was conducted according to EN ISO 6270-2 for five days [29]. To test the coating stability at low temperatures, the samples were placed in the climatic chamber for five days. The test cycle was set up in ten steps in which the temperatures changed from −5 °C, with 0% humidity, to 10 °C, with 70% humidity. The samples were left at each temperature for 8 h. The accelerated testing was performed on two samples per coating. After the humidity test, and the icing/deicing process, the physical properties of the samples were tested after they reached room temperature.

The corrosion protection performance of the epoxy coating and Al NP–epoxy nanocomposite were studied in 3.5 wt.% NaCl solution by EIS (VersaSTAT 3, AMETEK Scientific 131 Instruments, Princeton applied research, Berwyn, PA, USA). Open-circuit potential (OCP) was first obtained over a period of 20 min to study the changes in the corrosion potential of the coatings. The electrochemical cell consisted of a graphite rod as the auxiliary electrode, a saturated calomel electrode as the reference, and the epoxy coating and/or Al NP–epoxy nanocomposite specimen as the working electrode. The frequency range of 0.1 to 10^5 Hz with an amplitude of 10 mV was used. The impedance data were fitted using the ZSimpWin software. The Intermittent Contact-Scanning Electrochemical Microscopy (ic-ac-SECM) (M470, BioLogic, France) test of epoxy resin, epoxy paint, and epoxy nanocomposite with 1% of Al NP was used to observe real impedance distribution over a certain area. The 3DIsoPlot program was used to obtain the topography of the samples in 3D view. A three-electrode system, immersed in tap water, was above the surface of the nanocomposite. This three-electrode system consisted of the Ag/AgCl/KCl (saturated) reference electrode, the platinum sheet as the counter electrode, and of the UltraMicroElectrode (UME) probe for measuring the local electrochemical activity. The UME probe with 10 µm diameter platinum wire was used.

Pseudomonas aeruginosa and *Bacillus subtilis* were taken as the test bacteria. The antibacterial activity of the Al NP was performed using the well-diffusion method [30]. The final densities of the bacterial suspensions' cells were about 10^8 and 10^7 CFU/mL, respectively. The bacterial culture was inoculated from fresh colonies on agar plates into 20 mL Muller Hinton culture medium. The nanoparticle samples dissolved in distilled water (100 µg/mL) were added from the stock into each well. The zone of inhibition was measured using a ruler. After 24 h, the appearance of the inhibition zone was observed.

3. Results and Discussion

3.1. Characterisation of Al NP

The SEM microstructure of Al NPs is shown in Figure 2a. The surface of nanoparticles is a fluffy powder with spherical properties. To determine the composition of Al NP, the EDS analysis was carried out.

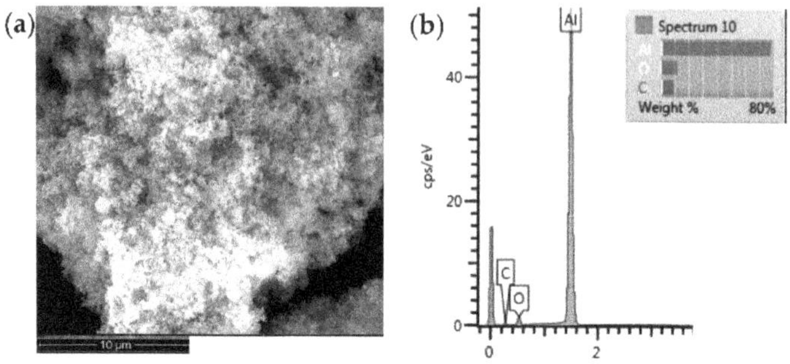

Figure 2. SEM image (**a**) and EDS profile (**b**) of pure Al NP.

The EDS analysis of Al NPs (Figure 2b) determined the mass fraction of the aluminium nanoparticles of about 80%. Spectrum minor elements such as O and C were due to contamination.

3.2. *Evaluation of Mechanical Properties*

Epoxy paint and epoxy paint nanocomposites with 0.50%, 0.75%, and 1.0% of Al NP were exposed to the humidity (ISO 6270-2) and climatic chambers. According to ISO 12944-6, nanocomposites were classified as corrosivity category C4 and durability up to 15 years [31]. After exposure of the samples to the corrosive media, mechanical properties were observed (Table 2).

Table 2. Results of changing thickness, hardness, adhesion force, and colours of epoxy paint and nanocomposite paint samples in different corrosion conditions.

	Samples	**Epoxy Paint**	**0.5%_Al**	**0.75%_Al**	**1.0%_Al**
thickness, µm	unexposed	249.9	278	256.4	275.5
	humidity chamber	275.9	263	263.2	285.6
	climatic chamber	248.8	259.9	271	292.3
Hardness, Shore D	unexposed	83.4	83.8	82.8	82.0
	humidity chamber	83.2	80.3	80.0	80.2
	climatic chamber	81.0	83.0	81.4	81.8
adhesion, MPa	unexposed	8.34 8.31	9.59 9.42	12.10 10.30	13.47 14.18
	humidity chamber	9.85 10.20	10.87 11.52	11.89 12.56	13.24 12.12
	climatic chamber	10.11 10.52	16.57 13.56	13.15 13.16	12.43 11.41
RAL colours	unexposed	3013	8015	8016	8017
	humidity chamber	3013	8015	8016	8017
	climatic chamber	3013	8015	8016	8017

According to Table 2, it is evident that there was no change in the thickness and hardness of the coating on the samples despite the addition of Al NPs and the exposure of the coating to corrosive conditions. The adhesion of the coating to the metal substrate showed a small increase with the addition of Al NPs. The results showed that the direct addition of Al NPs to the epoxy paint reduced the stress in the coating. The particle effect, the uniform dispersion of the filler in the epoxy paint, and the strength of the bond with the paint contribute to a better stress transfer. The strengthening mechanism was due to the tight bond between the nanoparticles and the epoxy paint, which can limit the mobility of the polymer chains [32]. Samples that were exposed to the humid and air-conditioned chambers showed a smaller increase in adhesion. An increase in the adhesion force could be caused by increasing or decreasing temperature and moisture in the humidity and climatic chambers. As shown in the paper, the addition of Al NPs to the epoxy matrix significantly changed the colour of the coating [24].

3.3. Evaluation of Anticorrosion Properties

The impedance data were analysed using the equivalent electrical circuits (EEC) shown in Figure 3. The model's outer and inner circles corresponded to the high and low-frequency loops, respectively. Indeed, the outer circle was used to characterise the coating properties. The inner circle of EEC with two-time constants, was the feature of the double layer formed at the coating/metal interface [33]. Figure 3 shows that the first resistance encountered by the alternating current was the electrolyte resistance, R_s, which had a negligibly small value. The coating resistance, R_{coat}, provided another resistance to the passage of current. At the phase boundary between solid and liquid, an electrical double layer appeared, representing the capacity of the coating, CPE_{coat}. After a long time of exposure of the coating to an aggressive medium, damage occurred to the epoxy paint and thus the electrolyte could more easily reach the surface of the grey cast iron the alternating current encountered a charge transfer resistance, R_{ct}, which also had its capacity, CPE_{dl} [34].

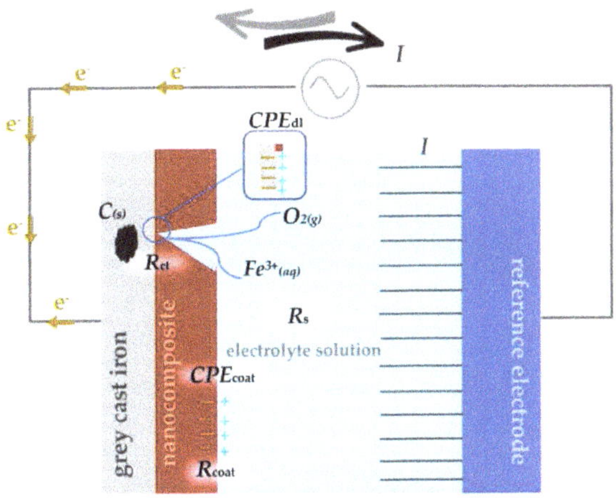

Figure 3. The equivalent electrical circuits used for describing the impedance response of epoxy coating and nanocomposite with 0.5, 0.75, and 1.0% Al NP.

Figure 3 shows the electrochemical mechanism of corrosion destruction on the surface of grey cast iron, which is called graphitization. When the aggressive medium reaches the surface of grey cast iron, it will initiate the anodic dissolution of iron by leaving free graphite [35]. To slow down this process, the epoxy resin and the epoxy paint were modified with Al NPs, and tests were performed in a 3.5 wt.% NaCl solution using alternating current.

The EIS results for the epoxy paint and nanocomposite samples with different concentrations of Al NPs (0.50, 0.75, 1.0%) were presented in Nyquist and Bode diagrams (Figure 4a,b). The measurement was made immediately after immersion in 3.5 wt.% NaCl solution. The results showed coating resistance, R_{coat}, which reflected the anti-penetrating ability of the coating to electrolyte solution [36]. All samples showed a single capacitive loop, meaning that there was no electrolyte penetration. Pure epoxy showed a lower coating resistance value, while the samples with 0.50 and 0.75% Al NPs had slightly higher resistances. The increase in resistance was significantly more pronounced in the sample containing 1.0% Al NPs. Nanoparticles embedded in the coating resisted the passage of current and provided high resistance. The large resistance values in the Nyquist plot were confirmed by the Bode phase angle in Figure 4b. The Bode phase angle plots were further analysed to explain the effectiveness of the coating. All tested samples achieved a constant phase angle value extending from the medium frequency range (100 Hz) to the high frequency range (10^5 Hz) during the entire exposure time, which indicated that the coating had not started to degrade. In the case of short-term immersion in the electrolyte, the properties of the protective coating remained intact.

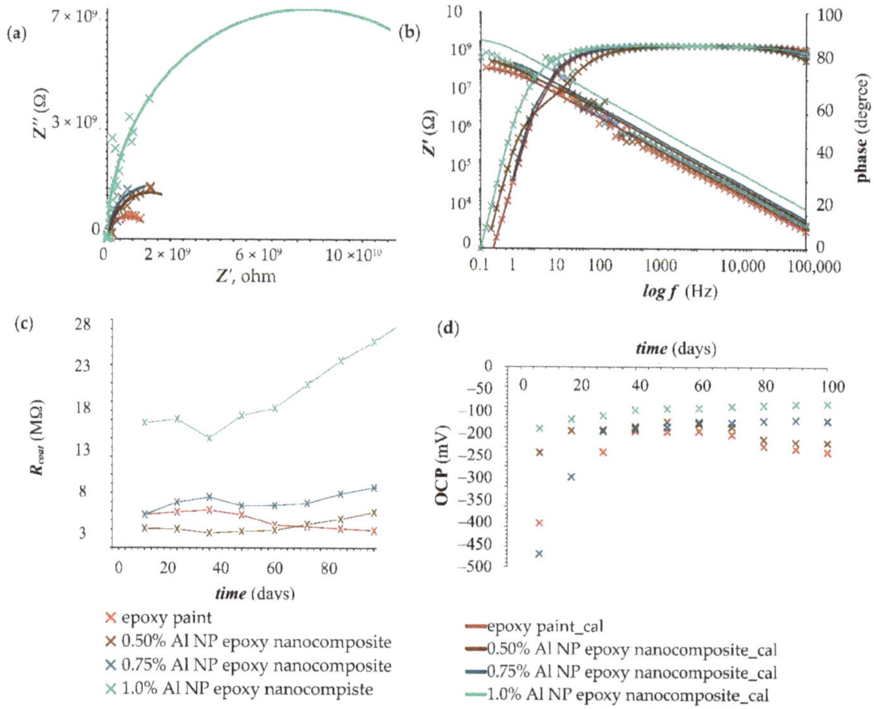

Figure 4. Nyquist and Bode plots of (**a**,**b**) epoxy paint and epoxy paint with different concentrations of Al NP immediately immersed in 3.5 wt.% NaCl solution. Corrosion resistance values (**c**) for samples in 3.5 wt.% NaCl solution during a 100-day immersion. Variations of OCP values (**d**) for samples during a 100-day immersion.

The value of R_{coat} for the blank epoxy coating (paint) without nanoparticles and for the coatings with different concentrations (0.50, 0.75, and 1.0%) of Al NP in the epoxy paint at different interval times over a 100-day immersion in 3.5 wt.% NaCl solution is shown in Figure 4c,d.

After 10 days, the epoxy paint showed the same resistance as the nanocomposite sample with 0.75% Al NPs. With a longer exposure time (after 50 days), the resistance of the blank epoxy gradually began to decrease (Figure 4c). Constant resistance values in all nanocomposite samples were maintained for 50 days, after which there was a slight increase in resistance in samples with 0.50% and 0.75% Al NPs, while the sample with 1.0% Al NPs showed a larger increase (Figure 4c). The obtained results indicated that it took 50 days to form an oxide film in the epoxy paint. The higher the concentration of nanoparticles, the greater the formation of the oxide film and the better the coating protection. Figure 4c shows that the highest resistance was achieved with the nanocomposite sample containing 1.0% Al NPs. This sample showed the highest resistance immediately at the beginning of the measurement. The increase in corrosion resistance could be attributed to a higher concentration of nanoparticles that created better adhesion and compatibility with the epoxy matrix, reduced the transport paths for the passage of corrosive electrolyte through the system coating, led to a decrease in the capacitance of the coating, and reduced corrosion reactions [16,30].

Open-circuit potential (OCP) is the potential of a working electrode when no current is applied to the cell [37]. In Figure 4d, all samples showed a constant OCP value after 50 days. After 100 days of exposure to an aggressive medium, the epoxy paint showed the highest negative potential value, which indicated weaker protective properties. Nanocomposites showed a more positive potential value that balanced out after 50 days indicating the beginning of the formation of a protective oxidative film. The values of the parameters shown on the equivalent circuit are listed in Table 3.

Table 3. EIS parameters values after fitting with equivalent circuit for epoxy paint and epoxy paint with different concentrations of Al NPs immediately immersed in 3.5 wt.% NaCl solution.

Parameters	Epoxy Paint	0.50% Al NP–Epoxy Nanocomposite	0.75% Al NP–Epoxy Nanocomposite	1.0% Al NP–Epoxy Nanocomposite
R_s, Ω	479.0	495.0	481.0	313.4
R_{coat}, Ω	$1.78{\cdot}10^8$	$1.80{\cdot}10^9$	$1.93{\cdot}10^9$	$1.87{\cdot}10^{10}$
CPE_{coat}, S·sec^n	$4.22{\cdot}10^{-9}$	$1.47{\cdot}10^{-10}$	$2.60{\cdot}10^{-10}$	$2.76{\cdot}10^{-11}$
n	1	1	1	1
R_{ct}, Ω	$1.17{\cdot}10^9$	$8.38{\cdot}10^8$	$1.30{\cdot}10^9$	$2.87{\cdot}10^9$
CPE_{dl}, S·sec^n	$3.65{\cdot}10^{-10}$	$2.05{\cdot}10^{-10}$	$3.01{\cdot}10^{-9}$	$7.22{\cdot}10^{-11}$
n	1	1	0.84	0.8

Considering the results reported in Table 3, CPE is a constant phase element of the double layer showing its capacitive properties, which depend on the empirical constant n. The CPE_{coat} value of the nanocomposite decreased by adding nanoparticles to the epoxy paint. The lower value of CPE_{coat} indicates superior corrosion resistance due to low electron storage.

To detect the presence of the aluminium oxide film, EDS microanalysis was performed at a point on the surface of the powder sample obtained by the reaction between Al NPs and distilled water (Figure 5).

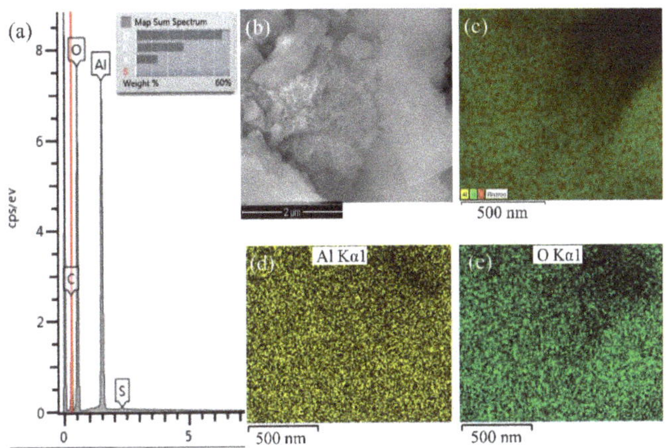

Figure 5. Results of (**a**) Surface spectrum, (**b**) SEM image, (**c**) collective colouring of surface maps of prepared aluminium oxide particles and partially coloured maps of the distribution of individual chemical elements in aluminium oxide particles: (**d**) Al K series, and (**e**) O K series.

Observing the size of the formed particles, an increase in the volume of the particles was visible in Figure 5a. The display of partial colours for aluminium and oxygen elements indicated that aluminium oxidation had occurred (Figure 5b,c). The EDS analysis cannot detect the presence of hydrogen in the compound, so according to the literature, the reactions that occur at the interface between Al NPs and a neutral aqueous medium are shown in Figure 6 [38].

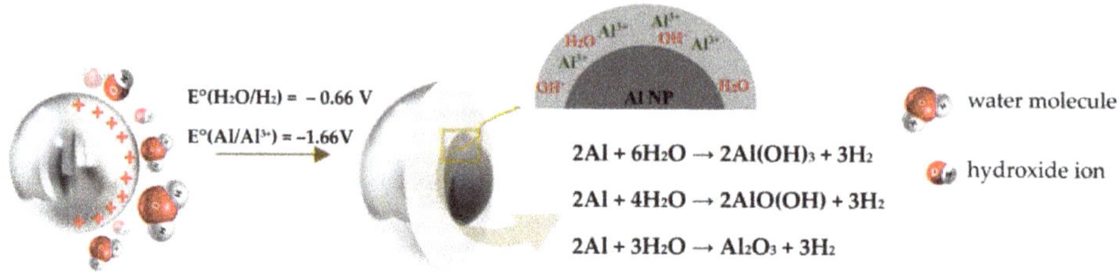

Figure 6. Mechanism of aluminium nanoparticle oxidation.

The oxidation process of a metal nanoparticle included two mechanisms, as shown in Figure 6. A nanoparticle in contact with an aqueous medium became electrically positively charged and, using van der Waals interactions, began to attract OH^- ions and the negative polar part of the H_2O molecule. After physical adsorption, due to the existence of a driving force (potential difference), Al NP oxidation and H_2O reduction occurred. The resulting oxidized products together with the metal surface created a chemisorbed monolayer [39,40]. The aqueous solution was a weak electrolyte, which means that it consisted more of water molecules than ions, and thus the formed aluminium ions could react with water molecules (H_2O) and form hydroxide ions (OH^-) [38]. According to the literature, the first possible reaction product is $Al(OH)_3$ (bayerite). The second possible reaction product is AlO(OH) (boehmite). The third possible reaction product is Al_2O_3 (alumina). These reactions are all thermodynamically favourable over a wide temperature range from room temperature to temperatures far above the melting point of aluminium (660 °C). In addition, all these reactions are highly exothermic [41]. The outer layer consisted of a mixture of Al_2O_3 and a hydrated layer, mostly in the form of amorphous $Al(OH)_3$, while the inner part was mostly made of Al_2O_3 and a small amount of aluminium oxyhydroxide in the form of AlO(OH). Such a coating (Al_2O_3-AlO(OH)) was characterised by continuity and resistance to certain electrolytes ($3.91 < pH < 8.64$) [39]. The layers may have different thicknesses and chemical compositions depending on the methods used for the production and passivation of the powders [42]. The oxide layer thickness remained stable for at least 30 days, showing that the native oxide acted as a passivation layer preventing further oxidation [43]. Consolidation of inorganic fillers into the polymeric coating formulation can diminish porosity and draw out the lifetime of the composite coating [44].

The appearance of a protective oxidative film on the surface of nanoparticles led to an increase in corrosion resistance. The mechanism of formation, the connection of Al NPs with epoxy paint, and the degradation of Al NPs in epoxy paint are shown in Figure 7.

According to literature data [26,45,46], by adding a hardener, the aliphatic amine started hardening by nucleophilic attack of the amine group on the epoxy ring of the monomer (Figure 7). At the same time, the nanoparticle could react with both the hardener and the epoxy paint due to its small size. When the epoxy ring was opened, the hydroxyl groups that serve as bonding sites became strong electromagnetic bonds between the epoxy and the metal molecules. At the beginning of exposure to an aggressive medium, the nanoparticles were well embedded in the epoxy paint, which protected them from rapid oxidation. To prevent the oxidation of Al NPs, layers of organic and inorganic coatings could be applied to their surface [42]. The protective barrier properties of the epoxy paint decreased over time. Organic polymeric coatings mostly protect against corrosion by forming a barrier to disengage the metal from the surrounding environment. Nonetheless, all polymeric coatings are porous to destructive species, for example, oxygen, water, and chloride particles [44]. The used aluminium nanoparticles in contact with the aqueous medium formed an aluminium oxide and smaller amounts of aluminium hydroxide and oxyhydroxide. Due to good dispersion, and interfacial compatibility of composites, a "maze effect" can be formed to prevent the penetration of corrosive media, prolong the penetration path of corrosive media in the coating, and ultimately delay the occurrence of corrosion [47]. This whole mechanism of protection of grey cast iron relied on a cathodic passivation process in which the aluminium in the coating acted as the sacrificial anode and protected the iron. According to the obtained EIS data (Figure 4c), an aluminium oxide film was formed after the sample had been immersed in an aggressive medium for 50 days. As a result, we can conclude that as the aggressive medium started to diffuse through the epoxy coating, more and more nanoparticles were completely and/or partially oxidized and the corrosion resistance increased (Figure 6).

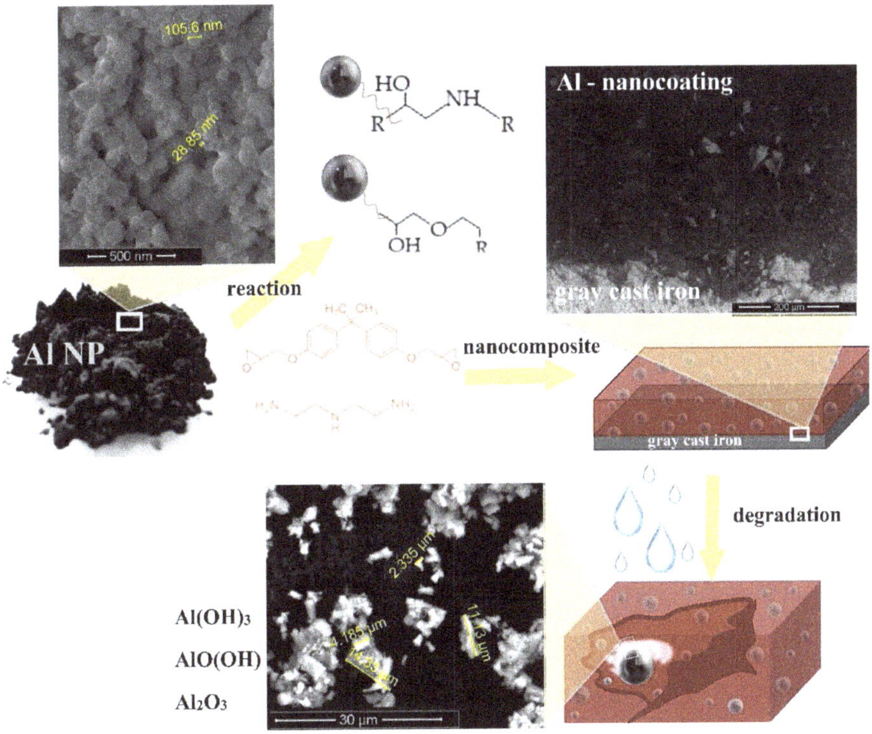

Figure 7. The mechanism of nanoparticle interaction with the epoxy paint and hardener, the degradation of the nanocomposite, and the formation of an oxidation film on the Al NP.

3.4. Surface Characterisation of 1.0% Al NP–Epoxy Nanocomposite

3.4.1. SEM/EDS Analysis

As 1.0% Al NP–epoxy nanocomposite was found to be the formulation with the best anticorrosion properties, the surface of the same sample was tested. The cross-sectional SEM image of the nanocomposite with 1.0% Al NPs prepared in epoxy paint is shown in Figure 8a,b.

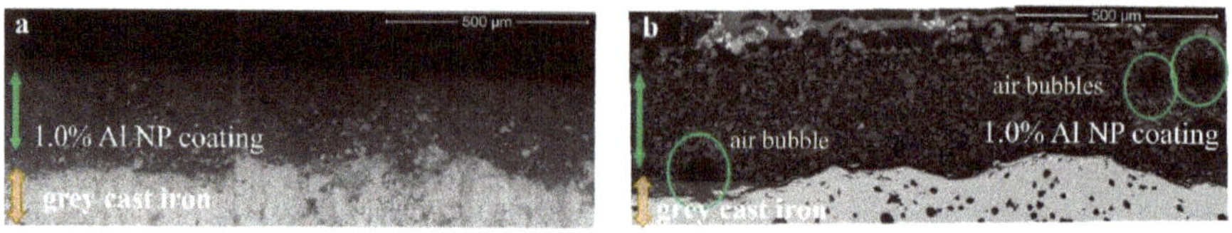

Figure 8. SEM cross-sectional morphology of nanocomposite with 1% Al NP prepared using (**a**) ultrasonic homogeniser and (**b**) mechanical mixing.

Figure 8 shows the result of two methods of preparing a nanocomposite containing 1.0% Al NPs. The preparation of the nanocomposite using ultrasonic mixing is shown in Figure 8a, while Figure 8b shows the nanocomposite prepared using a mechanical mixer. The results of mechanical mixing were published in the paper [24] and showed that the anticorrosive properties of the coating had improved in a short period of time. The reason for the short application of this nanocomposite was the appearance of air bubbles, as shown in Figure 8b. By applying ultrasonic mixing, better anticorrosive properties were achieved over a longer period. Figure 8a shows a cross-section of the nanocomposite, without air bubbles. Table 4 shows the advantages and disadvantages of both methods for the preparation of the nanocomposite material.

Table 4. Advantages and disadvantages of using mechanical and ultrasonic mixing in the preparation of nanocomposite coating.

	+	−
Mechanical stirring	Faster preparation of the sample It is not necessary to cool the Sample quiet mode of operation	Complicated handling of the device Complicated cleaning of the mechanical mixer The appearance of trapped air bubbles The appearance of nanoparticle agglomerates at higher concentrations
Ultrasonic stirring	Easy handling of the device Simple probe cleaning Easy air removal Better dispersion of nanoparticles	Heating the sample Longer sample preparation As the viscosity decreases, the device makes more noise

Better corrosion resistance and stability of the sample were obtained by ultrasonic preparation of nanocomposites due to better dispersion of nanoparticles and the absence of air bubbles. During sample preparation, air bubbles were released due to the heating of the coating. As the viscosity of the liquid decreased with the increase in temperature, air bubbles were released more easily. A risk of degradation of the base polymer, however, existed during processing [48]. To avoid this unwanted process, the epoxy paint must be heated to the temperature recommended by the manufacturer. Loss of solvent from the nanocomposite during preparation did not affect the reduction of anticorrosive properties.

Figure 9a shows the EDS spectra and chemical composition of the cross-section of the nanocomposite with 1% Al NP, which was prepared using an ultrasonic homogeniser in epoxy paint. SEM analysis of the sample is shown in Figure 9b. Figure 9d,e show the EDS distribution map of aluminium and oxygen in the cross-section of the nanocomposite with 1% Al NPs. The incorporation of Al NPs showed a uniform and homogeneous distribution in the entire measurement area of the nanocomposite cross-section. Additionally, according to the partial staining, we see that there was no formation of aluminium agglomerates. In the pure epoxy resin, as shown by the EDS analysis in Figure 10, the Al NPs were evenly distributed and remained at the nanoscale, which confirms that the nanoparticles did not start to form agglomerates.

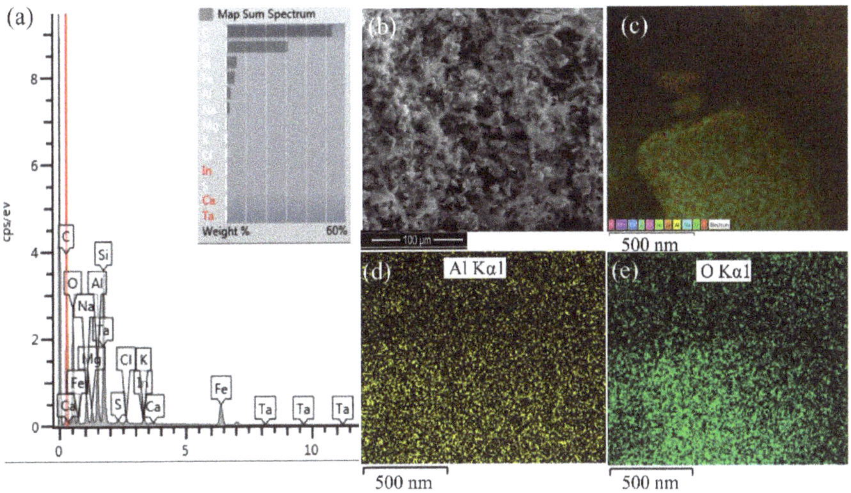

Figure 9. Results of (**a**) cross-sectional spectrum of the sample surface (**b**) SEM image of the sample surface (**c**) EDS cross-section map, and distribution of (**d**) aluminium and (**e**) oxygen in 1% Al NP–epoxy paint nanocomposite coating.

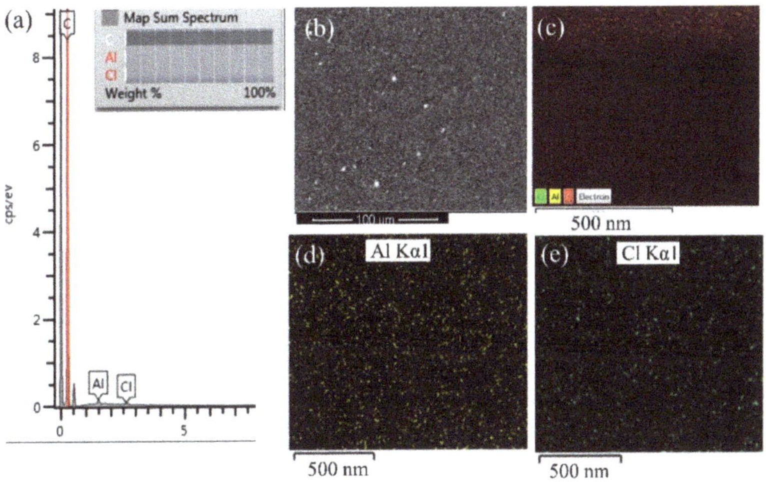

Figure 10. Results of (**a**) cross-sectional spectrum of the sample surface (**b**) SEM image of the sample surface (**c**) EDS cross-section map, and distribution of (**d**) aluminium and (**e**) chlorine in 1% Al NP–epoxy resin coating.

3.4.2. ic-ac-SECM Analysis

SECM provides electrochemical activity and topographic information about the surface reactions at the micrometre scale in aqueous environments [49]. We used this technique for the characterisation of the influence of the nanoparticles in the epoxy resin and epoxy paint. In the first experiment, it was necessary to find the appropriate vibration frequency so that the ultramicroelectrode (UME) could react with the surface (Figure 11). The piezo sensor was used to determine the distance of the UME from the surface of the sample. During the AC characterisation, the frequency applied to the piezo was automatically swept up to 600 Hz and the signal gain was recorded. The amplitude was 0.1.

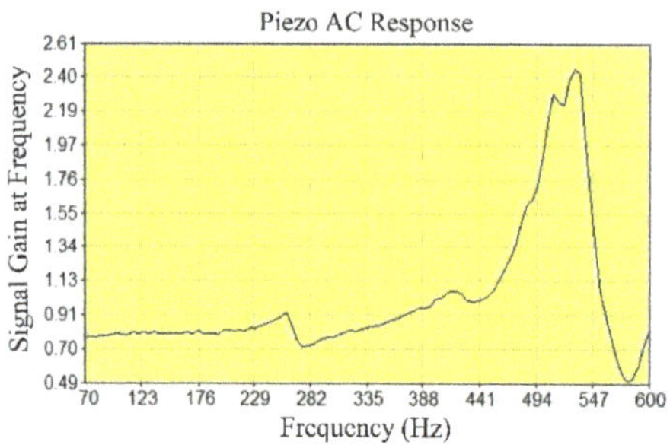

Figure 11. Determination of the vibration frequency of the UME.

The real impedance measured was the result of the surface activity and constant distance between the UME and the sample surface. The measurements were performed in tap water (Figure 11). The topography of the surface was determined on samples that did not contain nanoparticles and on samples obtained by ultrasonic mixing of epoxy resin and epoxy paint with 1.0% Al NPs.

The distribution of real impedance resistance values on the surface of pure epoxy resin, epoxy paint, 1.0% Al NP–epoxy resin, and 1.0% Al NP–epoxy paint nanocomposite coating is shown in Figure 12. The UME that recorded the topography moved along the x-axis from 0 to 0.5 mm, the y-axis from -0.25 μm to 0.25 μm, and the z-axis was set to the constant distance determined by the piezo sensor.

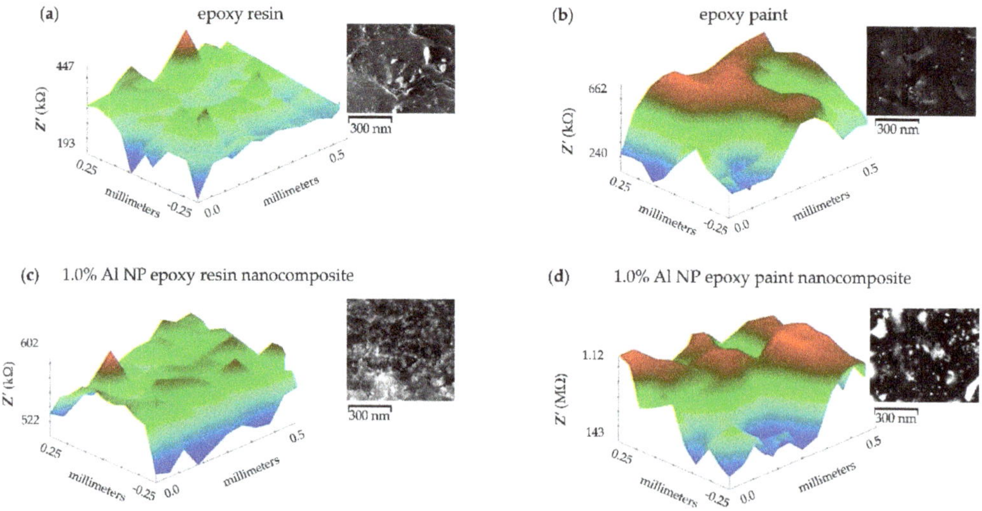

Figure 12. ic-ac-SECM distribution of real impedance per surface for samples: (**a**) epoxy resin, (**b**) epoxy paint, (**c**) 1.0% Al NP–epoxy resin nanocomposite, (**d**) 1.0% Al NP–epoxy paint nanocomposite coating immersed in tap water.

There is a small variation in the measured values of the impedance (from 522 to 602 kOhms) of the epoxy resin indicating a homogeneous structure (Figure 12a). A larger difference in real impedance distribution was observed in the epoxy paint sample (Figure 12b). The occurrence of localised peaks was surrounded by boundaries of higher impedance. The reason for the increase could be interpreted as the result of the effect of added coating components, such as additives in the epoxy paint. In the case of the sample containing 1.0% Al NPs in the epoxy resin, the impedance distribution was uniform, which resulted in a homogeneous structure (Figure 12c). The sample containing 1.0% Al NPs in the epoxy paint showed the greatest resistance but also a large distribution of resistance (Figure 12d). A deviation in the impedance resistance values appeared due to the densely distributed aluminium nanoparticles located right next to the surface and inside the coating itself. The aluminium particles, located right next to the surface, reacted with the UME and allowed the current to flow, thus increasing the value of the current. Particles that were further from the surface were not able to have a complete impact on the UME, but the epoxy resin came to the fore, and the current flow decreased.

The obtained values which are shown in Figure 12 for coating resistance were used to calculate the nanocomposite protection efficiency (CPE) using the following equation [24]:

$$CPE = \frac{Z'_{coatwithAlNP} - Z'_{coatwithoutAlNP}}{Z'_{coatwithAlNP}} \quad (1)$$

The calculated nanocomposite efficiency obtained by ic-ac-SECM measurement of epoxy, epoxy paint, 1.0% Al NP–epoxy resin nanocomposite, and 1.0% Al NP–epoxy paint nanocomposite coating immersed in tap water, is shown in Table 5.

Table 5. Calculated coating protection efficiency (CPE, %) for non-modified and modified epoxy coating immersed in tap water.

Samples	Epoxy Resin	1.0% Al NP–Epoxy Resin Nanocomposite	Epoxy Paint	1.0% Al NP–Epoxy Paint Nanocomposite
Z'_{max} (kΩ)	447	602	662	1120
CPE (%)	-	25.75	-	40.89

Nanoparticles in pure epoxy resin increase the protection efficiency by 25.75% and in the epoxy paint by 40.89%.

3.5. Evaluation of Antibacterial Properties

To develop a nanocoating with antibacterial properties, we investigated the antibacterial activity of Al NPs. Inorganic nanoparticles with antimicrobial activity are emerging as a new class of additives to coating materials to fulfil the increasing general demands for achieving dual properties of nanocomposites [21]. In the literature, there is almost no data on the toxicity of pure Al NPs, most of the literature studies aluminium oxide nanoparticles (alumina).

The antibacterial activity of Al NP nanoparticles against *P. aeruginosa* (Gram-negative) and *B. subtilis* (Gram-positive) was measured using the well-diffusion method (Figure 13a,b).

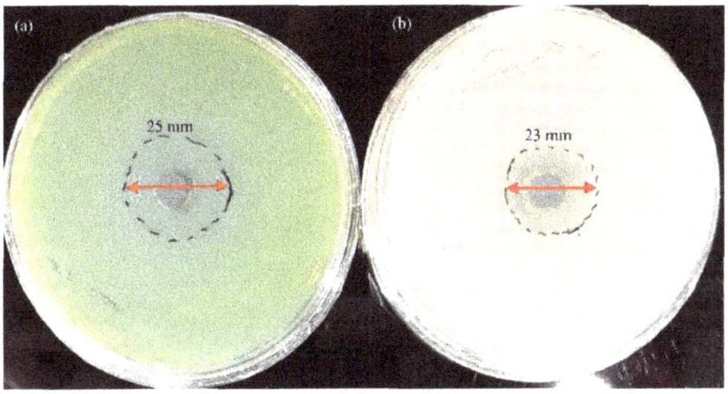

Figure 13. Antibacterial activity of pure Al NPs. Zone of inhibition of Al NP against (**a**) *P. aeruginosa* and (**b**) *B. subtilis* after 24 h.

The inhibition zone diameter indicated that Al NPs had a great antimicrobial effect (Figure 13a,b). Pure Al NPs produced zones of inhibition of 25 mm against *P. aeruginosa* and 23 mm against *B. subtilis*. In general, differences in the cell wall structure between Gram-positive and Gram-negative bacteria could affect the interaction between NPs and bacteria. Gram-positive bacteria had a thick outer cell wall formed by a thick peptidoglycan layer with hard polysaccharide chains linked by peptides. The thick outer cell wall could hinder NP penetration into the thick peptidoglycan layer [50]. It is known that, in an aerobic environment, pure aluminium reacts with oxygen and forms a thin oxide layer on its surface [51]. Factors that may be responsible for the antibacterial effect of aluminium oxide are (i) zeta potential and size of nanoparticles and (ii) formation of reactive oxidizing species (ROS) [52,53].

4. Conclusions

(1) The Al NP nanocomposite was successfully dispersed in the epoxy resin and epoxy paint, without agglomerates and the appearance of air bubbles, which was confirmed by SEM, EDS, and SECM analyses.

(2) Electrochemical tests confirm that the addition of Al NPs to the epoxy paints significantly increases the impedance values. The enhanced corrosion protection performance in the presence of spherical Al NPs was attributed to the cathodic passivation process. After a longer exposure time, oxides form on the Al NP surface, close the pores, and provide greater resistance to the aggressive medium. When the added content of Al NPs was 1.0 wt.%, the nanocomposite displayed the best mechanical and corrosion protection properties.

(3) The tests of antimicrobial properties against two typical bacteria of *P. aeruginosa* and *B. subtilis* indicate that the surface antibacterial layer of Al NPs possesses excellent antimicrobial properties.

(4) According to the obtained results, we believe that this Al NP–epoxy nanocomposite represents a multifunctional coating with excellent anticorrosive and antibacterial properties.

Author Contributions: Conceptualization, V.A. and M.S.; methodology, M.S. and M.V.D.; software, M.S. and M.K.; validation, M.S. and M.K.; formal analysis, M.S. and M.K.; investigation, M.S.; resources, V.A.; data curation, M.S.; writing—original draft preparation, M.S.; writing—review and editing, M.S.; visualization, M.S., M.K., M.V.D. and V.A.; supervision, V.A.; project administration, V.A. and M.K.; funding acquisition, V.A. All authors have read and agreed to the published version of the manuscript.

Funding: This research was funded by "Development of anticorrosion protection system for multipurpose pipe use", grant number KK.01.1.1.07.0045. This work was supported by the European Regional Development Fund under the Operational Program Competitiveness and Cohesion 2014–2020.

Institutional Review Board Statement: Not applicable.

Informed Consent Statement: Not applicable.

Data Availability Statement: Not applicable.

Conflicts of Interest: The authors declare no conflict of interest.

References

1. Melchers, R.E. Post-perforation external corrosion of cast iron pressurised water mains. *Corros. Eng. Sci. Technol.* **2017**, *52*, 541–546. [CrossRef]
2. Heikh, A.H.; Sarkar, A.; Singh, J.K.; Sohail, M.A.K.; Alharthi, N.; Ghosh, M. Corrosion Characteristics of Copper-Added Austempered Gray Cast Iron (AGCI). *Materials* **2019**, *12*, 503.
3. Umoren, S.A.; Solomon, M.M. Recent Developments on the Use of Polymers as Corrosion Inhibitors—A Review. *Open Mater. Sci. J.* **2014**, *8*, 39–54. [CrossRef]
4. Jain, P.; Patidar, B.; Bhawsar, J. Potential of Nanoparticles as a Corrosion Inhibitor: A Review. *J. Bio-Tribo-Corros.* **2020**, *6*, 1–12. [CrossRef]

5. Koch, G.; Varney, J.; Thompson, N.; Moghissi, O.; Gould, M.; Payer, J. International Measures of Prevention, Application, and Economics of Corrosion Technologies Study. *NACE Int.* **2016**, *2016*, 2–3.
6. Koch, G.H.; Brongers, M.P.H.; Thompson, N.G.; Virmani, Y.P.; Payer, J.H. Cost of Corrosion Study Unveiled. *NACE Int.* **2015**, 95–126.
7. Kania, H. Corrosion and Anticorrosion of Alloys/Metals: The Important Global Issue. *Coatings* **2023**, *13*, 216. [CrossRef]
8. Ferreira, E.S.; Giacomelli, C.; Giacomelli, F.C.; Spinelli, A. Evaluation of the inhibitor effect of L-ascorbic acid on the corrosion of mild steel. *Mater. Chem. Phys.* **2004**, *83*, 129–134. [CrossRef]
9. Ammar, A.U.; Shahid, M.; Ahmed, M.K.; Khan, M.; Khalid, A.; Khan, Z.A. Electrochemical study of polymer and ceramic-based nanocomposite coatings for corrosion protection of cast iron pipeline. *Materials* **2018**, *11*, 332. [CrossRef]
10. Foorginezhad, S.; Dargah, M.M.; Firoozirad, K.; Aryai, V.; Razmjou, A.; Abbassi, R.; Garaniya, V.; Beheshti, A.; Asadnia, M. Recent Advances in Sensing and Assessment of Corrosion in Sewage Pipelines. *Process. Saf. Environ. Prot.* **2020**, *147*, 192–213. [CrossRef]
11. Lv, M.; Du, M. A review: Microbiologically influenced corrosion and the effect of cathodic polarization on typical bacteria. *Rev. Environ. Sci. Biotechnol.* **2018**, *17*, 431–446. [CrossRef]
12. Slekovec, C.; Plantin, J.; Cholley, P.; Thouverez, M.; Talon, D.; Bartrand, X.; Hocquet, D. Tracking Down Antibiotic-Resistant Pseudomonas aeruginosa Isolates in a Wastewater Network. *PLoS ONE* **2012**, *7*, e49300. [CrossRef] [PubMed]
13. Perelomov, L.; Sizova, O.; Rahman, M.M.; Perelomova, I.; Minkina, T.; Sokolov, S.; Atroshchenko, Y. Metal-Tolerant Bacteria of Wastewater Tretment Plant in a Large City. *Sustainability* **2022**, *14*, 11335. [CrossRef]
14. Cyprowski, M.; Stobnicka-Kupiec, A.; Ławniczek-Wałczyk, A.; Bakal-Kijek, A.; Gołofit-Szymczak, M.; Górny, R.L. Anaerobic bacteria in wastewater treatment plant. *Int. Arch. Occup. Environ. Health* **2018**, *91*, 571–579. [CrossRef]
15. Roulová, N.; Mot'ková, P.; Brožková, I.; Pejchalová, M. Antibiotic resistance of Pseudomonas aeruginosa isolated from hospital wastewater in the Czech Republic. *J. Water Health* **2022**, *20*, 692–701. [CrossRef] [PubMed]
16. Zhao, X.; Yan, L.; Xu, X.; Zhao, H.; Lu, Y.; Wang, Y.; Jiang, C.; Shao, D.; Zhu, J.; Shi, J. Synthesis of silver nanoparticles and its contribution to the capability of Bacillus subtilis to deal with polluted waters. *Appl. Microbiol. Biotechnol.* **2019**, *103*, 6319–6332. [CrossRef]
17. Balakrishnan, A.; Jena, G.; Pongachira George, R.; Philip, J. Polydimethylsiloxane–graphene oxide nanocomposite coatings with improved anti-corrosion and anti-biofouling properties. *Environ. Sci. Pollut. Res.* **2021**, *28*, 7404–7422. [CrossRef] [PubMed]
18. Saravanan, P.; Jayamoorthy, K.; Ananda Kumar, S. Design and characterization of non-toxic nano-hybrid coatings for corrosion and fouling resistance. *J. Sci. Adv. Mater. Devices* **2016**, *1*, 367–378. [CrossRef]
19. Bertani, R.; Bartolozzi, A.; Pontefisso, A.; Quaresimin, M.; Zappalorto, M. Improving the antimicrobial and mechanical properties of epoxy resins via nanomodification: An overview. *Molecules* **2021**, *26*, 5426. [CrossRef]
20. Yuan, H.; Qi, F.; Zhao, N.; Wan, P.; Zhang, B.; Xiong, H.; Liao, B.; Ouyang, X. Graphene oxide decorated with titanium nanoparticles to reinforce the anti-corrosion performance of epoxy coating. *Coatings* **2020**, *10*, 129. [CrossRef]
21. Manjumeena, R.; Venkatesan, R.; Duraibabu, D.; Sudha, J.; Rajendran, N.; Kalaichelvan, P.T. Green Nanosilver as Reinforcing Eco-Friendly Additive to Epoxy Coating for Augmented Anticorrosive and Antimicrobial Behavior. *Silicon* **2016**, *8*, 277–298. [CrossRef]
22. Talabi, H.K.; Adewuyi, B.O.; Olaniran, O.; Ayekusibe, P.O. Reinforcement Efficiency of Copper Nanoparticles on Epoxy Matrix Bactericide Composite. *J. Chem. Technol. Metall.* **2021**, *56*, 321–326.
23. Jamkhande, P.G.; Ghule, N.W.; Bamer, A.H.; Kalaskar, M.G. Metal nanoparticles synthesis: An overview on methods of preparation, advantages and disadvantages, and applications. *J. Drug Deliv. Sci. Technol.* **2019**, *53*, 101174. [CrossRef]
24. Samardžija, M.; Alar, V.; Špada, V.; Stojanović, I. Corrosion Behaviour of an Epoxy Resin Reinforced with Aluminium Nanoparticles. *Coatings* **2022**, *12*, 1500. [CrossRef]
25. *Aalco Aluminium Alloy Specifications Datasheet*; Aalco Metals Ltd.: Wednesbury, UK, 2016; pp. 1–6.
26. Pesiri, D.; Aumann, C.; Bilger, L.; Booth, D.; Carpenter, R.D.; Dxe, R.; O'Neill, E.; Shelton, D.; Walter, K.C. Industrial scale nano-aluminum powder manufacturing. *J. Pyro.* **2007**, *714*, 1–16.
27. Kumar, A.M.; Khan, A.; Suleiman, R.; Qamar, M.; Saravanan, S.; Dafalla, H. Bifunctional CuO/TiO_2 nanocomposite as nanofiller for improved corrosion resistance and antibacterial protection. *Prog. Org. Coatings* **2017**, *114*, 9–18. [CrossRef]
28. *ISO 868*; Plastic and Ebonite—Determination of Indentation Hardness by Means of a Durometer (Shore Hardness). International Organization for Standardization: Geneva, Switzerland, 2003.
29. *ISO 6270-2*; Paints and Varnishes—Determination of Resistance to Humidity—Part 2: Condensation (In-Cabinet Exposure with Heated Water Reservoir). International Organization for Standardization: Geneva, Switzerland, 2017.
30. Chaudhary, R.G.; Tanna, J.A.; Gandhare, N.V.; Rai, A.R.; Juneja, H.D. Synthesis of nickel nanoparticles: Microscopic investigation, an efficient catalyst and effective antibacterial activity. *Adv. Mater. Lett.* **2015**, *6*, 990–998. [CrossRef]
31. *ISO 12944-6*; Paints and Varnishes-Corrosion Protection of Steel Structures by Protective Paint Systems—Part 6: Laboratory Performance Test Methods. International Organization for Standardization: Geneva, Switzerland, 2018.
32. Wei, H.; Xia, J.; Zhou, W.; Zhou, L.; Hussian, G.; Li, Q.; Ostrikov, K.K. Adhesion and cohesion of epoxy-based industrial composite coatings. *Compos. B Eng.* **2020**, *193*, 108035. [CrossRef]
33. Sherif, E.S.M.; Alam, M.A.; Al-Zahrani, S.M. Fabrication of Different Protective Coatings and Studying their Mechanical Properties and Corrosion Behavior in Sodium Chloride Solutions. *Int. J. Electrochem. Sci.* **2015**, *10*, 373–387.

34. Margarit-Mattos, I.C.P. EIS and organic coatings performance: Revisiting some key points. *Electrochim. Acta* **2020**, *354*, 136725. [CrossRef]
35. Chilkoor, G.; Sarder, R.; Islam, J.; Kumar, K.E.A.; Ratnayake, I.; Star, S.; Jasthi, B.K.; Sereda, G.; Koratkar, N.; Meyyappan, M.; et al. Maleic anhydride-functionalized graphene nanofillers render epoxy coatings highly resistant to corrosion and microbial attack. *Carbon N. Y.* **2020**, *159*, 586–597. [CrossRef]
36. Shankar, A.R.; Anandkumar, B.; Thinaharan, C.; George, R.P.; Rooby, J.; Philip, J.; Kamachi Mudali, U. Corrosion Evaluation of Buried Cast Iron Pipes Exposed to Fire Water System for 30 years. *Trans. Indian Inst. Met.* **2020**, *73*, 9–21. [CrossRef]
37. Liu, X.; Xiong, J.; Lv, Y.; Zuo, Y. Study on corrosion electrochemical behavior of several different coating systems by EIS. *Prog. Org. Coatings* **2009**, *64*, 497–503. [CrossRef]
38. Dave, P.N.; Chopda, L.V.; Sahu, L. Applications of Nanomaterials in Corrosion Protection Inhibitors and Coatings. *ACS Symp. Ser.* **2022**, *1418*, 189–212.
39. Trentin, A.; Pakseresht, A.; Duran, A.; Castro, Y.; Galusek, D. Electrochemical Characterization of Polymeric Coatings for Corrosion Protection: A Review of Advances and Perspectives. *Polymers* **2022**, *14*, 2306. [CrossRef]
40. Niroumandrad, S.; Rostami, M.; Ramezanzadeh, B. Effects of combined surface treatments of aluminium nanoparticle on its corrosion resistance before and after inclusion into an epoxy coating. *Prog. Org. Coatings* **2016**, *101*, 486–501. [CrossRef]
41. Kurtela, M.; Šimunović, V.; Stojanović, I.; Alar, V. Effect of the cerium (III) chloride heptahydrate on the corrosion inhibition of aluminum alloy. *Mater. Corros.* **2019**, *71*, 125–147. [CrossRef]
42. Chernavskii, P.A.; Peskov, N.V.; Mugtasimov, A.V.; Lunin, V.V. Oxidation of metal nanoparticles: Experiment and model. *Russ. J. Phys. Chem.* **2007**, *1*, 394–411. [CrossRef]
43. Petrovic, J.; Thomas, G. Reaction of Aluminum with Water to Produce Hydrogen, a Study of Issues Related to the Use of Aluminum for On-Board Vehicular Hydrogen Storage. *U.S. Dep. Energy* **2008**, *1*, 1–26.
44. Gromov, A.A.; Strokova, Y.I.; Teipel, U. Stabilization of metal nanoparticles—A chemical approach. *Chem. Eng. Technol.* **2009**, *32*, 1049–1060. [CrossRef]
45. Gerard, D.; Gray, S.K. Aluminium plasmonics. *J. Phys. D. Appl. Phys.* **2015**, *48*, 184001. [CrossRef]
46. Schmidt, R.G.; Bell, J.P. Epoxy Adhesion to Metals. *Adv. Polym.* **1986**, *75*, 33–71.
47. Feichtenschlager, B.; Pabisch, S.; Svehla, J.; Peterlik, H.; Sajjad, M.; Koch, T.; Kickelbick, G. Epoxy Resin Nanocomposites: The Influence of Interface Modification on the Dispersion Structure—A Small-Angle-X-ray-Scattering Study. *Surfaces* **2020**, *3*, 664–682. [CrossRef]
48. Taylor, P.; Halder, S.; Ghosh, P.K.; Goyat, M.S.; Ray, S. Ultrasonic dual mode mixing and its effect on tensile properties of SiO_2-epoxy nanocomposite. *J. Adhes. Sci. Technol.* **2013**, *27*, 37–41.
49. Xavier, J.R.; Nallaiyan, R. Application of EIS and SECM Studies for Investigation of Anticorrosion Properties of Epoxy Coatings containing ZrO_2 Nanoparticles on Mild Steel in 3.5% NaCl Solution. *J. Fail. Anal. Prev.* **2016**, *16*, 1082–1091. [CrossRef]
50. Slavin, Y.N.; Asnis, J.; Häfeli, U.O.; Bach, H. Metal nanoparticles: Understanding the mechanisms behind antibacterial activity. *J. Nanobiotechnol.* **2017**, *15*, 1–20. [CrossRef]
51. Karlsson, P.; Palmqvist, A.E.C.; Holmberg, K. Surface modification for aluminium pigment inhibition. *Adv. Colloid Interface Sci.* **2006**, *128–130*, 121–134. [CrossRef]
52. Prashanth, P.A.; Raveendra, R.S.; Krishan, R.H.; Ananda, S.; Bhagya, N.P.; Nagabhushana, B.M. Synthesis, characterizations, antibacterial and photoluminescence studies of solution combustion-derived α-Al_2O_3 nanoparticles. *J. Asian Ceram. Soc.* **2015**, *3*, 345–351. [CrossRef]
53. Gudkov, S.V.; Burmistrov, D.E.; Smirnova, V.V.; Semenova, A.A.; Lisitsyn, A.B. A Mini Review of Antibacterial Properties of Al_2O_3 Nanoparticles. *Nanomaterials* **2022**, *12*, 2635. [CrossRef]

coatings

Article

Epoxy Coating Modification with Metal Nanoparticles to Improve the Anticorrosion, Migration, and Antibacterial Properties

Marina Samardžija [1,*], Ivan Stojanović [2], Marija Vuković Domanovac [3] and Vesna Alar [2]

1 Department of Chemistry, Faculty of Mining-Geology-Petroleum Engineering, University of Zagreb, 10110 Zagreb, Croatia
2 Department of Welded Structures, Faculty of Mechanical Engineering and Naval Architecture, University of Zagreb, 10000 Zagreb, Croatia; ivan.stojanovic@fsb.unizg.hr (I.S.); vesna.alar@fsb.unizg.hr (V.A.)
3 Department of Industrial Ecology, Faculty of Chemical Engineering and Technology, University of Zagreb, 10000 Zagreb, Croatia; mvukovic@fkit.unizg.hr
* Correspondence: marina.samardzija@rgn.unizg.hr; Tel.: +385-1-5535-912

Abstract: Nanoparticles are capable of making more durable and stronger materials with better chemical resistance. They are used for a wide range of applications. Likewise, the potential of metal nanoparticles as antimicrobial agents has been widely studied. In this work, we investigate various nanoparticles (Al, Ni, Ag) incorporated into epoxy coating. The anticorrosion and antibacterial properties of the unmodified and modified coatings were evaluated. According to the SEM and EDS analyses, the coating did not contain agglomerates, which confirms the quality of the dispersion of inorganic nanoparticles in the coating. After 24 h and 10days immersions in a 3.5 wt.% NaCl solution, the corrosion behaviour for all nanocomposite was studied by means of EIS investigations. The study included the evaluation of the inhibition zone of the nanoparticles and the antimicrobial properties of the nanocomposite. It was found that the nanoparticles of Al and Ag provide excellent antibacterial properties. The epoxy nanocomposite with Al NP showed the migration of ions in the range from 0.75 to 1 mg/L in a wastewater solution for 30 days, indicating a potential for antimicrobe activity. The 1% Al NP epoxy nanocomposite showed good anticorrosion and antibacterial properties and demonstrated great potential for applications in pipelines.

Keywords: corrosion protection; nanoparticles; epoxy coating; antibacterial activity

1. Introduction

The 21st century is marked by the study of nanotechnology and the production of nanostructured materials. Nanoparticles (NPs) are a wide class of materials that includes particulate substances with sizes between 1 and 100 nm [1].The use of nanoparticles in different fields such as molecular biology, physics, organic and inorganic chemistry, medicine, and material science is of growing interest in future applications [2]. In recent years, scientists and researchers have been motivated to develop coatings with new features, such as the possibility of the migration of substances due to the improvement of anticorrosive and antibacterial properties. As a result, nanoparticles began to be incorporated into the polymer material. The material obtained was defined as a nanocomposite, which implies a composition of at least two immiscible phases, and one of them in the nanometer scale [3]. The ability of nanocomposite material to release nanoparticles is considered very harmful if applied in the food packaging industry [4], but its application on the surface of drainage pipes is a new idea for protection against microorganisms. Cast iron pipes have been widely used in water distribution systems formore than 150 years due to their high mechanical strength and cost effectiveness [5]. With the development of urbanization and industrialization, the amount of wastewater produced and discharged increased significantly year by year. Consequently, problems such as corrosion, damage, a reduction in the water transport capacity, and an increase in the habitat of pathogenic and opportunistic bacteria are becoming serious [6,7]. To achieve corrosion protection, organic coatings and corrosion inhibitors are widely used, but the limitations of these materials, such as high cost, heavy contamination, and operational difficulties, has not been completely overcome [8]. Polymer nanocomposite coatings are a new generation of coatings which have recently been used for the protection of drainage pipes from corrosion due to their superior mechanical strength, stiffness, improved barrier properties against oxygen and moisture, increased heat, wear, and use in the simple repair of damaged structures [9,10]. Similarly, a new class of "nanometallo-antibiotics" consisting of numerous metal NPs has appeared, whichinvolves investigating their antimicrobial properties [11]. It is clear that some metal nanoparticles are effective antimicrobial agents against several pathogenic microorganisms, and their action depends on size, shape, and surface charge [12,13]. Mejía and co-authors [14] developed a new thin antibacterial coating with long-term effectiveness based on silver release that has an antibacterial effect. According to the research of Tahir and co-authors, the incorporation of silver nanoparticles (Ag NP) into the epoxy resin reduces the occurrence of bubbling and delamination of the coating and improves the durability of the coating [15]. Epoxy antimicrobial coatings are of great interest for the protection of surfaces, since the survival of microorganisms on the surface environment can be detrimental to materials [16]. Microbially induced corrosion (MIC) can be defined as the process by which biological agents (live organisms) cause changes in the material properties, leading to the structural lowering in quality or value [17]. Considering the proven antibacterial efficacy of metal and metal oxide nanoparticles in an organic coating with clear known mechanisms of action against bacteria for silver, silver oxide, titanium dioxide, iron oxide, and zinc oxide, potential materials with the same effect as nickel and aluminum remain poorly researched [16,18,19].

Nickel is a metal relatively resistant to atmospheric corrosion, but it also has huge potential for the development of antibacterial properties [20,21]. The bactericidal action of nickel nanoparticles (Ni NP) has shown a strong effect with particle sizes in the range of 10–100 nm [21]. Aluminum nanoparticles (Al NP) also have high corrosion resistance, but due to the ability to create a passive oxidation layer their antibacterial potential increases [4]. It can be concluded that certain metal powder nanoparticles have a cytotoxic and genotoxic effect on bacteria, and their intercalation in the epoxy coating creates the potential for antibacterial protection. In this study, the effect of the Al NP, Ni NP, and Ag NP in an epoxy coating on the anticorrosion, migration, and the antibacterial property was investigated. In our current work, the composition and distribution of particles for these coatings were characterized using SEM and EDS investigations. The EIS technique was used to characterize the anticorrosion properties of the modified epoxy coating and epoxy nanocomposite. The migration of nanoparticles from the epoxy coating into the wastewater was observed for 30 d, and the possibility of antibacterial action of nanocomposite was tested according to ISO 22196.

2. Materials and Methods

2.1. Materials

The metallic substrate used during this study was gray cast iron with chemical composition 2.5 C, 1.5 Si, 1.05 Mn, 0.5 P, 0.07 S, and Fe in balance (wt.%). The epoxy coatings that we used in this work were based on diglycidyl ether of biophenol A, and the polyamine hardener was obtained from Hempel (Umag, Croatia). Aluminum, nickel, and silver nanoparticles with an average particle size of about 100 nm were provided by Guangzhou Hongwu Material Technology Co., Ltd., Guangzhou, China. The morphology of the powder nanoparticles was observed from the SEM micrographs, EDS, and XRF (Figure 1).

According to Figure 1a–c, the Al and Ni NPs showed spherical particles and the Ag NPs within irregular shapes. All nanoparticles showed different sizes. The small nanoparticles had a very large surface area to volume ratio, which gave them huge energy and high reactivity. The agglomerates with average particle sizes from 223.4 to 620.7 nm were observed in Figure 1a,b. The XRF analysis of the elemental composition of the nanoparticle powder showed a high proportion of Al, Ag, and Ni nanoparticles (Table 1).

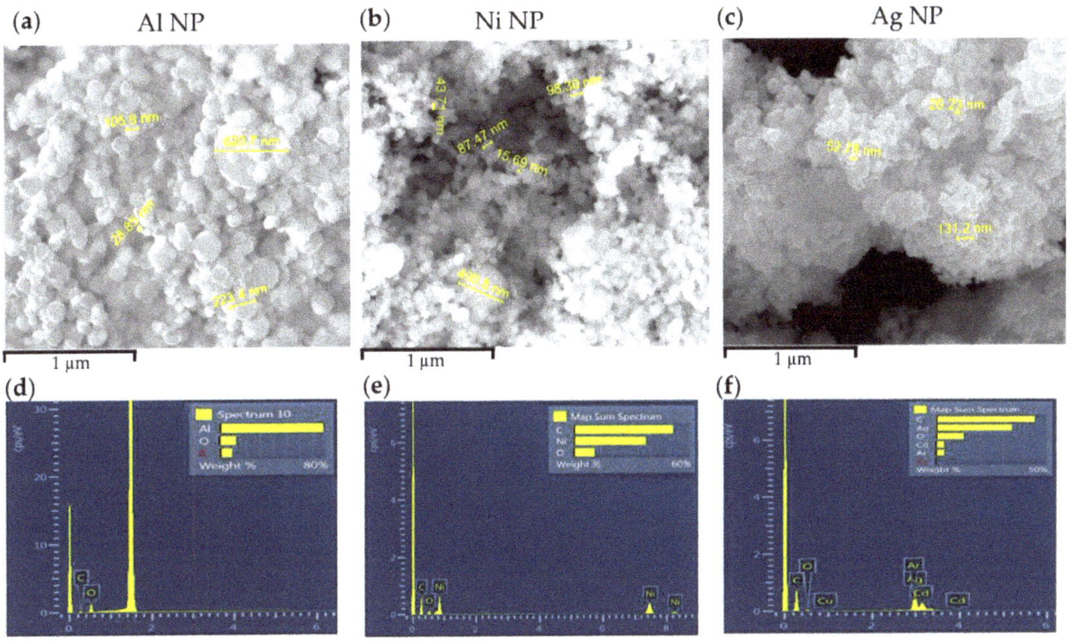

Figure 1. SEM image and EDS analysis of the (**a**,**d**) Al, (**b**,**e**) Ni, and (**c**,**f**) Ag nanoparticles.

Table 1. The XRF analysis of the elemental composition of the nanoparticle powder.

Al NP			**Ni NP**			**Ag NP**		
Elements	**wt.%**	**+/−**	**Elements**	**wt.%**	**+/−**	**Elements**	**wt.%**	**+/−**
Al	98.85	0.24	Ni	99.21	0.25	Ag	98.94	0.32
Cr	0.08	0.01	Cu	0.28	0.05	Cr	0.41	0.10
Fe	0.03	0.43	Si	0.26	0.05	Co	0.33	0.04
Residue	1.04	0.01	Co	0.11	0.02	Si	0.33	0.02
Total	100	-	Residue	0.14	0.01	Total	100	-
			Total	100				

2.2. *Preparation of the Epoxy Coating/Nanocomposite*

The epoxy coating was formulated by taking epoxy paint and polyamine hardener in the mass ratio of 4:1. Four formulations of prepared coatings are shown in Table 2.

At ambient temperature, various nanoparticles were incorporated into the epoxy coating under sonicate agitation for 20 min with a delay in the process due to the cooling of the nanocomposite. Polyamine hardener was added to the nanocomposite and mechanically stirred using a glass stirrer.

Before applying the coating, the panels of 9.5 cm × 0.9 cm × 15 cm gray cast iron were abrasively blasted and cleaned with ethanol (70 wt.%). The nanocomposite mixture was applied using a film applicator with a wet film thickness of 150 µm. The samples were kept at room temperature for 24 h, and then another layer of nanocomposites was applied in the opposite direction. The samples were left at room temperature (25 °C) for 7 days.

Table 2. The chemical composition of the tested samples.

Sample	Epoxy Coating (g)	Hardener (g)	m (Al NP) (g)	m (Ni NP) (g)	m (Ag NP) (g)	Thickness (μm)
Epoxy Coating	30	7.5	-	-	-	249.1
1% Al NP Nanocomposite	30	7.5	0.4545	-	-	256.1
1% Ni NP Nanocomposite	30	7.5	-	0.4545	-	266.5
1% Ag NP Nanocomposite	30	7.5	-	-	0.4545	264.9

2.3. Characterization of the Epoxy Coating/Nanocomposite

To evaluate the size of the nanoparticles, the quality of dispersion, and the appearance of the agglomerate of nanoparticles in the nanocomposite, scanning electron microscopy (SEM) (TESCAN Brno, Brno, Czech Republic) and energy-dispersive X-ray spectroscopy (EDS) analyses were used.

To investigate the anticorrosion behavior of Al, Ni, and Ag nanocomposites, electrochemical impedance spectroscopy (EIS) (VersaSTAT 3, AMETEK Scientific 131 Instruments, Princeton applied research, Berwyn, PA, USA) was used. These measurements were performed in a three-electrode electrochemical cell in a 3.5 wt.% NaCl solution. The coated gray cast iron was used as the working electrode, with a test area of 19.75 cm^2, a saturated calomel electrode was used as a reference electrode, and the graphite electrode was used as an auxiliary electrode. The frequency scan range of 0.1 to 10^5 Hz with an amplitude of 10 mV was used. The impedance data were fitted using the ZSimpWin software(AMETEK, Berwyn, PA, USA) (Version 3.2).

The samples of epoxy coating, 1% Al NP epoxy nanocomposite,1% Ni NP epoxy nanocomposite, and 1% Ag NP epoxy nanocomposite were placed in contact with simulation wastewater according to the DIN EN 877 [22]. For the assessment of Al^{3+}, Ni^{2+}, and Ag^{+} ions' migration from the epoxy coating, 10 g of each sample was immersed in 100 mL simulated wastewater for 30 days. In this study, samples were kept in the dark at a temperature of 40 °C. The temperature for this study was chosen according to the real conditions in drainage pipes. The mass concentration, electrical conductivity, and pH value of the wastewater were determined after 10, 20, and 30 days.The mass concentration of Al^{3+}, Ni^{+}, and Ag^{+} ions in wastewater was determined with the flame atomic absorption spectrometry (F-AAS) technique (Perkin Elmer Analyst 700, Waltham, MA, USA).The conductivity and pH values were measured with a pH/EC meter (HI5521–02, HANNA instruments, Woonsocket, RI, USA).

The antibacterial properties of the Al, Ni, and Ag were investigated with a well-diffusion method [23]. For this purpose, *P. aeruginosa* was used as a Gram-negative bacterial strain, and *B. subtills* as a Gram-positive bacterial strain. The 24 h aged active bacterial cultures were poured into the Muller Hinton culture medium. The nanoparticle samples dissolved in distilled water (100 g/mL) were added from the stock into each well. The zone of inhibition was measured using a ruler. After 24 h, the appearance of the inhibition zone was observed.

The antibacterial activities of nanocomposites were tested according to ISO 22196:2011 against *P. aeruginosa* and *B. subtilis* [24]. The surface of the epoxy coating, 1% Al epoxy nanocomposite, 1% Ni epoxy nanocomposite, and 1% Ag epoxy nanocomposite were cut to size 50 mm × 50 mm and inoculated with 0.4 mL of 10^5 CFU bacterial suspensions. All samples were covered with polyethylene foil dimensions of 40 mm × 40 mm (± 1 mm) and a thickness of 0.06 mm. All samples were placed in a Petri dish and incubated at 35 °C (±1 °C) and 90% relative humidity (RH) (±5%) for 24 h (±1 h). The number of viable bacterial cells was determined by plating different dilutions on plate count agar, incubating the plates for 24 h at 37 °C, and then visually counting the colonies [25]. All experiments are representative of repeated trials. Sample error bars on plots represent ± SD. The measurement of the inhibition zone was carried out in four sections (rotation by 45°) using a ruler with a length of 15 cm. The measurements were repeated three times.

3. Results and Discussion

3.1. SEM and EDS Analysis of Epoxy Coating/Nanocomposite

The morphology of epoxy coating and 1 wt.% of Al, Ni, and Ag nanocomposites were investigated using SEM analysis. Figure 2a shows the SEM images of epoxy coating without nanoparticles, which indicates a rough surface morphology with a large addition of other coating components such as fillers, additives, pigments and similar. The EDS analysis confirms the presence of aluminum particles in the initial sample (Figure 2b,c). It was assumed that these microparticles of aluminum were used as a pigment in the epoxy coating. The presence of these particles in the initial epoxy coating showed small irregularities and the appearance of agglomerates (Figure 2b). The EDS analysis of the initial epoxy coating (Figure 2c) determined the percentage of the aluminum microparticles, and this was 0.85%.

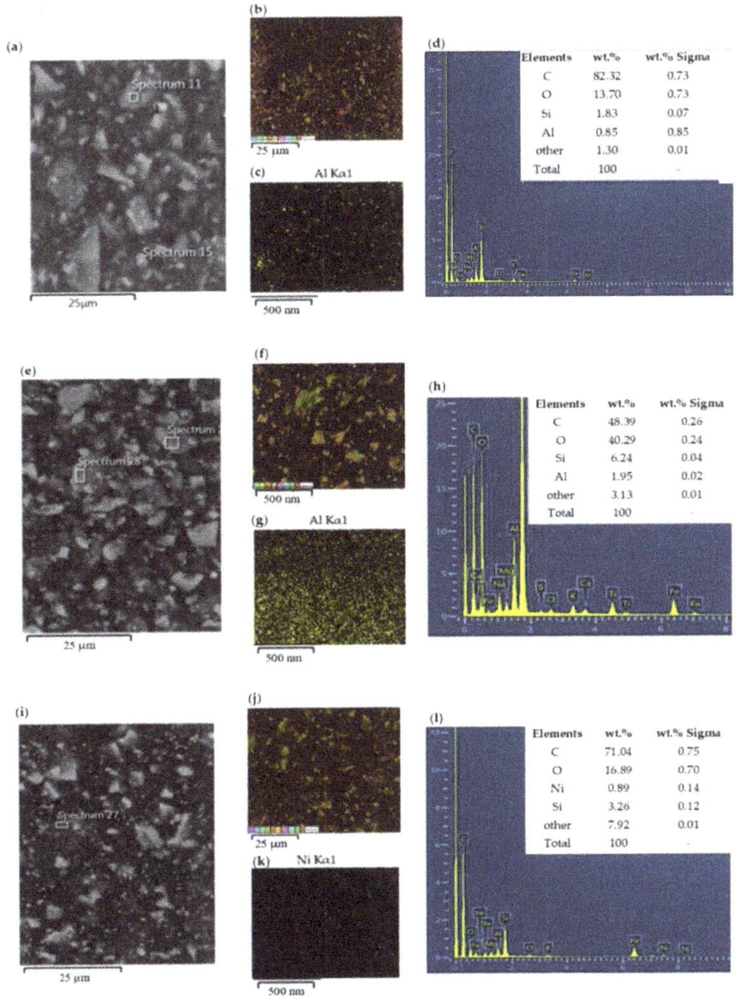

Figure 2. *Cont.*

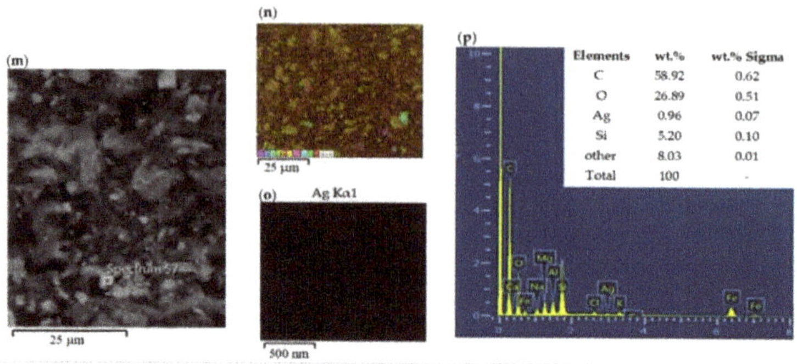

Figure 2. The SEM and EDS analyses of (**a**,**b**) epoxy coating, (**e**,**f**) 1% Al NP epoxy nanocomposite, (**i**,**j**) 1% Ni NP epoxy nanocomposite, and (**m**,**n**) 1% Ag NP epoxy nanocomposite. The EDS mapping and element weight percentages of distribution of microparticles of (**c**,**d**) Al in epoxy coating, and nanoparticles of (**g**,**h**) Al, (**k**,**l**) Ni, and (**o**,**p**) Ag in the nanocomposite.

With the addition of nanoparticles, the rough surface morphology of nanocomposites remained the same (Figure 2e,i,m).The action of the sonication process improved the distribution of the added aluminum particles (Figure 2g). The EDS analysis confirmed the uniform distribution of Ni and Ag nanoparticles within the epoxy coating, shown in Figure 2k,o. Furthermore, EDS analysis determined that all nanocomposite samples contained a certain percentage of nanoparticles that deviated very slightly from the added 1% nanoparticles (Figure 2d,h,l,p).

3.2. EIS Analysis of Epoxy Coating/Nanocomposite

The EIS measurements were performed to investigate the effects of the addition of Al, Ni, and Ag nanoparticles on the corrosion protection properties of the epoxy coating in a 3.5 wt.% NaCl solution with different immersion times. According to the electrochemical behavior of the various prepared nanocomposites, the model of the equivalent circuit with three resistance elements was chosen (Figure 3). This model demonstrated electrolyte resistance, R_e, coating capacitance, CPE_{coat}, coating resistance, R_{coat}, charge transfer resistance, R_{ct}, and double layer capacitance, CPE_{dl} [26]. The behavior of the CPE was shown by the parameter phase shift, n ($-1 \leq n \leq 1$); when $n = 0$, the CPE represented a pure resistor, if $n = -1$, the CPE stood for an inductor, and if $n = +1$, the CPE represented a pure capacitor [27].

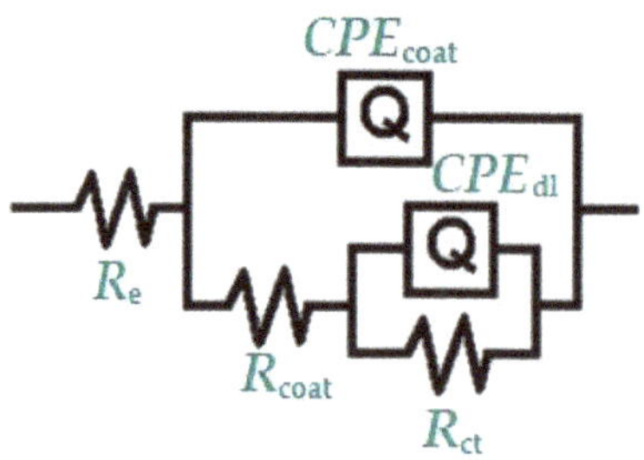

Figure 3. The equivalent circuits used for fitting the Nyquist and Bode plots for coating.

The Nyquist and Bode plots for the epoxy coating, 1% Al epoxy nanocomposite, 1% Ni epoxy nanocomposite, and 1% Ag epoxy nanocomposite after 24 h and 10 days of immersion in 3.5 wt.% NaCl solution are illustrated in Figure 4a–d. The calculated values of the EIS parameters were reported in Table 3. The coating protection efficiency (%*CPE*) also listed in the table was calculated using the equation [28]:

$$\%CPE = \frac{R_{\text{coat with NP}} - R_{\text{coat without NP}}}{R_{\text{coat with NP}}} \quad (1)$$

where $R_{\text{coat with NP}}$ is the resistance of Al, Ni, or Ag nanocomposites, and $R_{\text{coat without NP}}$ is the resistance of the epoxy coating.

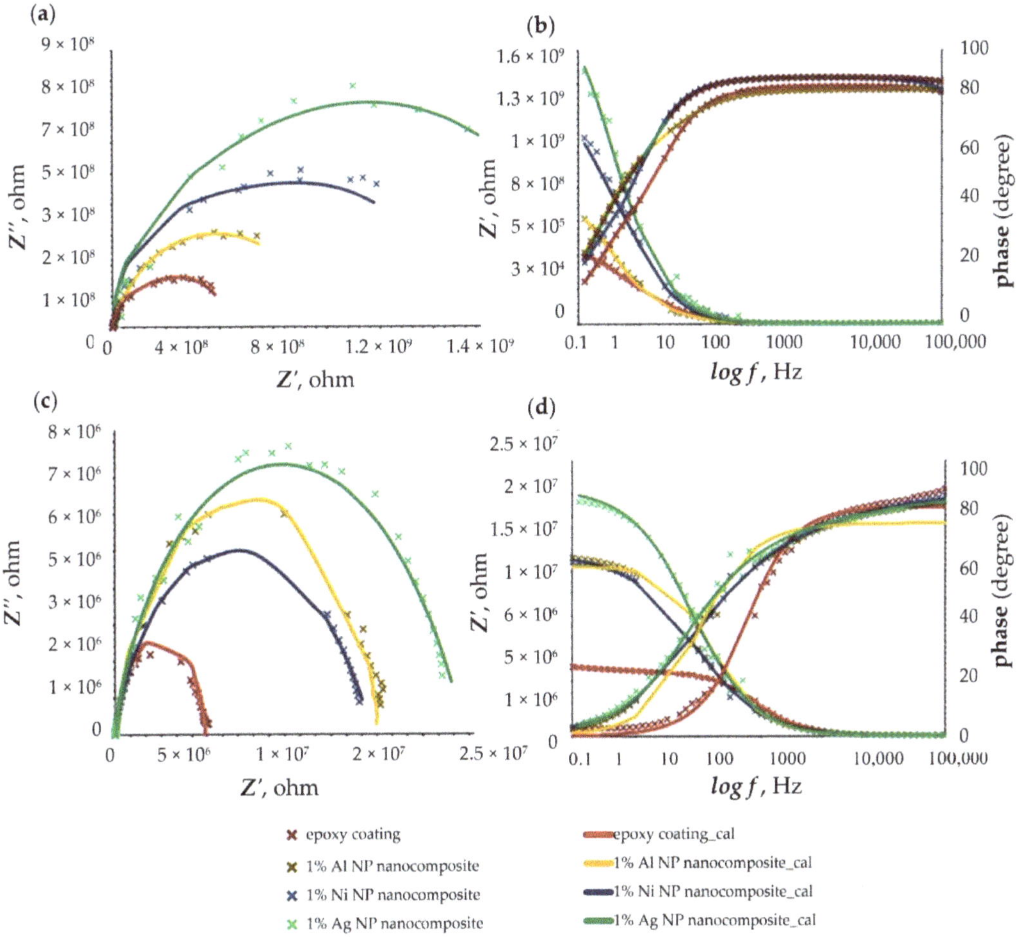

Figure 4. The (**a**,**c**) Nyquist and (**b**,**d**) Bode plots for epoxy coating, 1% Al epoxy nanocomposite, 1% Ni epoxy nanocomposite, and 1% Ag epoxy nanocomposite after (**a**,**b**) 24 h and (**c**,**d**) 10 days of exposure to 3.5 wt.% NaCl solution.

After immersion, as shown in Figure 4, the electrochemical response of the prepared nanocomposites showed one-time constants. The Nyquist diagrams for epoxy coating had the lowest value of R_{coat}, implying that the presence of nanoparticles prolonged the use of the gray cast iron. The nanocomposite with 1% of Ag NP showed the best corrosion resistance, followed by the 1% Ni NP nanocomposite and 1% Al NP nanocomposite (Figure 4a). All samples achieved a constant phase angle value extending from the medium frequency range (100 Hz) to the high frequency range (10^5 Hz) after 24 h exposure in

3.5 wt.% NaCl solution (Figure 4b). With increasing immersion time (after 10 days), the resistance decreased due to the penetration of the electrolyte through the nanocomposite layer, and the corrosion resistance for 1% Ni and 1% Al nanocomposites became equal (Figure 4a,c). Figure 4c shows the deviation of the Nyquist diagrams from a perfect semicircle. This phenomenon is referred to as frequency dispersion, and has been attributed to the roughness and non-uniformity of a working electrode, fracture structures, and the formation of porous layers [29]. This behavior can be observed in the Bode diagram, where the constant area of the phase angle value decreased, indicating that all samples had started to degrade (Figure 4d).

Table 3. Electrochemical parameters of the EIS experiments for all sample after 24 h, and 10 days of exposure to 3.5 wt.% NaCl solution at room temperature.

Samples	R_e, Ω	R_{coat}, Ω	CPE_{coat}, Ssec^n	n	R_{ct},Ω	CPE_{dl}, Ssec^n	n	CPE, %
				After 24 h				
Epoxy Coating	195.9	2.55×10^8	4.8×10^{-10}	0.9641	1.76×10^8	2.32×10^{-9}	0.9305	-
1% Al NP Nanocomposite	116.8	5.68×10^8	6.97×10^{-11}	0.9379	2.034×10^8	5.67×10^{-10}	0.9505	54.87
1% Ni NP Nanocomposite	204.1	7.90×10^8	2.72×10^{-12}	1	4.42×10^8	1.69×10^{-9}	1	67.72
1% AgNP Nanocomposite	217.1	8.89×10^8	5.95×10^{-12}	1	8.81×10^8	2.12×10^{-10}	1	71.32
				After 10 days				
Epoxy Coating	153.9	5.87×10^6	7.35×10^{-10}	0.924	4.87×10^4	4.19×10^{-9}	0.9951	-
1% Al NP Nanocomposite	100.0	1.67×10^7	5.63×10^{-10}	0.9102	2.35×10^6	3.85×10^{-10}	0.9862	64.85
1% Ni NP Nanocomposite	165.3	2.34×10^7	1.32×10^{-11}	0.9589	3.34×10^6	7.53×10^{-10}	0.9357	74.91
1% AgNP Nanocomposite	135.5	2.57×10^7	3.64×10^{-11}	0.9525	4.73×10^6	9.23×10^{-10}	0.9768	77.16

Based on Table 3, the double layer capacitance between the coating surface/electrolyte solutions (CPE_{coat} value) increased with time. This showed that the absorption level of the electrolyte solution in the nanocomposite had increased, but its resistance was still better than that of the epoxy coating. Further inspection of the table revealed that the n values were close to unity, implying that the interface behaved nearly capacitively. The addition of nanoparticles increased the value of %CPE. In all nanocomposites, the efficiency improved with time. The reason for such behavior could be nanoscale inorganic particles that cause better barrier properties. When metals interact with their surroundings, they can be converted into a more chemically stable form such as oxide, hydroxide, or sulfide, and thus provide better corrosion resistance [30,31].

Figure 5 illustrates the reaction between Al, Ni, and Ag NP and electrolyte, and the protection mechanism of the Al, Ni, and Ag nanocomposites in contact with corrosive electrolyte. When metallic aluminum comes into contact with oxygen, it becomes very reactive. A thin layer of alumina (4 nm thickness) forms in about 100 picoseconds on any exposed aluminum surface, which acts as a protective covering for further oxidation [32]. The Al NPs form an aluminum oxide layer on its surface in contact with the electrolyte, according to the following equations [33]:

$$2Al + 6H_2O \rightarrow 2Al(OH)_3 + 3H_2 \quad (2)$$

$$2Al + 4H_2O \rightarrow 2AlO(OH) + 3H_2 \quad (3)$$

$$2Al + 3H_2O \rightarrow 2Al_2O_3 + 3H_2 \quad (4)$$

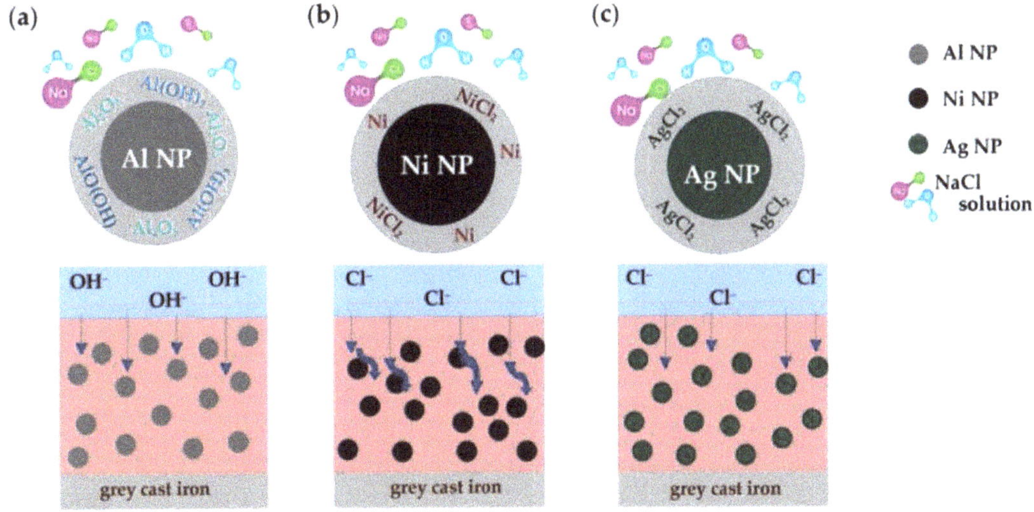

Figure 5. Protection mechanisms of (**a**) Al, (**b**) Ni, and (**c**) Ag nanoparticles in the epoxy coating in contact with NaClsolution.

The volume of formed aluminum oxides is greater than the volume of nanoparticles, which can prevent the further penetration of the corrosive medium, prolong the path of penetration of the corrosive medium into the coating, and ultimately delay the onset of corrosion (Figure 5a) [34,35].

The Ni NPs did not show passive behavior in the epoxy coating because the 3.5 wt.% NaCl solution had a pH value of7. Nickel did not react with water, and small amounts of chloride ions can be exposed on its surface. According to Pourbex's diagram, passive oxide films should appear at a pHvalue of 10 [20,36]. From these results, we can conclude that only Ni NPs and a small part of the formed $NiCl_2$ acted to increase the corrosion resistance of nickel nanocomposites. The nickel nanocomposite system did not have the ability to close the micropores formed in the epoxy coating during exposure to a corrosive medium. The effectiveness of this coating dropped, as can be seen in Figure 4c,d.

The Ag NPs have the property of releasing Ag^+ ions, which gives them the ability to react with other chemical species in the environment. Since silver is a noble metal ($E^\circ/V = 0.79$ vs. SHE), its affinity for oxygen will be low, so a layer of AgCl will form on its surface when exposed to 3.5 wt.% NaCl solution [36,37]. The generation of insoluble AgCl leads to the formation of a chemical barrier inside the epoxy coating. Ag NPs act as a chemical barrier against the further penetration of corrosive species from the electrolyte to the grey cast iron substrate [38].

3.3. Antibacterial Activity of the Al, Ni, and Ag Nanoparticles

The antibacterial activity of Al, Ni, and Ag nanoparticles was tested against various bacterial strains: *P. aeruginosa*, and *B. subtilise*. Figure 6 represents the antibacterial activity of nanoparticles for various bacteria in a well-diffusion technique. Arithmetic means of measurement results are given with measurement uncertainty U = 1.5 mm, k = 2, and P = 95%, where U is expanded measurement uncertainty, k the presented coverage factor, and P is confidence level.

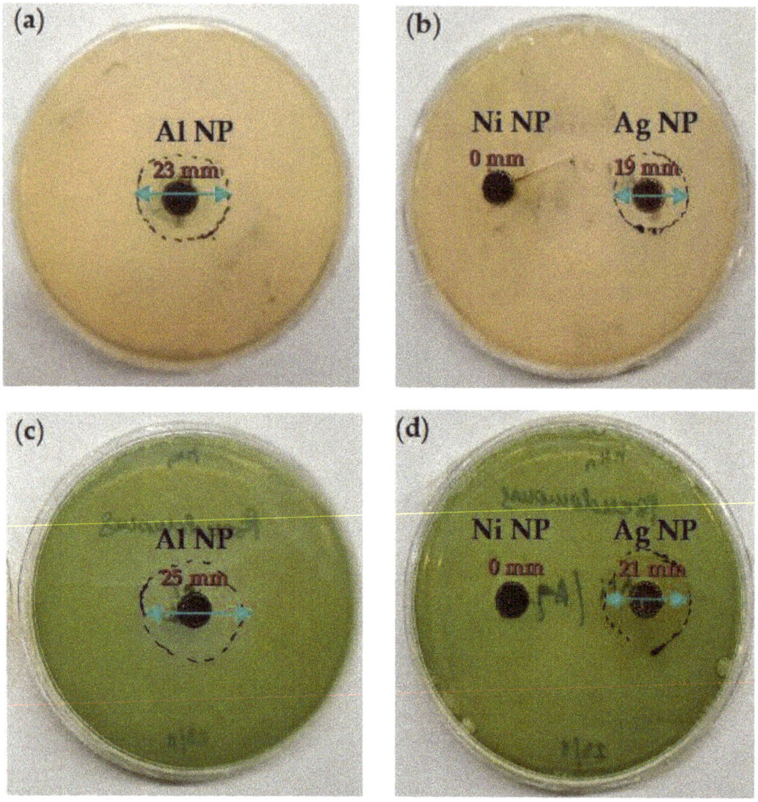

Figure 6. Zone of inhibition of (**a**) Al, and (**b**) Ni, and Ag nanoparticles against *B. subtilis*, and (**c**) Al, and (**d**) Ni, and Ag against *P. aeruginosa*.

The results indicated that Al and Ag nanoparticles showed effective antibacterial activity compared to Ni nanoparticles. The diameter of the inhibitory zone shows that Al NP had a larger zone of inhibition than Ag NP. In this respect, it was found that the Al and Ag NP were poisonous to both Gram-positive and Gram-negative bacteria. Nickel nanoparticles showed a very small inhibition zone, due to the large range of nanoparticle sizes. Typically, smaller NPs have better interaction with the bacteria due to their ability to penetrate a cell and inhibit bacteria growth [11,13]. Asghari and co-authors revealed that *P. aeruginosa* showed a high survival rate against Ni NPs and therefore can survive at quite high concentrations of Ni NPs [39]. It was assumed that due to their large size range and the possibility of agglomeration in aqueous media, Ni NPs could not enter the membrane of bacterial cells (*P. aeruginosa* and *B. subtilis*) to attach to functional groups of proteins to cause their denaturation [40].

3.4. Migration Ability and Antibacterial Activity of the Nanocomposites

A migration of nanoparticles from epoxy coating is a very important test because it provides a new property for drainage pipe protection. Figure 7 demonstrated the migration of Al and Ni nanoparticles from the epoxy coating into simulated wastewater. Released Ag^+ ions from the epoxy coating reacted with Cl^- ions from the wastewater to form a white precipitate of silver(I) chloride. This is the reason why the actual concentration of Ag^+ ions released could not be determined with F-AAS techniques. In Figure 7, the concentration of Ag^+ ions in the wastewater was below the detection limit of the instrument.

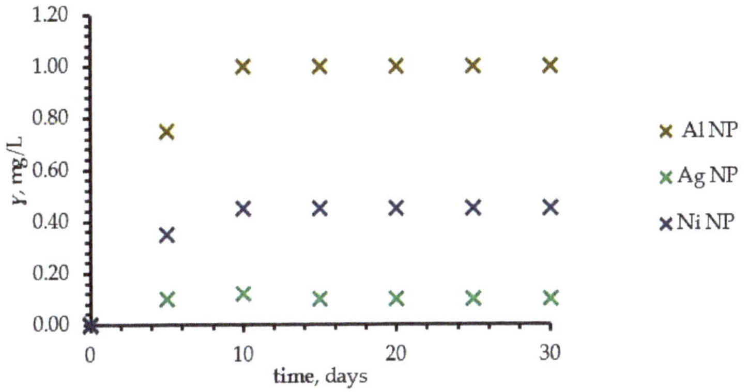

Figure 7. The mass concentration of the migration of Al, Ag, and Ni nanoparticles from epoxy coating to simulated wastewater.

The nanoparticles with 1% of Al NP showed significant migration, while nanocomposites with Ag and Ni nanoparticles showed the lowest value of migration from epoxy coating to the wastewater (Figure 7). The amount of migrated Al NP gradually increased in the beginning, but after 10 days of exposure the wastewater assumed a constant value of aluminum concentration (1 mg/L).For all samples, the pH-value increased very slightly, from 7.723 to 8.791, while the electrical conductivity showed a significant increase for the Al NP nanocomposite, from 1.085 to 2.083 mS/cm.

The migration of nanoparticles from the epoxy coating was stimulated by the action of the wastewater. Aluminum nanoparticles were the only ones capable of forming an oxide film on their surface. The resulting oxide film, with its passivity, closed the micropores of the epoxy and did not allow the oxidation and migration of other nanoparticles located further from the surface of the epoxy coating. The incorporation of certain metals and metal-oxides in the epoxy polymers also led to the development of 'active' materials that prevented the growth of microorganisms, and hence preserved the quality of water pipesduring the transportation of wastewater. According to ISO 22196, the antibacterial properties of the epoxy coating (control sample) and nanocomposites were determined by measuring the reduction in antibacterial activity. Consistent with the ISO guidelines, the reduction in antibacterial activity was estimated using the following equation [41]:

$$\text{reduction of antibacterial activity} = \frac{(CFU_{0h} - CFU_{24h})}{CFU_{0h}} \times 100 \quad (5)$$

where CFU_{0h} is the bacterial colony forming units obtained for the control samples before incubation, and CFU_{24h} is the bacterial colony forming units observed for the test conditions for the nanocomposites after incubation.

The antibacterial activity of the epoxy coating and nanocomposites against *P. aeruginosa* and *B. subtilis* after a 24 h exposition period is presented in Figure 8 and the reduction in antibacterial activity in Table 4.

Table 4. Average values of bacterial reductions of epoxy coating and 1% Al NP, 1% Ag NP, and 1% Ni NP in nanocomposite against *P. aeruginosa* and *B. subtilis* after 24 h.

		Nanocomposite		
	Epoxy Coating	**1% Al NP**	**1% Ag NP**	**1% Ni NP**
P. aerugionosa	4.17 ± 0.09	61.46 ± 2.64	45.83 ± 4.05	11.46 ± 0.92
B. subtilis	3.44 ± 0.11	55.31 ± 0.16	42.19 ± 1.41	11.25 ± 0.15

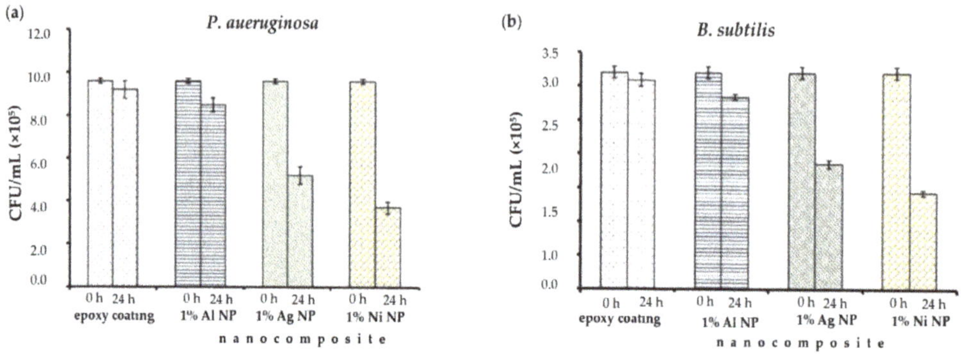

Figure 8. Antibacterial activity of epoxy coating, and 1% Al, 1% Ag, and 1% Ni epoxy nanocomposite on (**a**) *P. aerugionosa* and (**b**) *B. subtilis* during 24 h according to the ISO 22196 standard.

Based on the evaluation of bacteria (Figure 8), we concluded that after 24 h incubation the highest live count of *P. aeruginosa* and *B. subtilis* was observed on the epoxy coating samples and samples containing 1% of Ni NP in the nanocomposite (Table 4). The nanocomposite with 1% of Al NP exhibited the highest antibacterial effect, with a reduction of 61.46 ± 0.09% for *P. aeruginosa* and 55.31 ± 0.16% for *B. subtilis* (Figure 8, Table 4). The 1% of Ag NP in the epoxy coating showed the second-greatest antimicrobial potential, and achieved a 45.83 ± 4.05% cell reduction for *P. aeruginosa* and 42.19 ± 1.41% for *B. subtilis*, respectively, after 24 h incubation. However, the average leaching potential of Al NP was better than Ag NP in the epoxy coating (Figure 7). In contact with aqueous media, the Al NP formed an aluminum oxide and smaller amounts of aluminum hydroxide and oxyhydroxide [34]. The attraction between nanoparticles and bacteria depends on the surface charge of the nanoparticles and the bacteria [42]. Gram-positive bacteria (*B. subtilis*) consist of a thick layer of peptidoglycan, which is embedded in teichoic acid, while Gram-negative bacteria (*P. aeruginosa*) have a layer of lipopolysaccharide at the external surface. The teichoic acid and lipopolysaccharides impart a negative charge to the surface of bacterial cells [43]. The negatively charged bacterial surfaces of *P. aeruginosa* and *B. subtilis* strongly attract the resulting highly positively charged oxides, for example Al_2O_3 [44]. Better electrostatic attraction between bacteria and particles will result in better antibacterial action, which is visible in Figures 6c and 8a. Table 4 shows that less antibacterial activity was evident in samples exposed to the bacterium *B. subtilis* due to the thick outer cell wall, which can hinder NP penetration into the thick peptidoglycan layer.

In contact with corrosive media, Al NPs react with water molecules and the following redox reactions occur [45]:

$$Al(s) + 6H_2O(l) \rightarrow 2Al^{3+}(aq) + 6OH^-(aq) + 3H_2(g) \quad (6)$$

In the oxidation reaction, the Al^{3+} ion is formed, which is toxic to bacteria. Also, the resulting oxidation products can release the Al^{3+} ion, which will affect the life of the bacteria [41]. The mechanism of action of Al^{3+} ions has not been fully clarified yet. Cell penetration and the adsorption or diffusion of NPs at the cell surface is often the initial step in the stages involved in some microbial cell inhibition [46]. According to literature [47], an increased Al^{3+} concentration can stimulate reactive oxygen species (ROS), which can act as the main factor in DNA damage in bacterial cells.

4. Conclusions

The influence of nanoparticles in the epoxy coating leads to the modification of the internal defects in the coating due to the filling of voids and blocking of easy access to channels. Nanoparticles that could react with a corrosive medium prolong the path of penetration of corrosive species from the electrolyte to the metal substrate. Likewise, the resulting corrosion product on the nanoparticle is dangerous for lower organisms and can be used for the formation of a new property of the coating. These studies showed that Al and Ag nanoparticles embedded in nanocomposites can significantly improve the anticorrosive properties and inhibit bacteria, while Ni NPs have lower efficiency. Due to remarkable properties like high reactivity, the ability to form a homogeneous structure, bacterial resistance, cost-effectiveness, and unexplored literature, Al NPs attracted our enormous interest and will continue to serve for detailed analysis so that they can be applied as a protective additive in epoxy coating for the protection of pipelines.

Author Contributions: Conceptualization, M.S., V.A. and M.V.D.; methodology, M.S. and M.V.D.; software, M.S.; validation, M.S. and M.V.D.; formal analysis, M.S.; investigation, M.S. and M.V.D.; resources, V.A. and I.S.; data curation, M.S.; writing—original draft preparation, M.S.; writing—review and editing, M.S.; visualization, M.S.; supervision, V.A., I.S. and M.V.D.; project administration, V.A. and I.S.; funding acquisition, V.A. All authors have read and agreed to the published version of the manuscript.

Funding: This research was funded by "Development of anticorrosion protection system for multipurpose pipe use", Grant Number KK.01.1.1.07.0045. This work was supported by the European Regional Development Fund under the Operational Program Competitiveness and Cohesion 2014–2020.

Institutional Review Board Statement: Not applicable.

Informed Consent Statement: Not applicable.

Data Availability Statement: Not applicable.

Acknowledgments: We thank Michaela Hruškova Hasan (University of Zagreb, Faculty of Mining, Geology, and Petroleum Engineering) for the use of flame atomic absorption spectrometry techniquesand helpful advice.

Conflicts of Interest: The authors declare no conflict of interest.

References

1. Khan, I.; Saeed, K.; Khan, I. Nanoparticles: Properties, applications, and toxicities. *Arab. J. Chem.* **2019**, *12*, 908–931. [CrossRef]
2. Jamkhande, P.G.; Ghule, N.W.; Bamer, A.H.; Kalaskar, M.G. Metal nanoparticles synthesis: An overview on methods of preparation, advantages and disadvantages, and applications. *J. Drug Deliv. Sci. Technol.* **2019**, *53*, 101174. [CrossRef]
3. Nguyen-tri, P.; Nguyen, T.A.; Carriere, P.; Xuan, C.N. Nanocomposite Coatings: Preparation, Characterization, Properties, and Applications. *Int. J. Corros.* **2018**, *2018*, 19. [CrossRef]
4. Tanveer, M.; Farooq, A.; Ata, S.; Bibi, I.; Sultan, M.; Iqbal, M.; Jabeen, S.; Gull, N.; Islam, A.; Khan, R.U.; et al. Aluminum nanoparticles, chitosan, acrylic acid and vinyltrimethoxysilane based hybrid hydrogel as a remarkable water super-absorbent and antimicrobial activity. *Surf. Interfaces* **2021**, *25*, 101285. [CrossRef]
5. Zhang, G.; Li, B.; Liu, J.; Luan, M.; Yue, L.; Jiang, X.T.; Yu, K.; Guan, Y. The bacterial community significantly promotes cast iron corrosion in reclaimed wastewater distribution systems. *Microbiome* **2018**, *6*, 222. [CrossRef]
6. Wang, J.; Liu, G.; Wang, J.; Xu, X.; Shao, Y.; Zhang, Q.; Liu, Y.; Qi, L.; Wang, H. Current status, existent problems, and coping strategy of urban drainage pipeline network in China. *Environ. Sci. Pollut. Res.* **2021**, *28*, 43035–43049. [CrossRef]
7. Sun, H.; Shi, B.; Lytle, D.A.; Bai, Y.; Wang, D. Formation, and release behavior of iron corrosion products under the influence of bacterial communities in a simulated water distribution system. *Environ. Sci. Process. Impacts* **2014**, *16*, 576–585. [CrossRef]
8. Lou, Y.; Chan, W.; Cui, T.; Qian, H.; Huang, L.; Ma, L.; Hao, X.; Zhang, D. Microbiologically influenced corrosion inhibition of carbon steel via biomineralization induced by *Shewanella putrefaciens*. *Mater. Degrad.* **2021**, *5*, 59. [CrossRef]
9. Brindha, T.; Rathinam, R.; Dheenadhayalan, S.; Sivakumar, R. Nanocomposite Coatings in Corrosion Protection Applications: An Overview. *Orient. J. Chem.* **2021**, *37*, 1062–1067.
10. Kausar, A. A review of high performance polymer nanocomposites for packaging applications in electronics and food industries. *J. Plast. Film Sheeting* **2020**, *36*, 94–112. [CrossRef]
11. Zheng, K.; Setyawati, M.I.; Leong, D.T.; Xie, J. Antimicrobial silver nanomaterials. *Coord. Chem. Rev.* **2018**, *357*, 1–17. [CrossRef]
12. Rai, M.; Shegokar, R. Metal Nanoparticles in Pharma. In *Metal Nanoparticles in Pharma*; Springer International Publishing: Cham, Switzerland, 2017; pp. 1–493.
13. Franco, D.; Calabrese, G.; Guglielmino, S.P.P.; Conoci, S. Metal-Based Nanoparticles: Antibacterial Mechanisms and Biomedical Application. *Microorganisms* **2022**, *10*, 1778. [CrossRef]
14. Giraldo Mejía, H.F.; Yohai, L.; Pedetta, A.; Herrera Seitz, K.; Procaccini, R.A.; Pellice, S.A. Epoxy-silica/clay nanocomposite for silver-based antibacterial thin coatings: Synthesis and structural characterization. *J. Colloid Interface Sci.* **2017**, *508*, 332–341. [CrossRef]

15. Hussain, Z.; Tahir, S.A.; Mahmood, K.; Ali, A.; Arshad, M.I.; Ikram, S.; Nabi, M.A.U.; Ashfaq, A.; Rehman, U.U.; Muddassir, Y. Synthesis and characterization of silver nanoparticles with epoxy resin. *Dig. J. Nanomater. Biostructures* **2020**, *15*, 873–883. [CrossRef]
16. Manjumeena, R.; Venkatesan, R.; Duraibabu, D.; Sudha, J.; Rajendran, N.; Kalaichelvan, P.T. Green Nanosilver as Reinforcing Eco-Friendly Additive to Epoxy Coating for Augmented Anticorrosive and Antimicrobial Behavior. *Silicon* **2015**, *8*, 277–298. [CrossRef]
17. Chaudhari, B.; Panda, B.; Šavija, B.; Paul, S.C. Microbiologically Induced Concrete Corrosion: A Concise Review of Assessment Methods, Effects, and Corrosion-Resistant Coating Materials. *Materials* **2022**, *15*, 4279. [CrossRef]
18. Shen, W.; Zhang, T.; Ge, Y.; Feng, L.; Feng, H.; Li, P. Multifunctional AgO/epoxy nanocomposites with enhanced mechanical, anticorrosion and bactericidal properties. *Prog. Org. Coat.* **2021**, *152*, 106130. [CrossRef]
19. Shetty, P.; Bhushan, S.; Vidya, A.; Kodialbail, S. Biocorrosion Behavior of Epoxy—Based Multilayer Nanocomposite Coatings. *J. Bio- Tribo-Corrosion* **2023**, *9*, 45. [CrossRef]
20. Leygraf, C.; Wallinder, I.O.; Tidblad, J.; Graedel, T. Appendix I: The Atmospheric Corrosion Chemistry of Silver. In *Atmospheric Corrosion*, 2nd ed.; John Wiley & Sons, Inc.: Hoboken, NJ, USA, 2016; pp. 337–347.
21. Majeed, H.M.; Wadee, S.A. Antibacterial Activity and Mechanism of Nickel Nanoparticles against Multidrug Resistant *Pseudomonas aeruginosa*. *Ann. Trop. Med. Public Health* **2019**, *22*, 157–168. [CrossRef]
22. *EN 877*; Cast Iron Pipes and Fittings, Their Joints and Accessories for the Evacuation of Water from Buildings—Requirements, Test Methods and Quality Assurance. European Committee for Standardization: Brussels, Belgium, 2006.
23. *ISO/TS 16782*; Clinical Laboratory Testing—Criteria for Acceptable Lots of Dehydrated Mueller-Hinton Agar and Broth for Antimicrobial Susceptibility Testing. International Organization for Standardization: Geneva, Switzerland, 2016.
24. *ISO 22196*; Plastics—Measurement of Antibacterial Activity on Plastics Surfaces. International Organization for Standardization: Geneva, Switzerland, 2007.
25. *ISO 4833–1*; Microbiology of the Food Chain—Horizontal Method for the Enumeration of Microorganisms—Part 1: Colony Count at 30 °C by the Pour Plate Technique. International Organization for Standardization: Geneva, Switzerland, 2018.
26. Ghanbari, A.; Attar, M.M. A study on the anticorrosion performance of epoxy nanocomposite coatings containing epoxy-silane treated nano-silica on mild steel substrate. *J. Ind. Eng. Chem.* **2015**, *23*, 145–153. [CrossRef]
27. Obot, I.B.; Madhankumar, A. Synergistic effect of iodide ion addition on the inhibition of mild steel corrosion in 1 M HCl by 3-amino-2-methylbenzylalcohol. *Mater. Chem. Phys.* **2016**, *177*, 266–275. [CrossRef]
28. Samardžija, M.; Alar, V.; Špada, V.; Stojanović, I. Corrosion Behaviour of an Epoxy Resin Reinforced with Aluminium Nanoparticles. *Coatings* **2022**, *12*, 1500. [CrossRef]
29. Solomon, M.M.; Gerengi, H.; Kaya, T.; Umoren, S.A. Performance Evaluation of a Chitosan/Silver Nanoparticles Composite on St37 Steel Corrosion in a 15% HCl Solution. *ACS Sustain. Chem. Eng.* **2017**, *5*, 809–820. [CrossRef]
30. Anjum, M.J.; Ali, H.; Khan, W.Q.; Zhao, J.; Yasin, G. *Chapter 11—Metal/Metal Oxide Nanoparticles as Corrosion Inhibitors*, 1st ed.; Elsevier: Amsterdam, The Netherlands, 2020; pp. 181–201.
31. Jain, P.; Patidar, B.; Bhawsar, J. Potential ofNanoparticles asa Corrosion Inhibitor: A Review. *J. Bio-Tribo-Corros.* **2020**, *6*, 43. [CrossRef]
32. Geoprincy, G.; Nagendhra Gandhi, N.; Renganathan, S. Novel antibacterial effects of alumina nanoparticles on *Bacillus cereus* and *Bacillus subtilis* in comparison with antibiotics. *Int. J. Pharm. Pharm. Sci.* **2012**, *4*, 544–548.
33. Niroumandrad, S.; Rostami, M.; Ramezanzadeh, B. Effects of combined surface treatments of aluminium nanoparticle on its corrosion resistance before and after inclusion into an epoxy coating. *Prog. Org. Coat.* **2016**, *101*, 486–501. [CrossRef]
34. Samardžija, M.; Kurtela, M.; Vuković Domanovac, M.; Alar, V. Anticorrosion and Antibacterial Properties of Al NP—Epoxy Nanocomposite Coating on Grey Cast Iron. *Coatings* **2023**, *13*, 898. [CrossRef]
35. Feichtenschlager, B.; Pabisch, S.; Svehla, J.; Peterlik, H.; Sajjad, M.; Koch, T.; Kickelbick, G. Epoxy Resin Nanocomposites: The Influence of Interface Modification on the Dispersion Structure—A Small-Angle-X-ray-Scattering Study. Surfaces. *Surfaces* **2020**, *3*, 664–682. [CrossRef]
36. Pavapootanont, G.; Wongpanya, P.; Viyanit, E.; Lothongkum, G. Corrosion behavior of Ni steels in aerated 3.5-wt.% NaCl solution at 25 °C by potentiodynamic method. *Eng. J.* **2018**, *22*, 1–12. [CrossRef]
37. Keast, V.J. Atmospheric Corrosion of Silver and Silver Nanoparticles. *Corros. Mater. Degrad.* **2022**, *3*, 221–234. [CrossRef]
38. Bordbar, S.; Rezaeizadeh, M.; Kavian, A. Improving thermal conductivity and corrosion resistance of polyurea coating on internal tubes of gas heater by nano silver. *Prog. Org. Coat.* **2020**, *146*, 105722. [CrossRef]
39. Asghari, E.; Kaltschmidt, B.P.; van Merwyk, L.; Huser, T.; Kaltschmidt, C.; Hutten, A. *Pseudomonasaeruginosa* Clusters Toxic Nickel Nanoparticles to Enhance Survival. *Microorganisms* **2022**, *10*, 2220. [CrossRef] [PubMed]
40. Hussain, A.; Ahmad, M.N.; Jalal, F.; Yameen, M.; Falak, S.; Noreen, S.; Naz, S.; Nazir, A.; Iftikhar, S.; Soomro, G.A.; et al. Investigating the antibacterial activity of POMA nanocomposites. *Polish J. Environ. Stud.* **2019**, *28*, 4191–4198. [CrossRef]
41. Fonseca, S.; Cayer, M.; Ahmmed, K.M.T.; Khadem-mohtaram, N.; Charette, S.J.; Brouard, D. Characterization of the Antibacterial Activity of an SiO_2 Nanoparticular Coating to Prevent Bacterial Contamination in Blood Products. *Antibiotics* **2022**, *11*, 107. [CrossRef]
42. Mukherjee, A.; Mohammed Sadiq, I.; Prathna, T.C.; Chandrasekaran, N. Antimicrobial activity of aluminium oxide nanoparticles for potential clinical applications. *Sci. Microb. Pathog. Commun. Curr. Res. Technol. Adv.* **2011**, *4*, 245–251.
43. Li, Z.; Ma, J.; Ruan, J.; Zhuang, X. Using Positively Charged Magnetic Nanoparticles to Capture Bacteria at Ultralow Concentration. *Nanoscale Res. Lett.* **2019**, *14*, 195. [CrossRef]

44. Jiang, W.; Mashayekhi, H.; Xing, B. Bacterial toxicity comparison between nano- and micro-scaled oxide particles. *Environ. Pollut.* **2009**, *157*, 1619–1625. [CrossRef] [PubMed]
45. Yasakau, K.A.; Zheludkevich, M.L.; Lamaka, S.V.; Ferreira, M.G.S. Mechanism of corrosion inhibition of AA2024 by rare-earth compounds. *J. Phys. Chem. B* **2006**, *110*, 5515–5528. [CrossRef] [PubMed]
46. Sharmin, S.; Rahaman, M.M.; Sarkar, C.; Atolani, O.; Islam, M.T.; Adeyemi, O.S. Nanoparticles as antimicrobial and antiviral agents: A literature-based perspective study. *Heliyon* **2021**, *7*, 5515–5528. [CrossRef]
47. Gudkov, S.V.; Burmistrov, D.E.; Smirnova, V.V.; Semenova, A.A.; Lisitsyn, A.B. A Mini Review of Antibacterial Properties of Al_2O_3 Nanoparticles. *Nanomaterials* **2022**, *12*, 2635. [CrossRef] [PubMed]

coatings

Article

Experimental Study of Atmospherically and Infrared-Dried Industrial Topcoats

Ivan Stojanović [1], Mirta Logar [1,*], Ivan Fatović [1], Vesna Alar [1] and Daniela Rakela-Ristevski [2]

1 Chair of Materials Protection, Faculty of Mechanical Engineering and Naval Architecture, University of Zagreb, 10000 Zagreb, Croatia; ivan.stojanovic@fsb.hr (I.S.); ivan.fatovic@fsb.hr (I.F.); vesna.alar@fsb.hr (V.A.)
2 Končar Steel Structures Inc., 10000 Zagreb, Croatia; daniela.rakela@koncar-mk.hr
* Correspondence: mirta.logar@fsb.hr

Abstract: In this paper, five different solvent-borne industrial topcoats were dried with infrared (IR) radiation and under atmospheric conditions. A comparison of physical, mechanical, chemical, and electrochemical properties of differently dried topcoats was made. The results of differential scanning calorimetry (DSC), Fourier-Transform Infrared Spectroscopy (FTIR), and adhesion of a topcoat to the metal substrate (determined by the pull-off test) indicate a higher degree of crosslinking of examined topcoats, which improves the coating's protective properties. Scratch hardness was determined by the pencil hardness test. Impact resistance was examined with a falling-weight test. Changes in the shade of the coating were examined by visual inspection and using a gloss meter. The electrochemical measurements of open circuit potential (OCP) and electrochemical impedance spectroscopy (EIS) were conducted. The thermal stability of topcoats was tested by thermogravimetric analysis (TGA). The results show overall better properties of IR-dried topcoats. In addition, topcoats dry significantly faster when IR radiation is applied, which makes this drying method very interesting for industrial application.

Keywords: industrial topcoats; polyurethane; polysiloxane; infrared drying; corrosion protection

1. Introduction

Metal corrosion is one of the biggest problems worldwide, especially in the industrial environment [1]. According to the National Association of Corrosion Engineers (NACE International) and data from 2013, the global cost of corrosion is estimated at USD 2.5 trillion, equivalent to 3.4% of the global GDP [2]. In order to slow down corrosion processes, it is important to wisely choose the material and to ensure good surface preparation, but also to use additional protection methods, such as coatings, cathodic protection, corrosion inhibitors, etc. [3,4]. The application of polymer coatings is one of the most commonly used corrosion protection methods. In this method, the metal substrate is usually coated with a coating system, which often consists of three layers: primer, intermediate coat, and topcoat. Organic coatings provide the best protection to the metal substrate if the barrier properties of the coating are good, that is, in the absence of defects (pores, scratches, etc.) that extend to the very surface of the protected metal, which can initiate corrosion [5,6].

Competitive pressure in the metal industry is increasing. To survive, it is necessary for the steel manufacturers to shorten the finishing time, which especially refers to the coating protection process, so that they can put as many products as possible on the market in the shortest possible time. One of the solutions for faster drying/curing of protective coatings is using radiation, such as infrared radiation (IR) [7]. Infrared drying enables fast evaporation of the solvent, but should be used with caution since too-fast evaporation can cause blistering [8]. In the process of radiation curing, a coating is crosslinked as a result of interacting with incident radiation, directly on a substrate [9]. Any industry that requires drying, polymerization, or curing in manufacturing processes can use infrared radiation to speed up the process [10]. Currently, IR radiation is often used in the automotive industry to speed up the coating drying process, but more and more industries are turning to this solution [11]. For the industry, the simplest transition to this new technology is to apply IR radiation for drying coatings they already use. In order to enable a carefree transition to IR technology, it is necessary to conduct research on the IR-dried coatings' protective properties in comparison with the properties of conventionally dried coatings.

The curing or crosslinking of coatings is the formation of lateral connections of polymer chains. As a result of this process, the crosslinked polymer becomes harder, and its glass transition temperature shifts towards higher temperatures. Furthermore, mechanical strength and chemical resistance improve with a higher curing degree [12]. These properties reflect the degree of polymer crosslinking. However, making predictions about the coating curing process is a complex and expensive issue that requires a lot of experimental work in order to obtain a functional curing prediction model [10]. Because of that, experimental literature that describes this issue is rather scarce. Some researchers, like Saure et al. [13,14] and Geipel and Stephan [15], use FTIR to analyze the behavior of polymer coatings during drying. Others also use differential scanning calorimetry (DSC) and thermogravimetric analysis (TGA), like Vessot et al. [16] did, to study the drying and curing of polyurethane-based coatings. A fully cured coating should have good adhesion to the metal substrate and good corrosion resistance [17]. Many researchers, including Alsamuraee et al. [5], use electrochemical impedance spectroscopy (EIS) as a method to evaluate the state of the coating surface and its barrier properties, which are key to achieving the satisfactory corrosion resistance of the coating. Nevertheless, the impact of catalytic infrared radiation (CIR) on the top coating curing process and comparison to the conventional drying method by using comprehensive test methods have not been broadly explored.

The aim of this paper is to study the difference between atmospherically and infrared-dried/cured industrial topcoats. Liquid, solvent-borne industrial topcoats from two different manufacturers were examined. A comparison of physical, mechanical, corrosion, and chemical properties of differently dried topcoats was made. The adhesion of the topcoat to the metal substrate was tested using a pull-off adhesion test. The gloss of differently dried topcoats was compared using a gloss meter. The pencil hardness test was used to evaluate the hardness of the topcoats. Regarding mechanical properties, impact resistance was also tested. Topcoats' corrosion resistance was assessed using electrochemical impedance spectroscopy (EIS). The stabilization of the corrosion potential in 3.5% NaCl solution was examined by open circuit potential (OCP). Thermogravimetric analysis (TGA) was used to evaluate the thermal stability of differently dried topcoats. Differential scanning calorimetry (DSC) showed the glass transition temperature of the polymer on which the topcoat is based. The degree of crosslinking, which directly affects the mechanical properties of the topcoats, was examined by Fourier-Transform Infrared Spectroscopy (FTIR). This paper compares experimentally obtained properties of atmospherically and IR-dried coatings.

2. Materials and Methods

2.1. Materials and Drying Methods

In this paper, five different topcoats were tested. Their characteristics are presented in Table 1. The table shows the chemical composition, the time required for the topcoat to become dry-to-touch when dried under atmospheric conditions (according to the manufacturer's data sheet and confirmed experimentally), the gloss, and the manufacturer's origin. All the examined topcoats are based on polyurethane (PUR), except for topcoat 2, which is based on polysiloxane (PSX). The times required for the coating to become dry-to-touch vary from 5.5 to 8 h. Tested topcoats 1–4 had a glossy finish, while topcoat 5 had a matte finish. The first three topcoats are of Danish origin, while the other two are of German origin. The examined topcoats are used in the power transformer industry.

Table 1. Characteristics of examined topcoats.

Topcoat	1	2	3	4	5
Chemical composition	PUR	PSX	PUR	PUR	PUR
Dry-to-touch [h]	8	5.5	8	7	6
Gloss	Glossy	Glossy	Glossy	Glossy	Matte
Manufacturer's origin	Danish	Danish	Danish	German	German

The described topcoats were prepared according to the manufacturer's instructions in a mixing bowl. They were applied with a spiral applicator, with the required film thickness according to the manufacturer's technical instructions, on a steel substrate with dimensions 150 × 120 × 18 mm. The substrate was prepared by steel grit blasting to the required cleanliness Sa 2.5, according to ISO 8501-1 [18], and a medium degree of roughness, according to ISO 8503-1 [19]. Half of the applied samples were air-dried, while the other half were dried with infrared (IR) radiation in the cabin for IR drying of liquid coatings owned by the company Končar—Steel structures. The drying chamber consisted of three catalytic gas IR emitters operating at wavelengths from 2 to 10 μm. Substrates with applied topcoats were dried at a distance of 100 cm from the IR emitter. The end of the curing period was detected using a contact method, i.e., a stroke of a pencil. It is considered that the coating is crosslinked when the pencil no longer leaves a mark when lightly stroked over the coating. After the coating had cooled down, a fingernail test was performed to confirm the curing. A properly cured coating should be tack-free and should not scratch [20]. Figure 1 shows a representative view of the dried samples. Topcoats dried in atmospheric conditions are presented in the top row, while the same topcoats dried in an IR cabin are presented in the bottom row.

Figure 1. Substrates with topcoats.

2.2. *Characterization of Differently Dried Topcoats*

2.2.1. Dry Film Thickness, Adhesion, and Gloss

Dry film thickness (DFT) was determined using an Elcometer 456 Coating Thickness Gauge (Elcometer Limited, Edge Lane, Manchester, UK), according to ISO 2808 [21]. Measurements were made at ten different places per sample with instrument accuracy ±2.5 μm, and the mean value was calculated. The adhesion of differently dried coatings to the metal substrate was tested using an Elcometer 510 Automatic Pull-off Adhesion Gauge (Elcometer Limited, Edge Lane, Manchester, UK) with an accuracy of ±1% of full scale, according to ISO 4624 [22]. The influence of infrared drying on topcoat gloss was examined with an Elcometer 480 Glossmeter (Elcometer Limited, Edge Lane, Manchester, UK) at three different angles, according to ISO 2813 [23]. For both the adhesion and gloss value tests, three measurements were performed, and the mean value was calculated.

2.2.2. Mechanical Properties

The mechanical properties of topcoats were determined using a pencil test and a rapid-deformation (impact resistance) test. To perform the hardness tests, an Elcometer 501 Pencil Hardness Tester (Elcometer Limited, Edge Lane, Manchester, UK) apparatus was used, according to ISO 15184 [24]. The hardness of the hardest pencil that does not leave any mark on the topcoat was recorded as the topcoat scratch hardness. The impact resistance tests were performed according to ISO 6272 [25], using an Elcometer 1615 Variable Impact Tester (Elcometer Limited, Edge Lane, Manchester, UK). A falling-weight test with a large-area indenter and a weight of 1 kg was used. As a result, the drop height (in inches) for which the topcoat did not yet crack was recorded.

2.2.3. Electrochemical Characterization

Electrochemical characterization of the topcoats was performed by measuring the open circuit potential (OCP) and by the electrochemical impedance spectroscopy (EIS) method. The device used for electrochemical measurements was a VersaSTAT 3 Potentiostat/Galvanostat (AMETEK Scientific 131 Instruments, Princeton applied research, Berwyn, PA, USA). Changes in the corrosion potential of a metal substrate with a topcoat in contact with a 3.5% NaCl solution were tested using the open circuit potential [26]. Potential stabilization was carried out for 1000 s with a saturated calomel electrode (SCE) as a reference electrode, at room temperature (23 ± 2 °C). The corrosion potential of the coated sample was measured after 1, 168, and 240 h spent in the electrolyte. A 3.5% NaCl solution was chosen as the electrolyte because it simulates real conditions from nature, since this is the average concentration of salt in the oceans [26].

Electrochemical impedance spectroscopy was used to assess the chemical resistance of topcoats. Measurements were performed after 1 and 240 h in a 3.5% NaCl solution, at a frequency range of 0.1 to 105 Hz, at room temperature (23 ± 2 °C), recording 10 points per decade [27]. The tested electrochemical cell consisted of a metal substrate with a topcoat as a working electrode, a saturated calomel electrode as a reference electrode, and two graphite rods as counter electrodes. In order to check the repeatability of the data, each measurement was executed in two replications. Experimental data were analyzed with the AMETEK ZSimpWin program.

2.2.4. Thermogravimetric Analysis (TGA) and Differential Scanning Calorimetry (DSC)

The thermal stability of topcoats was determined with thermogravimetric analysis, using TGA Q500 (TA Instruments, New Castle, DE, USA). Samples weighing around 10 mg were analyzed in a stream of nitrogen (60 mL/min) in the temperature range from −100 to 600 °C with a heating rate of 10 °C/min [28].

A Mettler Toledo differential scanning calorimetry analyzer, DSC822e (Mettler Toledo, Greifensee, Switzerland), was used to determine phase transitions. Samples of about 10 mg were analyzed in a stream of nitrogen (40 mL/min) with a heating and cooling rate of 10 °C/min in the temperature range from −100 °C to 150 °C by a double heating/cooling cycle. Samples were first heated from 25 °C to 150 °C at a rate of 10 °C/min and held at that temperature for 5 min to erase the thermal history of the sample during the preparation process. Samples were then cooled from 150 °C to −100 °C at a cooling rate of 10 °C/min, held at −100 °C for 5 min, and reheated from −100 °C to 150 °C at a rate of 10 °C/min [29]. From the second heating cycle, values of the glass transition temperature Tg were obtained. Liquid nitrogen was used to cool the samples to low temperatures.

2.2.5. Fourier-Transform Infrared Spectroscopy (FTIR)

The chemical composition of topcoats was determined using Fourier-Transform Infrared Spectroscopy (FTIR). A Spectrometer Spectrum One (Perkin Elmer, Waltham, MA, USA) using an Attenuated Total Reflection (ATR) chamber (ZnSe) was used. Measurements were carried out at room temperature in the range of wave numbers 4000–650 cm^{-1} [30]. Topcoat samples were scraped from the metal substrate and recorded in their basic form, without prior preparation.

A flowchart of the described procedures and methods is presented in Figure 2.

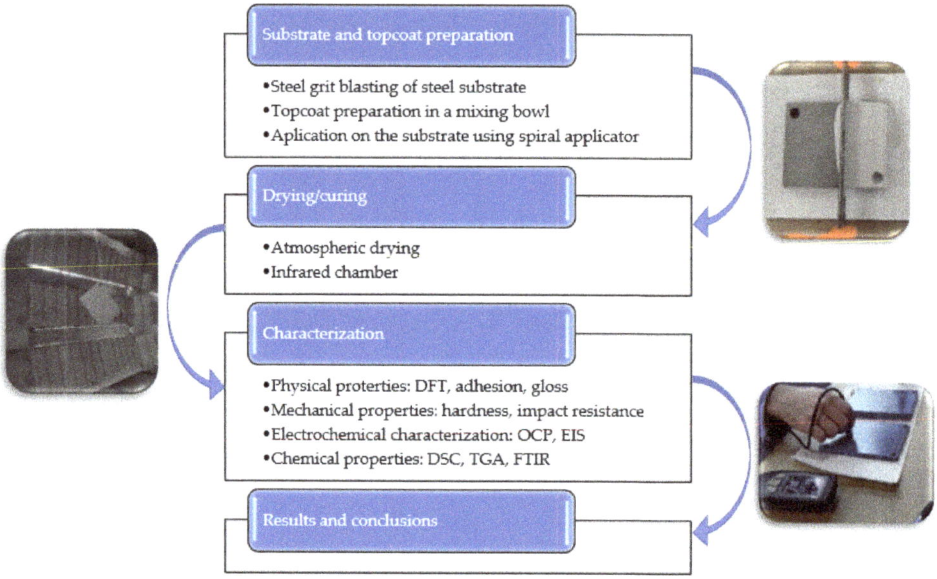

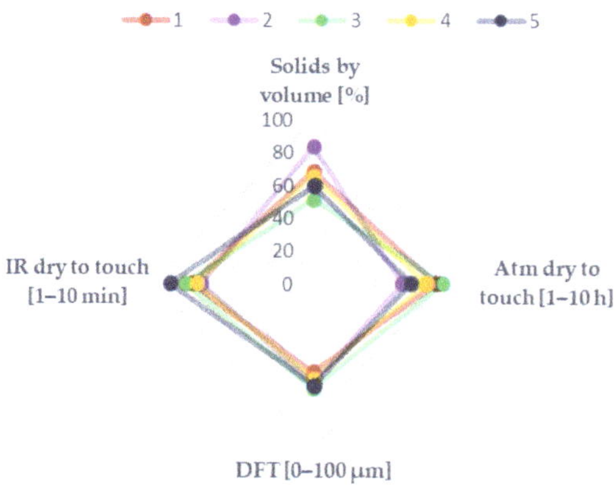

Figure 2. Flowchart of the procedures and methods used in this paper.

3. Results and Discussion

3.1. Drying Time, Dry Film Thickness, Adhesion, and Gloss

According to the manufacturer's data sheet and proven experimentally, topcoats need 5 to 8 h to become dry-to-touch when dried atmospherically. On the other hand, applying IR radiation significantly reduces the drying time, considering that the tested topcoats were dry-to-touch in just 10 min. Regarding dry film thickness, the goal was to obtain a roughly equal thickness of each film so that the topcoats would be comparable to each other. All samples are within 10 percent of the mean dry film thickness, as shown in Table 2.

Table 2. Type of drying (under atmospheric conditions—atm, applying infrared radiation—IR), time required for the topcoat to be dry-to-touch, dry film thickness, and pull-off values of differently dried topcoats.

Topcoat	Type of Drying	Drying Time [min]	DFT [μm]	Pull-Off [MPa]
1	atm	480	62.8	11.2
	IR	10	65.9	19.4
2	atm	330	62.4	16.1
	IR	10	60.3	19.2
3	atm	480	62.1	17.2
	IR	10	65.7	24.8
4	atm	420	58.1	11.4
	IR	10	60.8	8.2
5	atm	360	63.8	7.6
	IR	10	63.7	10.4

Pull-off adhesion values, which are also shown in Table 2, are higher when applying IR radiation. This means that the adhesion of the coating to the substrate is better when the coating is dried with IR radiation. The explanation of this effect lies in better crosslinking of the polymer when using IR radiation for curing, due to which the resistance to the tensile load applied in the pull-off adhesion test is better, and the topcoat has higher tensile strength [31]. A representative view of pull-off values is shown in Figure 3.

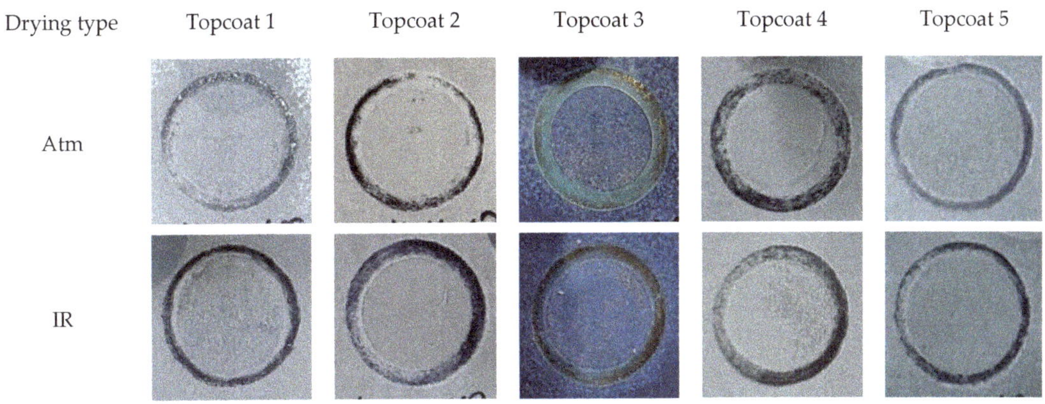

Figure 3. Representative pull-off values for each topcoat.

The gloss values of five tested topcoats, at three different angles, are shown graphically in Figure 4. Figure 4a shows gloss values of atmospherically dried topcoats, while graph Figure 4b shows gloss values of infrared-dried topcoats. Values obtained at different angles mostly match, regardless of the drying method. The only slightly larger deviation is with topcoat 4, whose gloss values are higher in the case of atmospheric drying. The lowest gloss values were recorded for topcoat 5, which is matte. Such low gloss values were expected.

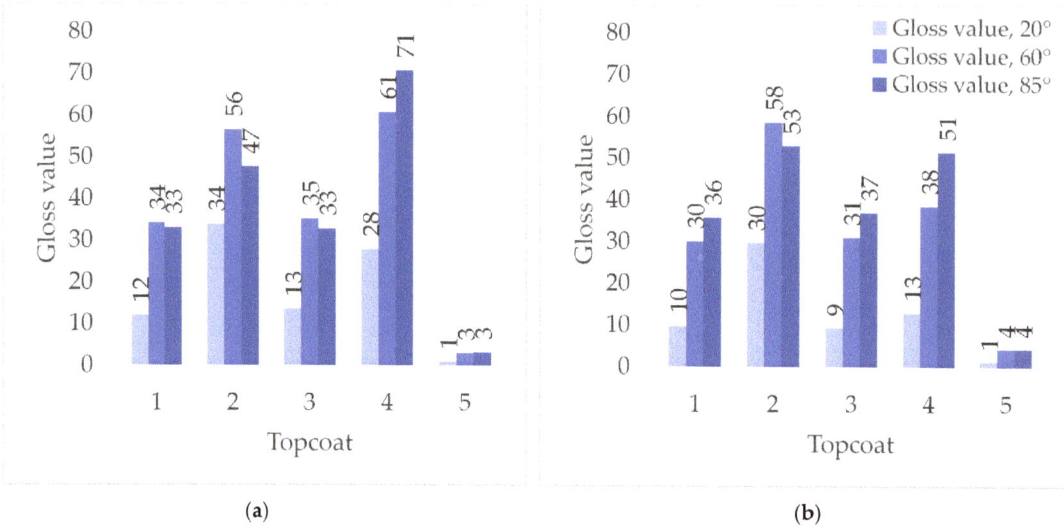

Figure 4. Gloss values at different angles for (**a**) atmospherically and (**b**) IR-dried topcoats.

3.2. Mechanical Properties

Coatings that have higher hardness resist wear better [32]. Also, good impact resistance is desirable, since it describes the coating's ability to absorb impact energy without breaking [33]. The scratch hardness of the examined topcoats was determined with a pencil test, and the results are shown in Table 3. All of the topcoats show similar scratch hardness, around HB. In two cases, IR-dried topcoats show higher hardness than the air-dried ones, while in other cases, the hardness is the same. The results of the impact test differ. Topcoats based on polysiloxane showed the lowest impact resistance. Polyurethane-based topcoats show better impact resistance. Some topcoats showed no plastic deformation, even when the drop height was at the highest level (which is limited by the apparatus). In half of the cases, the impact resistance was slightly better for atmospherically dried topcoats, while in the other half, it was somewhat better in the case of IR drying. That means that the drying method has no significant influence on the impact resistance of the topcoat.

Table 3. Pencil hardness and impact resistance of differently dried topcoats.

Topcoat	Type of Drying	Scratch Hardness	Impact Resistance [in]
1	atm	HB	33
	IR	HB	28
2	atm	HB	7
	IR	F	10
3	atm	HB	40 *
	IR	HB	40 *
4	atm	B	19
	IR	F	21
5	atm	HB	40 *
	IR	HB	36

* Topcoats did not crack even when the drop height was at the highest level.

3.3. Electrochemical Characterization

The results of measuring the open circuit potential after stabilization in 3.5% NaCl solution are shown in Table 4. Measurements after 1 hour differ significantly from the measurements after 168 h (7 days) and 240 h (10 days), which means that the corrosion potential did not stabilize after 1 hour spent in the electrolyte. The results of measuring the corrosion potential after 168 and 240 h are similar, which indicates the stabilization of the corrosion potential of the coated sample.

Table 4. Open circuit potential after stabilization in a 3.5% NaCl solution.

Topcoat	E_{corr} vs. SCE [mV]		
	1 h	168 h	240 h
1 IR	−464.9	−195.4	−202.7
1 atm	−513.4	−333.7	−288.3
2 IR	−46.0	−217.8	−216.3
2 atm	−540.1	−102.1	−258.6
3 IR	−138.9	−214.1	−151.8
3 atm	−562.2	−201.2	−241.4
4 IR	49.4	−182.5	−140.8
4 atm	200.9	−433.1	−360.7
5 IR	−432.1	−179.1	−169.6
5 atm	−559.7	−135.9	−127.0

Equivalent electrical circuits used to describe EIS results are shown in Figure 5a,b. Circuits consist of electrolyte resistance (R_s), coating resistance (R_c), coating capacitance (C_c), charge transfer resistance (R_{ct}), and double-layer capacitance between metal and electrolyte solution (C_{dl}). In some cases, an ideal capacitor model (C) was used to describe the capacitor, while in some other cases, the results were better described by a constant phase element (Q), which represents an imperfect capacitor. After 10 days in the electrolyte, with the 1 atm, 1 IR, and 2 atm topcoats, diffusion of the electrolyte to the metal substrate occurred, so the equivalent circuit in Figure 5b was used. This circuit contains an additional element W, i.e., Warburg impedance, which describes the phenomenon of diffusion [5,30].

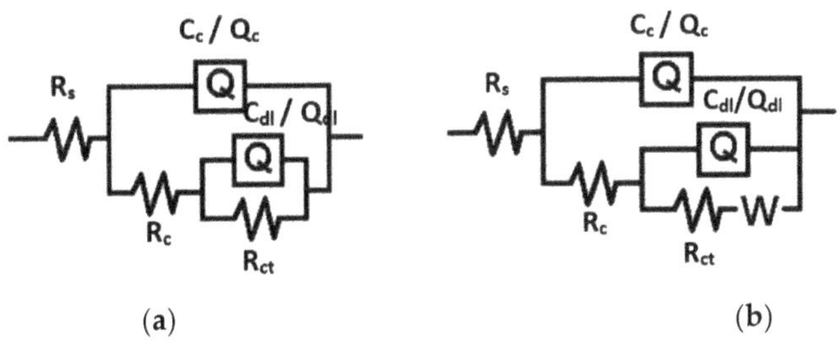

Figure 5. Equivalent electric circuit models: (**a**) electric circuit with three resistors and two capacitors, (**b**) the same circuit with the addition of element W, Warburg impedance [5].

Numerical values of coating resistance that were calculated using the ZSimpWin program are shown in Figure 6. Nyquist and Bode plots obtained by mathematical models, according to equivalent electric circuits, are shown in Figures 7 and 8. The IR-dried samples initially showed higher resistance than the air-dried samples. After exposure to the electrolyte, topcoats 1 and 2 show greater stability when IR-dried, while the other topcoats show greater stability in the case of atmospheric drying. Resistance of the coating decreased after ten days of exposure to the electrolyte, probably because of the penetration of the electrolyte into the micropores of the coating. Topcoat 1's protective properties deteriorated the most. This was probably due to a too-low film thickness, as the optimal thickness of the dry film is 100 μm, according to the manufacturer's recommendations. After one hour of immersion in the electrolyte, the PSX-based topcoat showed the best corrosion resistance. After ten days in the electrolyte, its resistance remained high. Topcoat 4 showed the highest resistance after 10 days. According to its technical data sheet, this topcoat has active pigments in its composition. Active pigments reduce or prevent corrosion of the metal substrate electrochemically by a galvanic or passivating mechanism or by building-up solid compounds with barrier properties that plug the damage in the coating [34].

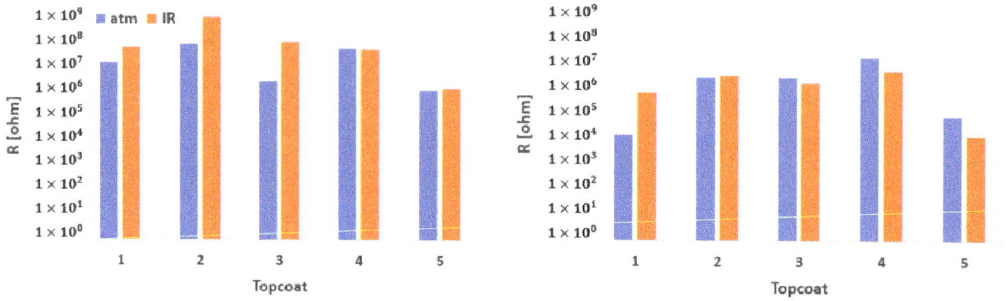

(a) (b)

Figure 6. Topcoats resistance after 1 h (**a**) and 240 h (**b**) in 3.5% NaCl solution.

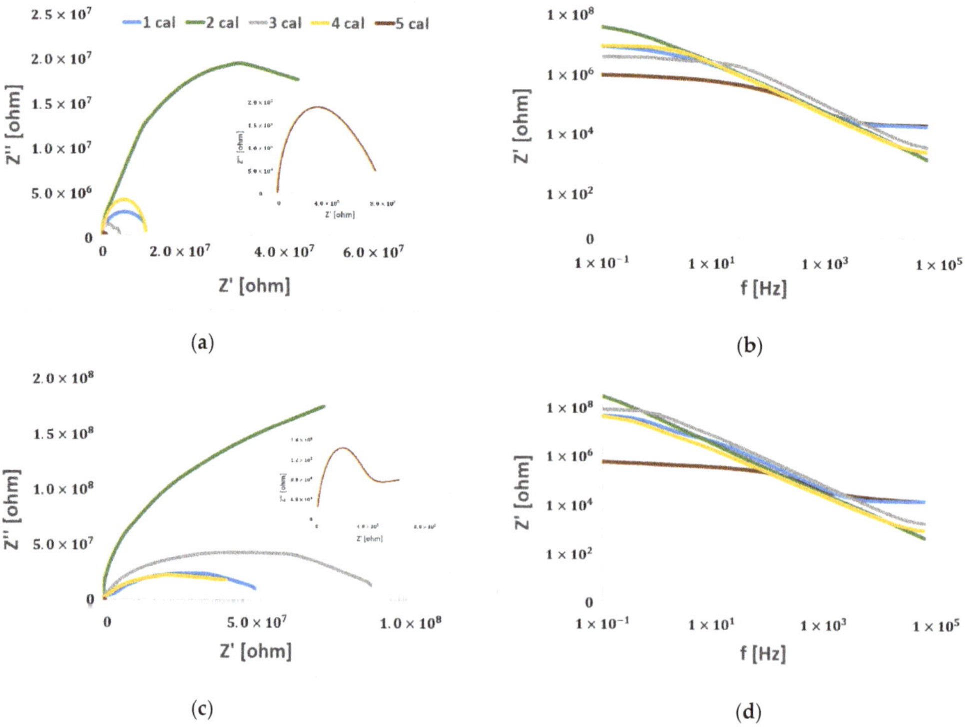

(a) (b)

(c) (d)

Figure 7. Nyquist and Bode plots for topcoats dried atmospherically (**a**,**b**) and with IR radiation (**c**,**d**) after 1 h in 3.5% NaCl solution.

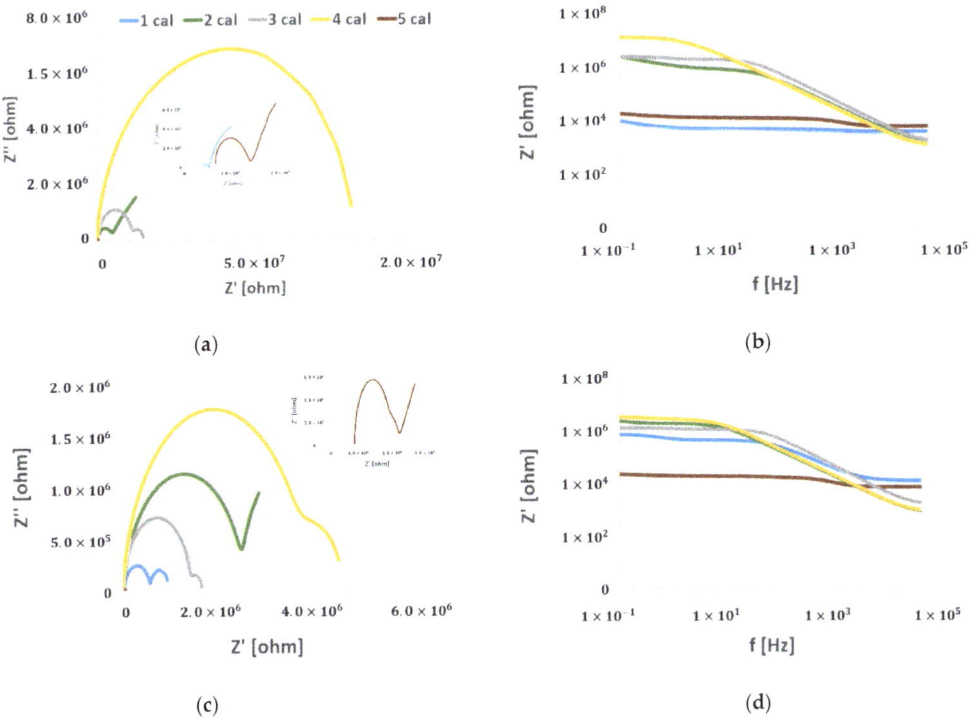

Figure 8. Nyquist and Bode plots for topcoats dried atmospherically (**a**,**b**) and with IR radiation (**c**,**d**) after 240 h in 3.5% NaCl solution.

3.4. Thermogravimetric Analysis (TGA) and Differential Scanning Calorimetry (DSC)

From TG and DTG curves, obtained by thermogravimetric analysis, the following features were determined: temperature at the beginning of decomposition ($T_{5\%}$) at 5% mass loss and temperature at the end of decomposition (T_f), as shown in Table 5. The initial decomposition temperature was, in most cases, higher in the case of IR-drying, which indicates the higher thermal stability of topcoats dried with IR radiation [35]. Only topcoat 5 showed better thermal stability when dried atmospherically. Topcoat 5's technical data sheet states that there is a presence of special barrier pigments in its composition. Some barrier pigments have reflective properties to avoid unnecessary heating [36]. Instead of absorbing the infrared energy, they reflect it back, reducing the amount of heat generated during the curing process, which can affect the curing kinetics and, consequently, the thermal stability. TGA thermograms of examined topcoats are shown in Figure A1.

The result of differential scanning calorimetry is the glass transition temperature, T_g, of the polymer (Table 5). Higher T_g values are desirable since they indicate better curing of the polymer [31,37]. As the curing of the polymer progresses, the molecular network mobility decreases; the degree of crystallinity increases; and the glass transition temperature of the polymer, T_g, increases with it. In the case of PUR-based topcoats, T_g is higher in the case of IR-drying, while a PSX-based topcoat (topcoat 2) exhibits a higher T_g value when dried atmospherically. Infrared radiation can have two effects on the glass transition of polymers. First is the heating effect, in which IR radiation transfers energy to the polymer, resulting in an increase in temperature. With higher curing temperature, the degree of crosslinking and the glass transition temperature increase [38]. That is the case with all examined PUR topcoats. On the other hand, the PSX-based topcoat (topcoat 2) has the shortest drying time, according to its technical data sheet. This latter was exposed to infrared radiation for the same amount of time as PUR-based topcoats. The PSX might require a shorter period of exposure to IR radiation because infrared radiation induces polymer scission, which increases the mobility of the polymer chains, and as a result of that, the glass transition temperature decreases [39]. DSC thermograms for the second heating cycle are shown in Figure A2.

Table 5. Results obtained by TGA ($T_{5\%}$, T_f) and DSC analyses (T_g).

Topcoat	$T_{5\%}$ [°C]	T_f [°C]	T_g [°C]
1 atm	252.9	437.6	39.4
1 IR	254.7	438.1	45.0
2 atm	257.6	536.0	85.7
2 IR	259.3	536.0	81.2
3 atm	226.4	391.3	39.8
3 IR	236.1	388.0	43.2
4 atm	263.8	474.3	37.6
4 IR	267.9	460.0	41.3
5 atm	268.1	413.3	45.5
5 IR	261.8	402.0	56.1

3.5. Fourier-Transform Infrared Spectroscopy (FTIR)

Table 6 gives an overview of the peak positions recorded by ATR-FTIR and a description of the characteristic absorption bands for polyurethanes. Walo et al. [28] state that absorption bands (recorded for atmospherically dried topcoats 1 and 3) at 3457.24 and 3457.58 cm^{-1} characterize free N-H stretching, while absorption bands (recorded for IR-dried topcoats 1 and 3) at 3382.27 and 3372.84 cm^{-1} characterize hydrogen-bonded N-H stretching. Similarly, Li et al. [29] state that absorption bands (of atmospherically dried topcoats 1 and 3) at 1737.06 and 1734.32 cm^{-1} characterize free urethano, while absorption bands (of IR-dried topcoats 1 and 3) at 1721.69 and 1722.90 cm^{-1} characterize hydrogen-bonded urethano. Hydrogen bonds explain the better mechanical properties of IR-cured topcoats, given that the intermolecular forces generated by them can improve the crosslinking density of polymers [29]. For topcoats 4 and 5, these wave numbers differ less, regardless of the drying method. FTIR spectra of PUR-based topcoats are shown in Figure A3, and the FTIR spectrum of the PSX-based topcoat is shown in Figure A4.

Table 6. Peak positions recorded on the ATR-FTIR spectrum and characteristic absorption bands for polyurethanes explained [40,41].

Wavenumber [cm^{-1}]								Assignment
Topcoat 1		Topcoat 3		Topcoat 4		Topcoat 5		
atm	IR	atm	IR	atm	IR	atm	IR	
3736.48	3738.29		3736.24	3677.26	3677.40			O-H stretching
3457.58	3382.27	3457.24	3372.84	3371.98	3367.21	3365.08	3380.38	N-H stretching
2955.58	2929.27	2941.43	2930.34	2928.97	2928.88	2929.56	2929.53	C-H asymmetric stretching
2864.64	2862.15	2869.03	2866.86	2860.18	2860.24	2859.81	2859.13	C-H symmetric stretching
1737.06	1722.90	1734.32	1721.69	1724.49	1723.25	1725.34	1724.99	C=O stretching vibrations interacting with H bond urethane
1443.11	1449.88	1448.57	1452.53	1461.47	1460.05	1452.89	1453.02	CH_2 scissoring and CH_3 deformation
		1115.85						Free C-O-C stretching
1082.86	1081.24	1071.56	1067.21	1073.40	1072.60	1073.10	1073.76	H bound C-O-C stretching

Table 7 gives an overview of the peak positions recorded by ATR-FTIR and a description of the characteristic absorption bands for polysiloxanes. Atmospherically and IR-dried topcoats show absorption peaks at similar wavelengths, with the exception of the absorption peak of the IR-cured topcoat at 1064.24 cm^{-1}. This peak indicates the asymmetric stretching of the Si-O-Si groups with a higher crosslinking degree [42]. The absence of a peak at that wavelength of the atmospherically dried topcoat indicates a lower degree of crosslinking of that topcoat, thus explaining the poorer mechanical properties of this coating, which was confirmed by the pencil and impact resistance tests.

Table 7. Peak positions recorded on the ATR-FTIR spectrum and characteristic absorption bands for polysiloxanes explained [42].

Wavenumber [cm^{-1}]		Assignment
Topcoat 2		
atm	IR	
1258.34	1259.31	CH_3 symmetric bending
	1064.24	Si-O-Si asymmetric stretching (high crosslinking degree)
1018.59	1021.55	Si-O-Si asymmetric stretching (surface groups, low crosslinking degree)
844.76	845.46	Si-CH_3 vibration
798.99	799.40	Si-C vibration

4. Conclusions

The results of this study showed overall better properties of infrared-dried industrial topcoats in relation to the same topcoats dried under atmospheric conditions. In addition to better crosslinking of polymers, which results in better protective properties of the topcoat, drying times are significantly shortened, which is especially important from an industrial point of view. Conclusions based on the conducted tests are as follows:

- Pull-off adhesion values are higher when applying IR radiation to cure the topcoat.
- Gloss values remain similar, regardless of the drying method. Furthermore, no changes in the shade of the topcoat are visible to the eye.
- Topcoat hardness is equal or higher in the case of IR-drying, while differently dried topcoats show a similar impact resistance.
- EIS initially showed a higher resistance of the IR-dried coatings. After 10 days in the electrolyte, topcoats from the Danish manufacturer show similar or better resistance in the case of IR drying, while IR-dried topcoats from the German manufacturer show a slightly lower resistance than the atmospherically dried ones.
- TGA showed a better thermal stability of IR-cured topcoats.
- DSC results showed that the glass transition temperature of the PUR-based topcoat increases when infrared radiation is applied, while the glass transition temperature of PSX decreases with the application of IR radiation, which indicates the sensitivity of PSX to IR radiation and the possible polymer scission effect.
- The results of FTIR indicate a higher curing degree of IR-dried topcoats.
- Although, in principle, infrared radiation improves the properties of the coating, it is necessary to pay attention to both the base polymer in the coating and the existence of special fillers or pigments that can affect the curing of the coating.

Author Contributions: M.L. and I.F. performed the experiments; V.A. analyzed the data; M.L. wrote the paper; I.S. designed the experiments and revised and edited the paper; D.R.-R. provided resources and the drying experimental setup. All authors have read and agreed to the published version of the manuscript.

Funding: This study is based on a project "Smart plant for drying liquid coatings", which is co-financed within the Operational Programme Competitiveness and Cohesion from the European Regional Development Fund under reference number KK.01.2.1.02.0030.

Institutional Review Board Statement: Not applicable.

Informed Consent Statement: Not applicable.

Data Availability Statement: Not applicable.

Conflicts of Interest: The authors declare no conflict of interest.

Appendix A

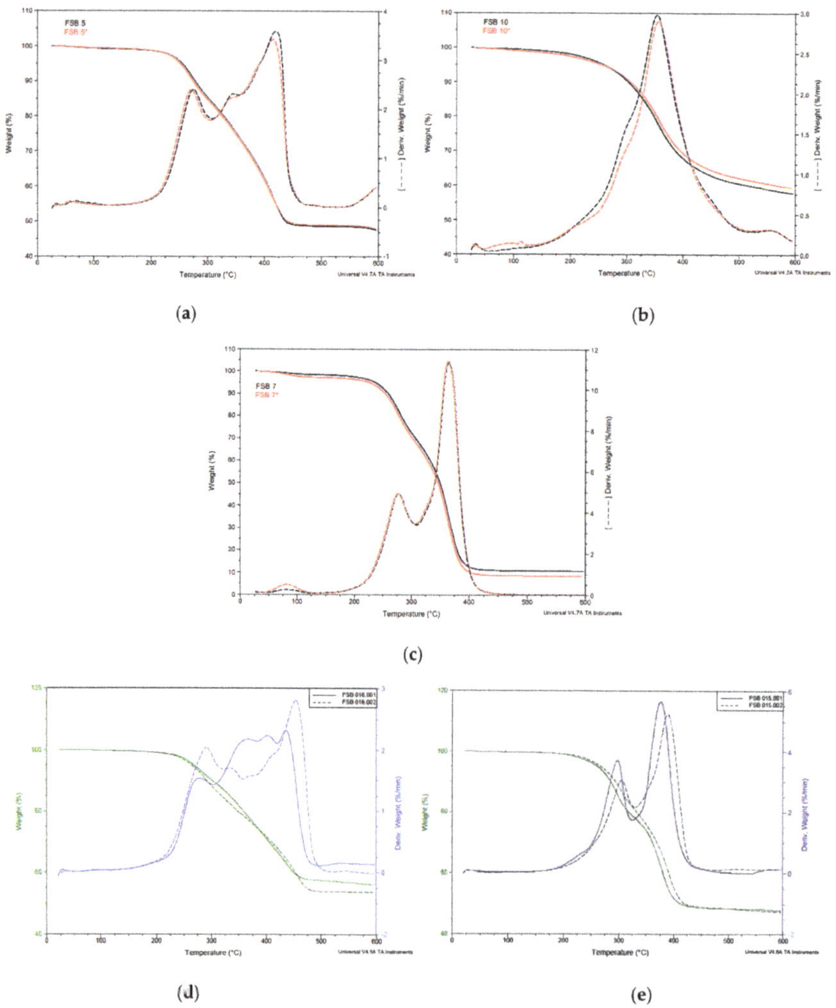

Figure A1. TGA thermograms of topcoats 1 (**a**), 2 (**b**), 3 (**c**), 4 (**d**), and 5 (**e**). Graphs obtained for topcoats from Danish manufacturer (1, 2, 3) that are dried atmospherically are colored red, and IR-dried topcoats are colored black. Graphs obtained for topcoats from German manufacturer (4, 5) that are dried atmospherically are marked on the graph with dashed lines, and IR-dried topcoats are marked with a solid line.

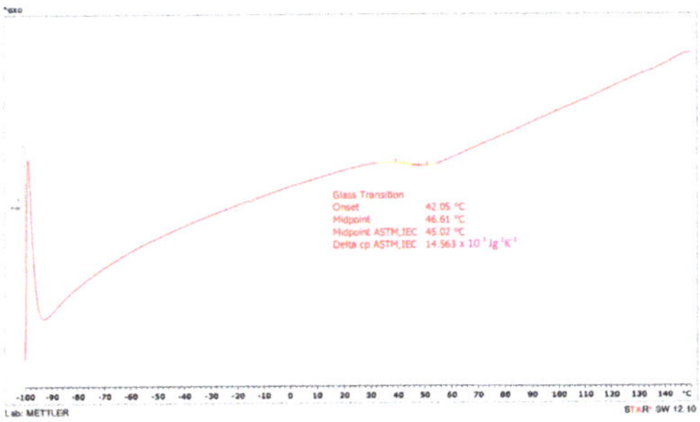

(a)

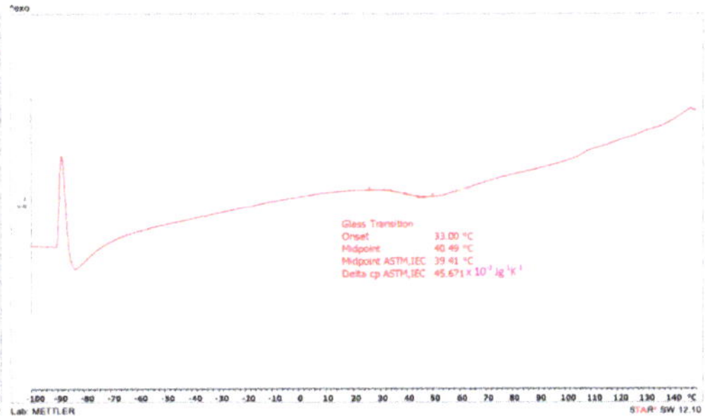

(b)

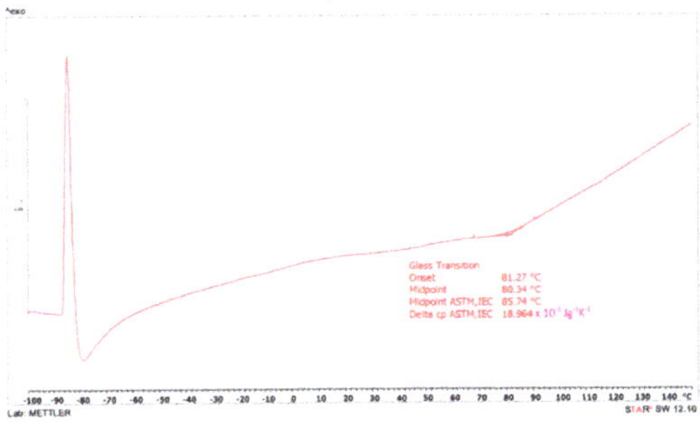

(c)

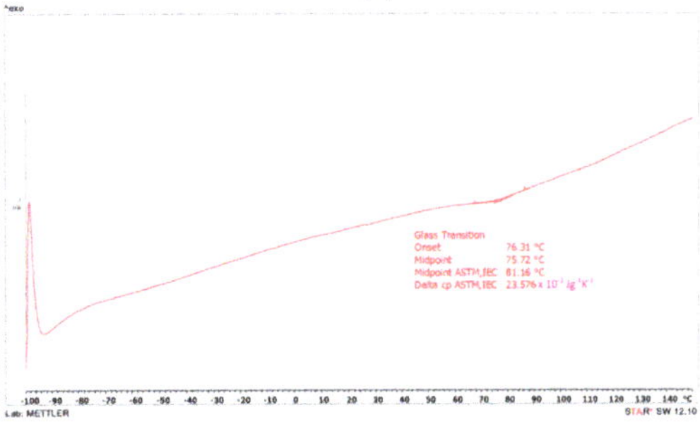

(d)

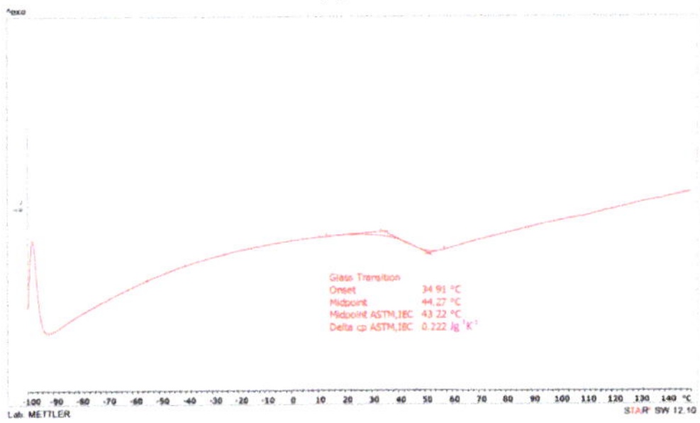

(e)

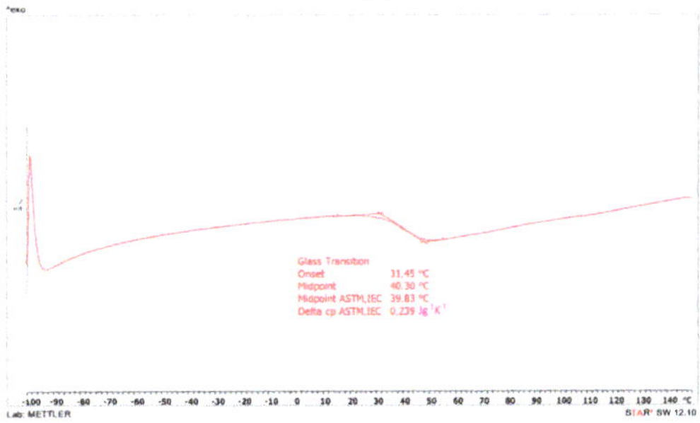

(f)

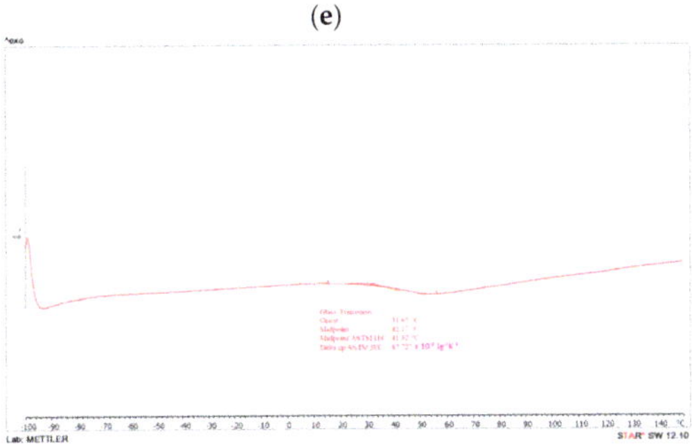

(g)

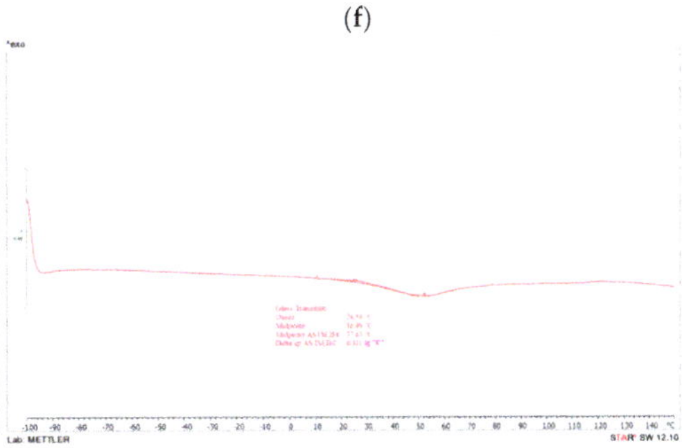

(h)

Figure A2. *Cont.*

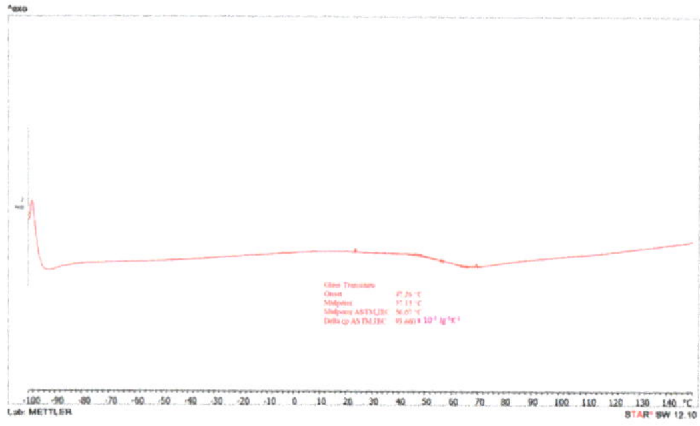

(i)

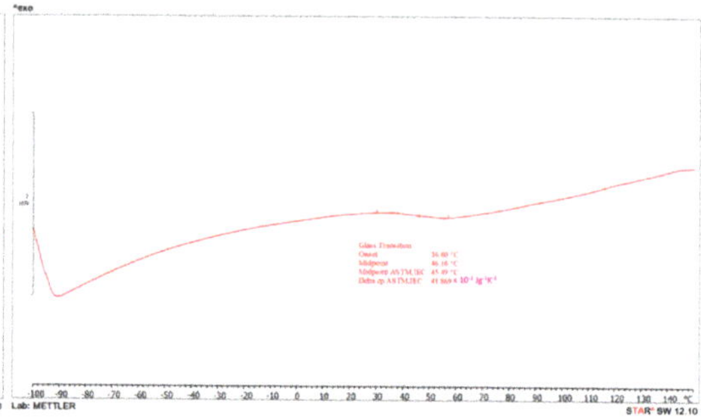

(j)

Figure A2. DSC thermograms for topcoat 1 IR-dried (**a**) and atm-dried (**b**), topcoat 2 IR-dried (**c**) and atm-dried (**d**), topcoat 3 IR-dried (**e**) and atm-dried (**f**), topcoat 4 IR-dried (**g**) and atm-dried (**h**), and topcoat 5 IR-dried (**i**) and atm-dried (**j**).

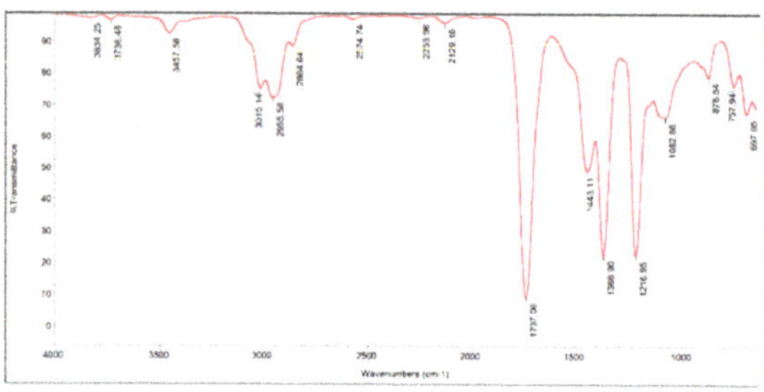

(**a**)

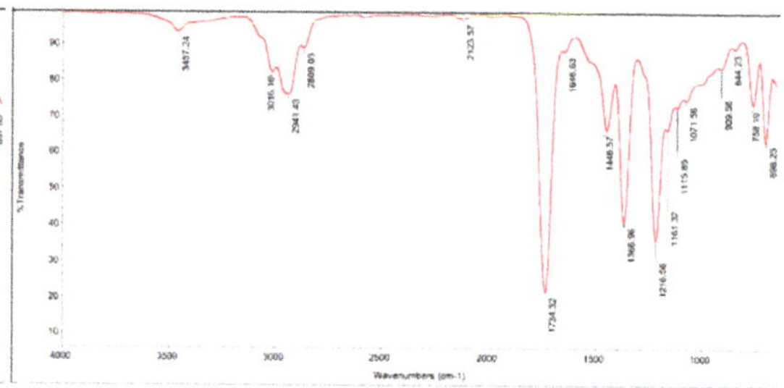

(**b**)

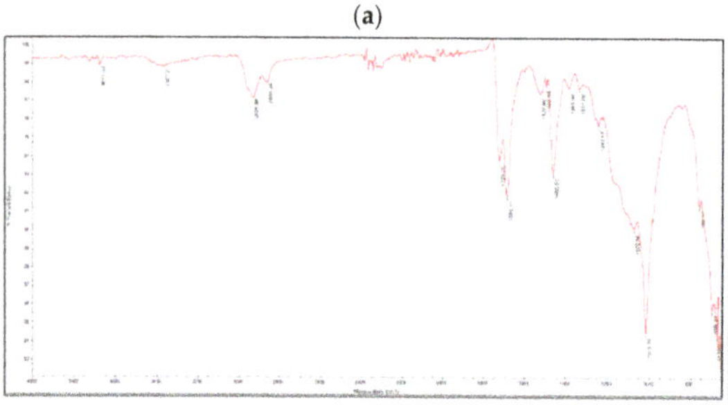

(**c**)

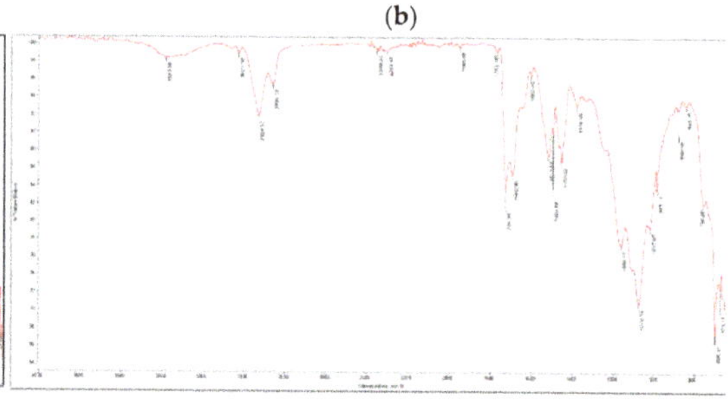

(**d**)

Figure A3. FTIR spectra of topcoats 1 (**a**), 3 (**b**), 4 (**c**), and 5 (**d**).

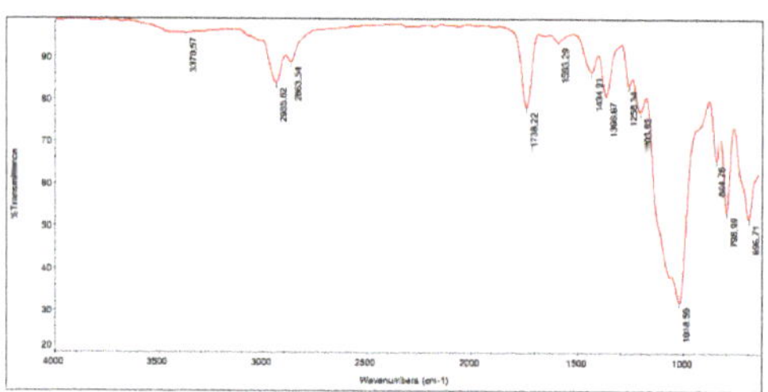

Figure A4. FTIR spectrum of topcoat 2.

References

1. Valdez, B.; Schorr, M.; Zlatev, R.; Carrillo, M.; Stoytcheva, M.; Alvarez, L.; Eliezer, A.; Rosas, N. Corrosion Control in Industry. In *Environmental and Industrial Corrosion—Practical and Theoretical Aspects*; Valdez, B., Schorr, M., Eds.; InTech: Rijeka, Croatia, 2012; Volume 1, pp. 19–51. [CrossRef]
2. NACE International. Economic Impact. Available online: http://impact.nace.org/economic-impact.aspx (accessed on 19 April 2023).
3. Hinton, B.R.W. Corrosion prevention and control. In *Handbook on the Physics and Chemistry of Rare Earths*; Gischneidner, K.A., Jr., Eyring, L., Eds.; North-Holland: Amsterdam, The Netherlands, 1995; Volume 21, pp. 29–35. [CrossRef]
4. Maaß, P. Corrosion and Corrosion Protection. In *Handbook of Hot-Dip Galvanization*; Maaß, P., Peißker, P., Eds.; Wiley-VCH: Weinheim, Germany, 2011; Volume 1, pp. 1–19. [CrossRef]
5. Alsamuraee, A.M.A.; Jaafer, I.H.; Ameen, A.H.; Abdullah, Q.A. Electrochemical impedance spectroscopic evaluation of corrosion protection properties of polyurethane/polyvinyl chloride blend coating on steel. *Am. J. Sci. Res.* **2011**, *2*, 761–768. [CrossRef]
6. Zheng, S.X. *Principles of Organic Coatings and Finishing*, 1st ed.; Cambridge Scholars Publishing: Newcastle upon Tyne, UK, 2020; pp. 1–14. Available online: https://www.cambridgescholars.com/resources/pdfs/978-1-5275-4119-1-sample.pdf (accessed on 7 June 2023).
7. Davidson, R.S.; Holman, R.J. Developments and trends in radiation curing. *Rev. Prog. Color.* **2003**, *33*, 46–58. [CrossRef]
8. Nienhuis, J.G. *Review on Drying and Curing Techniques of Coatings*; SHR: Wageningen, The Netherlands, 2004. Available online: http://virtual.vtt.fi/virtual/proj6/coste18/nienhuisdryingpaper.pdf (accessed on 7 June 2023).
9. Wali, A.S.; Kumar, S.; Khan, D. A review on recent development and application of radiation curing. *Mater. Today Proc.* **2023**, *82*, 68–74. [CrossRef]
10. Florencia, M.; López, A.; Barbosa, S.E.; Martini, R.E. Use of infrared radiation as curing agent for bioactive paper fabrication in times compatible with industrial process. *Chem. Eng. Process. Process Intensif.* **2020**, *157*, 108165. [CrossRef]
11. Pendar, M.-R.; Rodrigues, F.; Páscoa, J.C.; Lima, R. Review of coating and curing processes: Evaluation in automotive industry. *Phys. Fluids.* **2022**, *34*, 101301. [CrossRef]
12. Pathania, A.; Arya, R.K.; Ahuja, S. Crosslinked polymeric coatings: Preparation, characterization, and diffusion studies. *Prog. Org. Coat.* **2017**, *105*, 149–162. [CrossRef]
13. Saure, R.; Gnielinski, V. Moisture measurement by FT-IR-spectroscopy. *Dry. Technol.* **1994**, *12*, 1427–1444. [CrossRef]
14. Saure, R.; Wagner, G.R.; Schlunder, E.U. Drying of solvent-borne polymeric coatings. II. Experimental results using FTIR spectroscopy. *Surf. Coat. Technol.* **1998**, *99*, 257–265. [CrossRef]
15. Geipel, C.; Stephan, P. Experimental investigation of the drying process of automotive base paints. *Appl. Therm. Eng.* **2005**, *25*, 2578–2590. [CrossRef]
16. Vessot, S.; Andrieu, J.; Laurent, P.; Galy, J.; Gerard, J.F. Air convective drying and curing of polyurethane-based paints on sheet molding compound surfaces. *J. Coat. Technol.* **1998**, *70*, 67–76. [CrossRef]
17. Stojanović, I.; Cindrić, I.; Janković, L.; Šimunović, V.; Franjić, H. Performance Assessment of Differently Dried Coating Systems for Corrosion Prevention in Power Transformer Industry. *Coatings* **2021**, *12*, 331. [CrossRef]
18. *ISO 8501-1*; Preparation of Steel Substrates before Application of Paints and Related Products—Visual Assessment of Surface Cleanliness—Part 1: Rust Grades and Preparation Grades of Uncoated Steel Substrates and of Steel Substrates after Overall Removal of Previous Coatings. International Organization for Standardization: Geneva, Switzerland, 2007.
19. *ISO 8503-1*; Preparation of Steel Substrates before Application of Paints and Related Products—Surface Roughness Characteristics of Blast-Cleaned Steel Substrates—Part 1: Specifications and Definitions for ISO Surface Profile Comparators for the Assessment of Abrasive Blast-Cleaned Surfaces. International Organization for Standardization: Geneva, Switzerland, 2012.
20. Jonkers, G.; Nienhuis, J.G.; van de Velde, B. Climate Independent Painting: Is Infrared Heating a Solution for Professional Painters? *Macromol. Symp.* **2002**, *187*, 249–260. [CrossRef]
21. *ISO 2808*; Paints and Varnishes—Determination of Film Thickness. International Organization for Standardization: Geneva, Switzerland, 2019.
22. *ISO 4624*; Paints and Varnishes—Pull-Off Test for Adhesion. International Organization for Standardization: Geneva, Switzerland, 2016.
23. *ISO 2813*; Paints and Varnishes—Determination of Gloss Value at 20°, 60° and 85°. International Organization for Standardization: Geneva, Switzerland, 2014.
24. *ISO 15184*; Paints and Varnishes—Determination of Film Hardness by Pencil Test. International Organization for Standardization: Geneva, Switzerland, 2020.
25. *ISO 6272*; Paints and Varnishes—Rapid-Deformation (Impact Resistance) Tests—Part 1: Falling-Weight Test, Larne-Area Indenter. International Organization for Standardization: Geneva, Switzerland, 2002.
26. Wright, J.M.; Colling, A. *Seawater: Its Composition, Properties and Behaviour*, 2nd ed.; Butterworth-Heinemann: Oxford, UK, 1995; p. 5. [CrossRef]
27. Samardžija, M.; Alar, V.; Špada, V.; Stojanović, I. Corrosion Behaviour of an Epoxy Resin Reinforced with Aluminium Nanoparticles. *Coatings* **2022**, *12*, 1500. [CrossRef]
28. Walo, M.; Przybytniak, G.; Kavakli, P.A.; Güven, O. Radiation-induced graft polymerization of N-vinylpyrrolidone onto segmented polyurethane based on isophorone diisocyanate. *Radiat. Phys. Chem.* **2013**, *84*, 85–90. [CrossRef]

29. Li, S.; Liu, Z.; Hou, L.; Chen, Y.; Xu, T. Effect of polyether/polyester polyol ratio on properties of waterborne two component polyurethane coatings. *Prog. Org. Coat.* **2020**, *141*, 105545. [CrossRef]
30. Šoić, I.; Martinez, S.; Dubravić, M. Gel-Electrolyte EIS setup used for probing of IR Dried/Cured industrial coatings. *Prog. Org. Coat.* **2019**, *137*, 105331. [CrossRef]
31. Uday, M.; Kiran Kumar, P. Heat transfer studies for infrared radiation assisted curing in polymer composites. *J. Phys. Conf. Ser.* **2020**, *1473*, 12027. [CrossRef]
32. Bai, H.; Zhong, L.; Kang, L.; Liu, J.; Zhuang, W.; Lv, Z.; Xu, Y. A review on wear-resistant coating with high hardness and high toughness on the surface of titanium alloy. *J. Alloys Compd.* **2021**, *882*, 160645. [CrossRef]
33. DeArmitt, C. Functional Fillers for Plastics. In *Applied Plastics Engineering Handbook*, 2nd ed.; Kutz, M., Ed.; William Andrew Publishing: Norwich, NY, USA, 2017; Volume 3, pp. 517–532. [CrossRef]
34. Rammelt, U.; Reinhard, G. Characterization of active pigments in damage of organic coatings on steel by means of electrochemical impedance spectroscopy. *Prog. Org. Coat.* **1994**, *24*, 309–322. [CrossRef]
35. Kuźnia, M.; Magiera, A.; Zygmunt-Kowalska, B.; Kaczorek-Chrobak, K.; Pielichowska, K.; Szatkowski, P.; Benko, A.; Ziąbka, M.; Jerzak, W. Fly Ash as an Eco-Friendly Filler for Rigid Polyurethane Foams Modification. *Materials* **2021**, *14*, 6604. [CrossRef]
36. Coser, E.; Froes Moritz, V.; Krezinger, A.; Ferreire, C.A. Development of paints with infrared radiation reflective properties. *Polímeros* **2015**, *25*, 305–310. [CrossRef]
37. Acik, G.; Karabulut, H.R.F.; Altinkok, C.; Karatavuk, A.O. Synthesis and characterization of biodegradable polyurethanes made from cholic acid and L-lysine diisocyanate ethyl ester. *Polym. Degrad. Stab.* **2019**, *165*, 42–48. [CrossRef]
38. Michel, M.; Ferrier, E. Effect of curing temperature conditions on glass transition temperature values of epoxy polymer used for wet lay-up applications. *Constr. Build. Mater.* **2020**, *231*, 117206. [CrossRef]
39. Al-Qaraqawi, I.Y.; Abdulmalik, D.A.; Madi, N.K.; Almaadeed, M. Gamma irradiation effects on polymethyl methacrylate. *Phys. Stat. Sol.* **2007**, *4*, 3727–3730. [CrossRef]
40. Mora-Murillo, L.D.; Orozco-Gutierrez, F.; Vega-Baudrit, J.; González-Paz, R.J. Thermal-Mechanical Characterization of Polyurethane Rigid Foams: Effect of Modifying Bio-Polyol Content in Isocyanate Prepolymers. *J. Renew. Mater.* **2017**, *5*, 220–230. [CrossRef]
41. Rangel-Vazquez, N.-A.; Sánchez-López, C. Spectroscopy Analyses of Polyurethane/Polyaniline IPN Using Computational Simulation (Amber, MM+ and PM3 Method). *Polímeros* **2014**, *24*, 1–11. [CrossRef]
42. Steinbach, J.C.; Schneider, M.; Hauler, O.; Lorenz, G.; Rebner, K.; Kandelbauer, A. A Process Analytical Concept for In-Line FTIR Monitoring of Polysiloxane Formation. *Polymers* **2022**, *12*, 2473. [CrossRef] [PubMed]

coatings

Article

Wear Resistance Study of Bionic Pitted Ni Cladding Layer on 7075 Aluminum Alloy Drill Pipe Surface

Xu Li [1,2,3], Ke Gao [1,2,3], Yan Zhao [1,2,3,*], Xiaobo Xie [1,2,3], Xiaoshu Lü [1,4,5], Cong Zhang [1,2,3] and Hongxin Ai [1,2,3]

1 College of Construction Engineering, Jilin University, Changchun 130026, China
2 Engineering Research Center of Geothermal Resources Development Technology and Equipment, Ministry of Education, Jilin University, Changchun 130026, China
3 Key Laboratory of Drilling and Exploration Technology in Complex Conditions, Ministry of Natural Resources, Jilin University, Changchun 130026, China
4 Department of Electrical Engineering and Energy Technology, University of Vaasa, P.O. Box 700, 65101 Vaasa, Finland
5 Department of Civil Engineering, Aalto University, P.O. Box 12100, 02130 Espoo, Finland
* Correspondence: zhaoyan1983@jlu.edu.cn

Abstract: To enhance the lifespan of drill pipes and minimize wear, this study introduces a bionic structure model inspired by the pit shape structure found in the dung beetle's abdomen. The stress distribution and wear of bionic pitted structure and ordinary structure are simulated by finite element software. The findings revealed that the bionic structure significantly improves stress distribution, resulting in an impressive 81.3% increase in lifespan. Subsequently, the surface of the 7075 aluminum drill pipe was coated with Ni powder by a laser cladding system. Wear tests were conducted to analyze the wear and surface damage behavior of the cladding layer. The microstructure, composition, and microhardness of the cladding layer were measured and observed. The results showed that the cladding layer was mainly composed of Al_3Ni_2 and had high hardness. Additionally, a transition region exists between the cladding layer and the substrate, comprising relatively low hardness Al, thereby enhancing the drill pipe's ability to withstand alternating loads. Furthermore, the bionic structure possesses the capability to store particles, effectively reducing the occurrence of abrasive wear and increasing the lifespan by 70.0%.

Keywords: bionic structure; wear resistance; stress simulation; lifespan; laser cladding

1. Introduction

The performance of drill pipes has become increasingly important in recent years due to the growing number of complex well types, including deep wells, ultra-deep wells, and large displacement horizontal wells. Aluminum alloy drill pipes are popular due to their lightweight, good corrosion resistance, low bending stress, and non-magnetic properties [1–3]. However, the low hardness of the aluminum alloy drill pipe and the inevitable spiral bending downhole result in significant wear between the pipe and the wellbore and casing, leading to a reduced lifespan for the aluminum alloy drill pipe [4–6]. So, the demand for improving the wear resistance of aluminum alloy drill pipes has become urgent.

In order to improve the wear resistance of aluminum alloys, surface treatment methods such as liquid-phase metallurgy, solid-phase metallurgy, friction stir welding [7,8] micro-arc oxidation, spraying, and laser cladding are usually mentioned [9–11]. Laser cladding is a popular method for improving the wear resistance of aluminum alloy drill pipes due to its advantages of fast solidification, low substrate deformation, good metallurgical bonding, excellent performance, and good forming quality [12–15]. Ni-based powders have been studied and widely used in laser cladding materials for their good wettability, corrosion resistance, high-temperature self-lubrication, and moderate price [16]. However, applying laser cladding on the entire substrate surface may result in defects such as porosity, cracks, and material waste [17,18]. Bionics can effectively solve this problem.

Bionic research has found that the structure of some biological surfaces has high wear resistance, which is a feature gradually formed by biological evolution and optimization over hundreds of millions of years to adapt to the living environment [19]. The discovery of this phenomenon points out a new direction for the study of the wear resistance of friction pairs. Therefore, researchers explored the protective effects of various biological structures [20–22] on frictional wear surfaces, and then a bionic non-smooth surface was proposed. Jones et al. [23] and Minhaeng et al. [24] created dimple-like pits on cylindrical Co-Cr-Mo and 15CrMo alloy surfaces, respectively. Their research demonstrated the wear resistance of these bionic pits. Zhang [25] experimented with various bionic units on gray cast iron specimens and found that coupling these units improved the sliding wear resistance of the material. Mao C et al. [26] preliminarily explored the effects of dung beetle structure on the wear resistance pit tire tread compounds. Therefore, combining the bionic coupling functional surface with laser cladding can not only improve wear resistance but also reduce the formation of pores and cracks [27,28].

In this paper, the relationship between the diameter of the bionic pit shape structure and the center distance was summarized by observing the abdomen of multiple dung beetles. Then, a bionic pit shape structure model was established. The stress distribution and wear of bionic pitted structure and ordinary structure were simulated and compared using ANSYS, which is rarely studied in the existing literature. Subsequently, laser cladding was used to add bionic and normal structural cladding layers to the surface of the drill pipe, and wear tests were conducted using granite. Wear tests were conducted to analyze the wear and the surface damage behavior of the cladding layer. The microstructure, composition, and microhardness of the cladding layer were measured and observed, respectively, by SEM, XRD, and microhardness tester. The feasibility of the simulation results was verified, and a good foundation for subsequent studies was laid.

2. Measurement of Biological Surface Structure and Finite Element Analysis

2.1. Measurement of Biological Surface Structure

Dung beetles, as soil-dwelling animals, dig a storage chamber in the soil and push food to hide in it. The amount of earth to be moved to excavate a storage chamber is equivalent to 400 times the weight of the dung beetle itself [29]. In order to reduce the damage caused by soil, dung beetles evolved their pit shape structure on their abdomen over the years.

So as to investigate the features of the pit shape structures and furnish a basis for model establishment and test specimen preparation, three dung beetles were observed under a microscope. The aim was to determine the relationship between the diameter of the pit shape structure and the distance from the center of the circle.

The dung beetles used for observation were exclusively male and collected from the Yitong River area in Jilin Province, as depicted in Figure 1. After being thoroughly washed and disinfected with water and alcohol, they were made into samples and fixed onto a platform using playdough. The abdomen of the dung beetles was observed using a digital microscope. Figure 2 shows the abdomen of a dung beetle magnified 20 times under a microscope. It can be seen that the abdomen of the dung beetle has an obvious pit shape structure, and the pit units are circular and approximately distributed in a straight line, as shown in Figure 2a. In addition, the pit units have a local five-dot arrangement, as shown in Figure 2b. This structure can effectively reduce wear and greatly improve the wear resistance of the dung beetle's body surface.

Figure 1. (**a**) Dung beetles for observation. (**b**) The abdomen of a dung beetle.

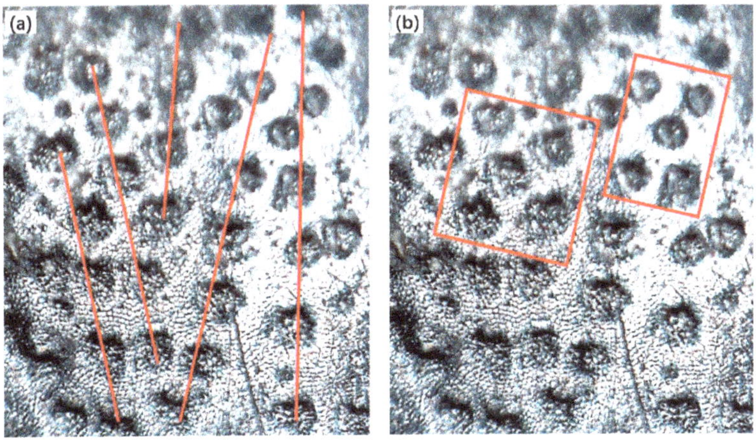

Figure 2. (**a**) Linear arrangement; (**b**) five-dot arrangement.

During the observation, two circular areas were marked on the abdomen of each dung beetle. The diameter of the pits within each area and the distance between the centers of two adjacent pits were measured. This process resulted in six groups of data for pit diameter and center distance. After calculating the average diameter and center distance for each group, Table 1 was obtained.

Table 1. Abdominal pit shape structure parameters of dung beetle.

Number	Region	S (mm)	D (mm)	$\overline{S}$ (mm)	$\overline{D}$ (mm)	$\overline{S}/\overline{D}$
1	1-1	0.28, 0.27, 0.28 0.29, 0.27, 0.28	0.14, 0.14, 0.15 0.13, 0.15, 0.14	0.278	0.142	1.95
	1-2	0.30, 0.31, 0.28 0.32, 0.31	0.15, 0.16, 0.16 0.14, 0.16	0.304	0.154	1.97
2	2-1	0.26, 0.27, 0.27 0.24, 0.26, 0.25	0.12, 0.13, 0.13 0.14, 0.13, 0.12	0.258	0.128	2.01
	2-2	0.24, 0.23, 0.24 0.23, 0.25, 0.26	0.12, 0.13, 0.13 0.11, 0.12, 0.13	0.242	0.123	1.96
3	3-1	0.25, 0.25, 0.27 0.26, 0.26	0.12, 0.12, 0.13 0.13, 0.11, 0.14	0.258	0.125	2.06
	3-2	0.21, 0.22, 0.23 0.22, 0.21, 0.21	0.11, 0.10, 0.12 0.11, 0.11, 0.12	0.217	0.112	1.93

In the table, S is the distance between the centers of two pits (mm); D is the pit diameter (mm); $\overline{S}$ is the average center distance of pits (mm); $\overline{D}$ is the average diameter of pits (mm).

$$S = \frac{S_1 + S_2 + \ldots S_n}{n} \tag{1}$$

$$D = \frac{D_1 + D_2 + \ldots D_n}{n} \tag{2}$$

It can be seen that the center distance of dung beetle abdominal pits ranges from 0.2 to 0.32 mm and the diameter from 0.1 to 0.16. The ratio between the average center distance and the average diameter in Table 1 is shown in Figure 3.

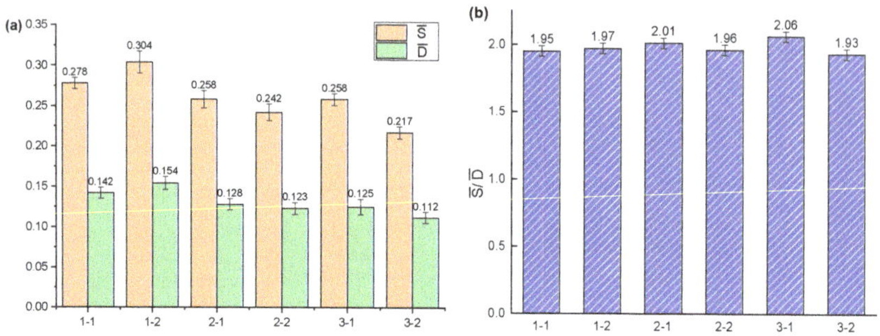

Figure 3. (**a**) $\overline{S}$ and $\overline{D}$; (**b**) the ratio of $\overline{S}/\overline{D}$.

It can be seen that the ratio of center distance to diameter is extremely close for different dung beetle abdominal pits despite their differences in size. The ratio of the average center distance to the average diameter of the six groups was averaged, and the result was 1.98.

2.2. Finite Element Analysis

2.2.1. Geometric Model and Material Properties

Taking 7075 aluminum drill pipe for drilling as the research object, the three-dimensional geometric model of the drill pipe was established. The model was coated with Ni on the exterior, with two different structures: pit shape and normal. These structures are depicted in Figure 4a. The pit shape structure was inspired by the abdominal structure of dung beetles, as shown in Figure 4b. The center distance between adjacent pits was 3.96 mm and the diameter of the pits was 2 mm. The ratio between them was 1.98, which is consistent with the result in Figure 3. The outer diameter of the model is 147 mm, with 2 mm of the cladding layer. At the same time, annular cross-grinding rock samples were established outside the coating, as shown in Figure 5. The material was granite, with a thickness of 6 mm.

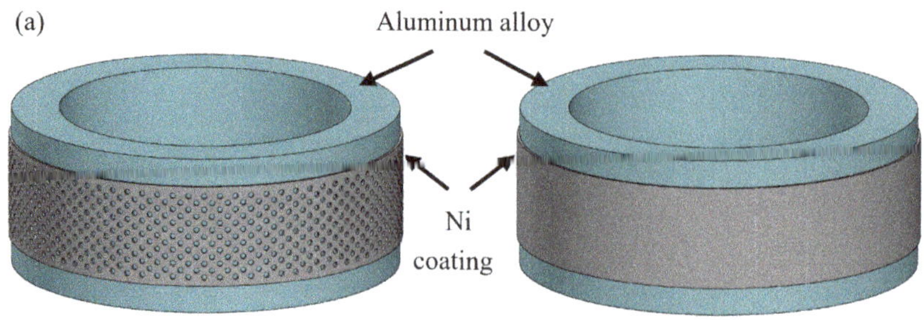

Figure 4. *Cont.*

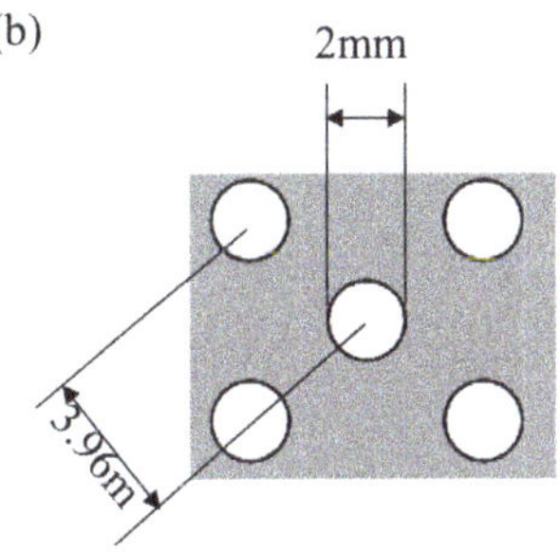

Figure 4. (**a**) Pit shape structure model and normal structure model; (**b**) pit shape structure parameters.

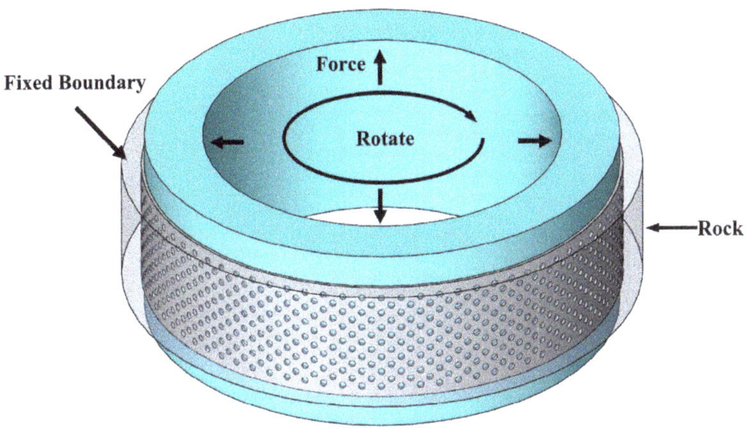

Figure 5. Rock sample diagram and boundary condition.

The boundary conditions are set in the load module. The motion path of the drill pipe could only rotate around the Y-axis. Pressure was applied on the inner side. For the rock sample, it is necessary to restrict its movement conditions in all directions, so fixed constraints were imposed. The boundary conditions are shown in Figure 5. The simulation was conducted for a duration of 2 s, with a lateral force of 100 N and a rotational speed of 30 rpm. The basic parameters of the materials used are outlined in Table 2 [30].

Table 2. Physical parameters for finite element analysis.

Model	Material	Density (kg·m^{-3})	Elastic Modulus (GPa)	Poisson's Ratio
Coating	Ni	8900	210	0.3
Drill pipe	Aluminum alloy	2810	72	0.33
Rock	Granite	2650	40	0.25

2.2.2. Wear Simulation

At the continuum mechanics scale, the wear phenomenon can use the phenomenological model to estimate. The model relates various state parameters of the contact surface to the amount of wear [31]. The finite element analysis software ANSYS can approximately calculate the wear caused by the material loss by repositioning the contact node at the contact surface. The Ansys Mechanical platform uses the Archard model to evaluate the wear.

In the field of continuum mechanics, wear can be estimated using a phenomenological model that correlates different state parameters at the interface to material loss [29]. To approximate the material loss caused by wear, the finite element analysis software Ansys repositions the contact nodes on the contact surface. The amount of wear is estimated using the Archard model in the Ansys Mechanical platform.

In 1953, Archard successfully built the Archard model based on Holm's model. It expresses the relationship between wear rate and contact pressure, material hardness and sliding speed, and its expression is as follows:

$$\frac{dV}{dt} = \frac{KP^a v_s^b}{\sigma_s} n \tag{3}$$

In the formula,

V—wear volume;
P—contact load;
V_s—slipping speed between the contacting objects;
a—contact stress index;
b—speed index;
n—in-plane normal;
K—the wear coefficient;
σ_s—plastic yield strength of softer materials.

The yield strength is assumed to be equal to the hardness value, and the yield pressure in Equation (3) can be replaced by hardness (H). Then, the Archard wear model expression can be rewritten as follows:

$$\frac{dV}{dt} = \frac{KP^a v_s^b}{H} n \tag{4}$$

H—The Brinell hardness of softer materials; the Brinell hardness of Ni is about 282.

Where P and v_s are determined by boundary conditions, whereas k, a, and b need to be input. In this paper, k is 1×10^{-6} and both a and b are 1.

The workflow of ANSYS for wear simulation is illustrated in Figure 6. It mainly consists of two steps: condition setting and wear simulation cycle. Before starting the simulation, the setup conditions inside the red box need to be defined, including the pressure, velocity, and friction coefficients. Since the purpose of the simulation is to compare the two structures, the coefficient of friction remains the same. In the wear simulation cycle, the contact nodes in each iteration are repositioned, each node along the X, Y, and Z direction of the relative displacement of node pressure and the speed is calculated, and then the material from the coating surface is removed to show wear and tear. The iterative process continues until the stable wear phase is reached. If the stable wear phase is not reached, the cycle proceeds to the next iteration, as shown in the blue box [32].

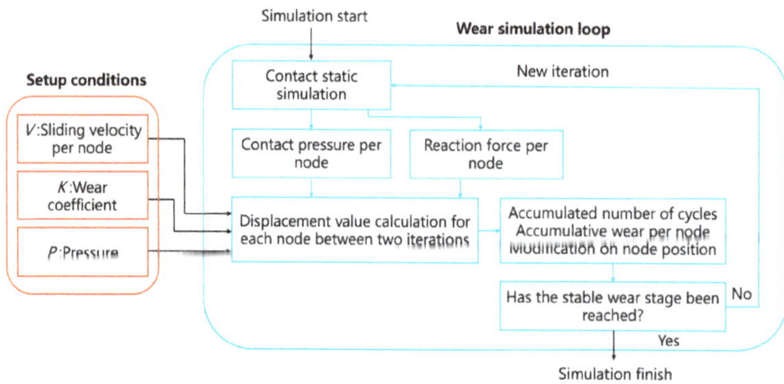

Figure 6. Workflow of the wear simulation method.

3. Experiment

3.1. Experimental Material and Laser Cladding Process

The experimental materials for this study consisted of a base material and a cladding material. The base material utilized was a 7075 aluminum alloy drill pipe, with ∅147 mm as outer diameter and 60 mm high. Prior to laser cladding, the oxide layer on the alloy surface was removed using a grinding wheel, followed by cleaning with acetone and coating with a light-absorbing material. The cladding material used was nickel powder, with a purity of 99.9% and a particle size of 50–100 μm (as shown in Figure 7). Prior to conducting the experiments, the powder was dried in a vacuum at 150 °C for 2 h.

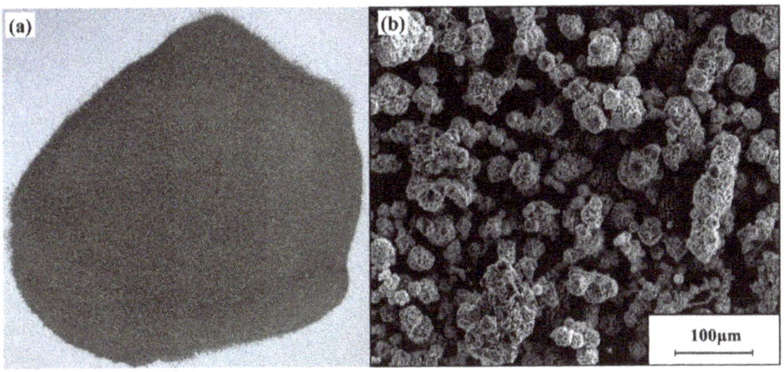

Figure 7. (**a**) Ni powder; (**b**) magnified by 20 times.

Laser cladding was carried out on an aluminum alloy tube using a CO_2 multifunctional CNC laser processing equipment. The equipment includes a fiber laser, a cooling system, and a control system. The laser power is 300 W and the speed is 30 mm/min. The laser wavelength is 1064 nm. The principle of laser cladding is illustrated in Figure 8. The surface of the aluminum alloy is heated by a laser-localized heating tube until it melts and by adopting the method of synchronous and rapid spraying of feeding powder in front of the laser beam to precast good Ni powder. When the beam is removed, the material goes through cooling consolidation in the outer layer of the aluminum alloy and forms a metallurgical bonding layer. The laser path to form different shapes of the cladding layer is controlled. The cladding thickness of the two structures is 2 mm, and the cladding is relatively uniform. During the laser cladding process, the flux of CO_2 is 15 L/min and the purity is greater than 99.9%.

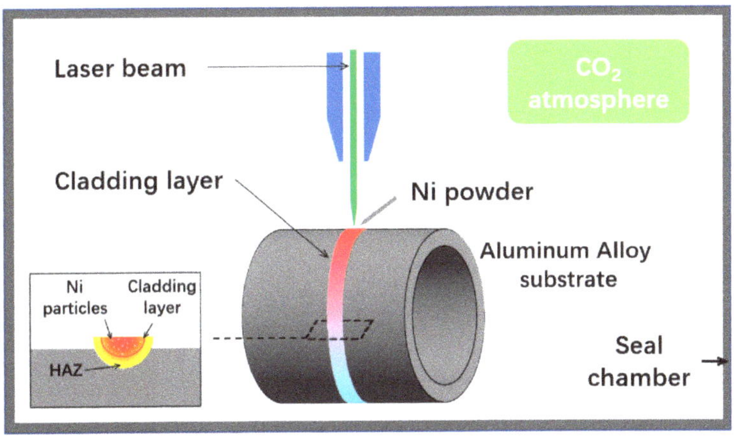

Figure 8. Schematic diagram of a laser cladding process.

3.2. Sliding Wear Tests

In order to simulate the actual drilling conditions of aluminum alloy drill pipe in the well and measure the wear resistance of the laser cladding layer, a drill pipe wear test bench was designed and machined (Figure 9a). Its working principle is shown in Figure 9b. The bench is capable of conducting wear tests on normal and biomimetic structures simultaneously. This ensures that the test conditions remain consistent and leads to improved reliability of the results.

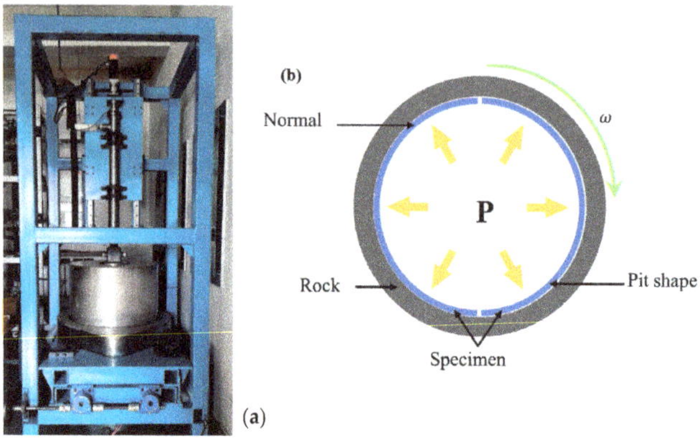

Figure 9. (**a**) Test bench; (**b**) schematic diagram of the test.

In this test, granite was selected as the rock sample, and the rock rotation speed ω was 40 r/min. The circumferential force was 500 N. In order to simulate the actual working environment of the drill pipe in the downhole, the contact part was continuously washed with drilling fluid during the test, and water was selected as the drilling fluid. Each test cycle was 30 min. The wear resistance was assessed by measuring the mass loss before and after the test. To minimize testing errors, the average value was determined based on three measurements conducted for each group.

3.3. Microhardness Testing

After conducting the wear test, the specimen was cut along the surface of the vertical cladding layer. The cut surface underwent a smoothing and polishing process using sandpaper with grit sizes of #600, #1000, #2000, and #5000, respectively. Subsequently, the specimen was subjected to a hardness test using an HV1000 micro-Vickers hardness tester with a 1 kg load and a loading time of 15 s.

3.4. Microstructural Observation and Component Testing

After polishing and cleaning, the specimens were titrated and etched using Keller reagent (1.0 mL HF + 1.5 mL HCl + 2.5 mL HNO_3 + 95 mL H_2O) for 15 s, followed by washing with anhydrous ethanol. Carl Zeiss-AxioImager A2m optical metallographic microscope (OM), Hitachi S-4800, and Hitachi JSM-IT500A scanning electron microscope (SEM) equipped with energy dispersive spectroscopy (EDS) were used to analyze the organization of the specimen, and Rigaku D/Max 2500C X-ray diffractometer (XRD) was used to analyze the physical phase of the cladding layer, with a scanning range of 20~90°.

4. Results and Discussion

4.1. Analysis of Simulation Results

First, the stress distribution was observed. Since the maximum stress appeared on the applied surface, which was not helpful for this study, the stress nephograms of the contact surface between two models at 2 s were extracted. Because the coating was circular and the two kinds of structures were symmetrical, part of the model was cut to show that the rotation direction was from left to right.

As shown in Figure 10, the surface stress distribution of the two kinds of structures is obviously dissimilar, but both show strong regularity. In the pit shape structure, the stress distribution around each pit unit is similar, with significantly higher stresses on both sides of the rotation direction compared to other areas. This stress concentration phenomenon is caused by the shape variation between the bionic structure and the surrounding coating. Additionally, it leads to a variation in stresses between two adjacent pit units in the circumferential direction, where the stresses initially decrease and then increase. On the other hand, the stress distribution in the horizontal direction of the normal structure is uniform without any concentration phenomenon. Moreover, the axial stress distribution demonstrates a clear decreasing trend from the middle to both sides.

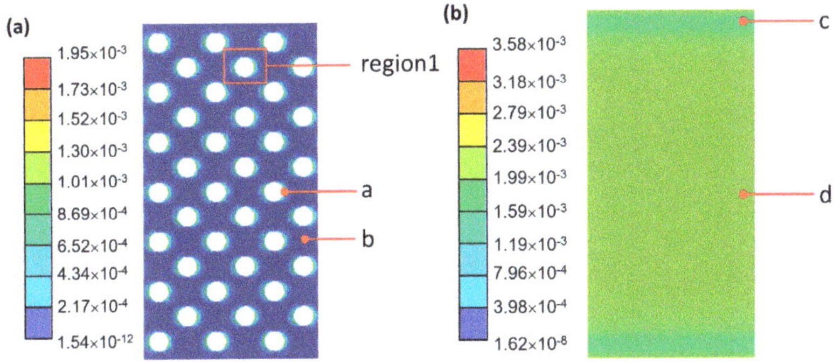

Figure 10. Stress nephogram of the contact surface between (**a**) pit shape structure and (**b**) normal structure and rock samples.

The stress value of every node in region 1 of Figure 10a was extracted. The coordinate origin was set at the center of the circle, with the positive X and Y axes representing the forward and vertical directions, respectively, establishing the coordinate system. We then created a position–stress scatter diagram (Figure 11a) and a three-dimensional diagram (Figure 11b). The stress values of points a–d are presented in Table 3.

Table 3. Stress at selected points.

Position	a	b	c	d
Stress (Mpa)	5.73×10^{-4}	1.39×10^{-4}	1.62×10^{-3}	2.29×10^{-3}

As can be seen from Figure 11, the stress is mainly concentrated on both sides of the rotation direction of the pit unit, forming a fan-shaped influence range that is concentrated within 0.5 mm. In contrast, the stress is relatively small in the vertical direction and beyond 0.5 mm. Additionally, the stress peak value is approximately 6–7 times greater than the minimum stress value. This stress concentration phenomenon can cause failure on both sides of the pit unit and then spread to other areas.

Points a and b were located in the middle position, whereas region 1 was situated at the edge. The stresses at points a and b exceeded the maximum stresses in region 1, indicating that the overall stress distribution of the pit shape structure also shows a decreasing trend from the middle to both sides. Moreover, the stress at point d was approximately four times higher than that of point a, suggesting that the peak stress of the normal structure was significantly greater than the peak stress of the pit shape structure, so the damage of the normal structure was greater than that of the crater structure. The macroscopic manifestation had greater wear than the cratered structure. This is because the lateral force applied causes elastic deformation in both Al alloy and Ni. However, the significant difference in elastic modulus between the two materials results in different deformation volumes. The normal structure, due to the continuous coating, causes the stress to be unreleasable and stored within the coating, whereas the bionic structure coating breaks the continuous surface, allowing for the release of stress. In addition, the stress at point d is approximately 1.4 times higher than that at point c, indicating that the normal structure has a more uniform stress distribution, with smaller gaps between the stress peaks and valleys.

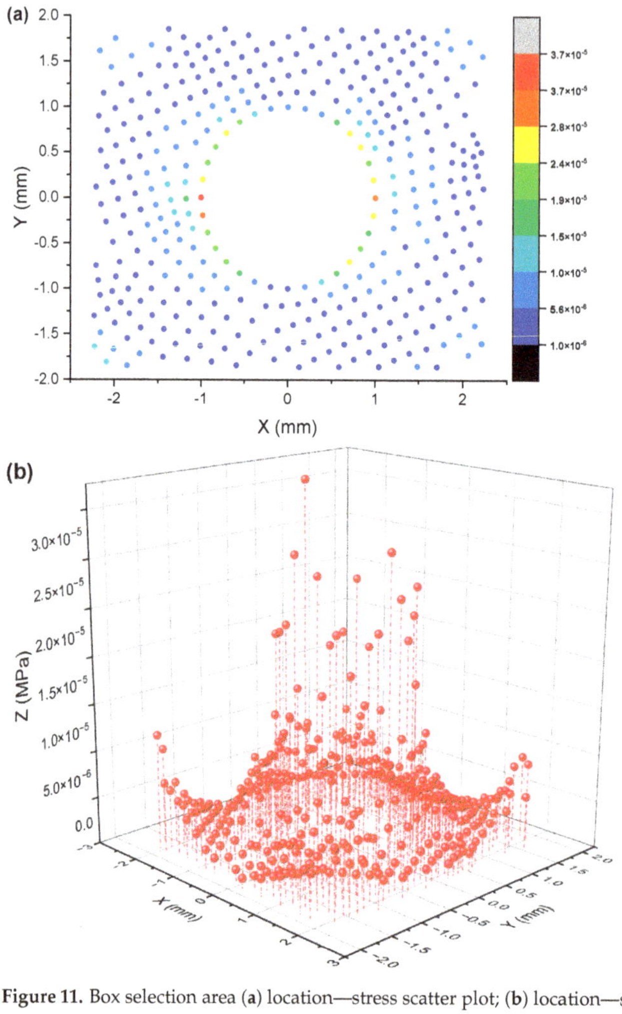

Figure 11. Box selection area (**a**) location—stress scatter plot; (**b**) location—stress 3D map.

After the wear of the coating and rock rotation for 2 s, the wear results of the two kinds of structures were compared in Table 4. Figure 12 depicts the comparison of the wear volume of the two models via a red part. The pit shape structure coating exhibited a reduction of 84.7% in wear volume compared to the normal structure coating, ignoring the change in the surface area caused by the variation in coating thickness. The wear depth of the two types of structures is depicted in the blue part in Figure 12. The results indicate that the lifespan of the pit shape coating increased by 81.3% compared to the normal coating, demonstrating a significant improvement effect.

Table 4. Wear result.

Model	Wear of 2 s (mm^3)	Comparison of Wear	Surface Area (mm^2)	Wear Depth (mm)	Comparison of Life Span
Normal structure	3.54×10^{-9}		17,203	2.06×10^{-13}	
Pit shape structure	5.42×10^{-10}	84.7% reduction	14,012	3.87×10^{-14}	81.3% improvement

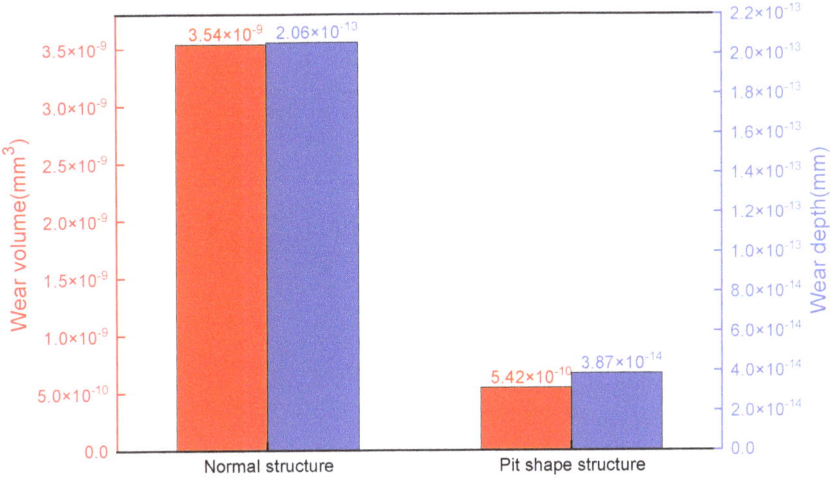

Figure 12. Comparison of wear and wear depth after simulation.

4.2. Analysis of Cladding Layer

Figure 13 presents the scanning electron microscope images taken before corrosion with the Keller reagent. The image clearly displays a distinct demarcation line between the molten cladding layer and the substrate, both before and after polishing. This demarcation line is indicated by the red arrows in Figure 13a,b. A good metallurgical bonding state exists between the cladding layer and the substrate, and no obvious cracks are observed at the interface bonding before and after polishing. In addition, a small number of pores is found at the bottom of the coating, but overall, the performance of the coating is not affected.

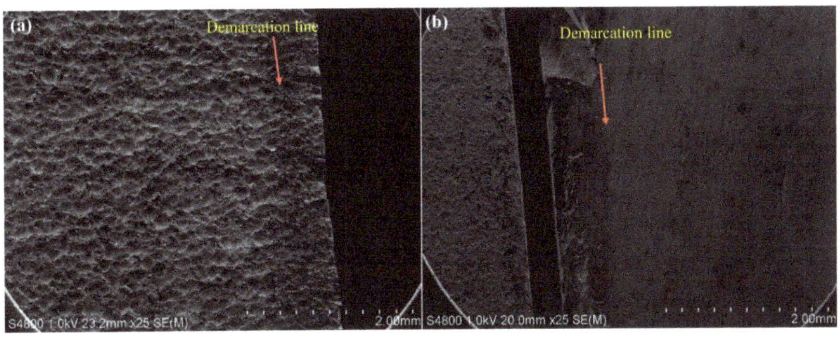

Figure 13. (**a**) Before polishing; (**b**) after polishing.

The specimen was then etched using the Keller reagent. The morphology of the cladding layer and the substrate after corrosion under an optical metallographic microscope are shown in Figure 14. It is evident that a transition region, approximately 140 μm in thickness, exists between the cladding layer and the substrate. Within this region, the coarse crystal region has a thickness of about 100 μm, whereas the fine crystal region has a thickness of about 40 μm. Due to the limited magnification of the metallographic microscope, a scanning electron microscope was used to further analyze the microstructure and products between the cladding layer and the substrate.

Figure 14. Magnified by 200 times under a metallurgical microscope.

Figure 15 shows the SEM micrographs between the substrate and the cladding layer. It shows that the grain size and morphology in the fine crystal zone are significantly different from those in the coarse crystal zone. The grain morphology in the fine crystal zone has an equiaxed crystal distribution with a grain size length from 5 to 10 μm, and the coarse crystal zone exhibits a columnar morphology with grain sizes between 10 and 20 μm. The EDS spectra of point A and point B illustrated in Figure 16 indicate that the composition of both points is primarily Al with some impurities. This suggests that the coarse and fine crystal regions are a result of recrystallization and grain growth during the process of melting and condensation of Al.

Figure 17a displays the SEM morphology of the cladding layer. It shows a uniformly distributed cast-equiaxed crystal organization without Ni powder particles in the cladding layer. At the same time, there are tiny gaps in the cladding layer. In Figure 17b, the EDS spectra of point C are shown. The primary composition of the clad layer is the Al-Ni phase, with minor amounts of other elements including C, O, Si, and Cr. This can be attributed to the laser cladding process being conducted in a CO_2 environment and the presence of impurities in the Ni powder.

In order to further determine the molten layer products, the cladding layer was analyzed using XRD. The results are shown in Figure 18. Figure 18 indicates that the cladding layer is mainly composed of Al_3Ni_2 and Al phase. It is worth noting that Ni only has a peak at 45 degrees that overlaps with the identified intermetallic phase, whereas the other angles do not show consistency. Secondly, both Al and Al_3Ni_2 also overlap with the identified intermetallic phase at the 45 degree peak, and the other angles show a high level of consistency. Lastly, there is no ferromagnetism on the surface of the cladding layer. Based on these observations, it can be concluded that the Ni powder completely reacted with Al to form Al_3Ni_2.

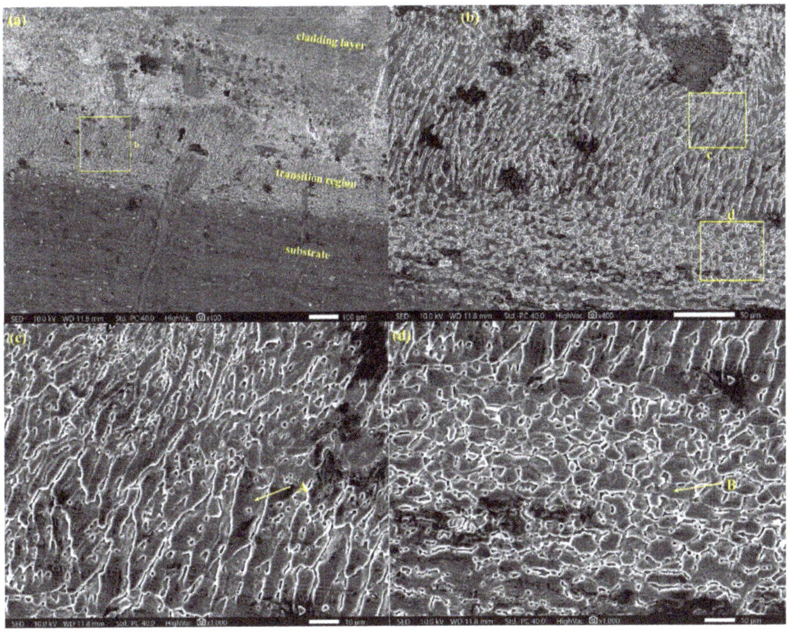

Figure 15. The SEM micrographs of (**a**) specimen, (**b**) transition region, (**c**) coarse crystal region, and (**d**) fine crystal region.

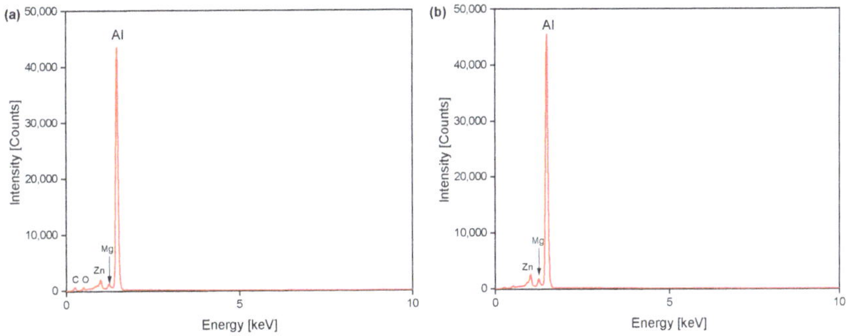

Figure 16. EDS analyses of (**a**) point A and (**b**) point B.

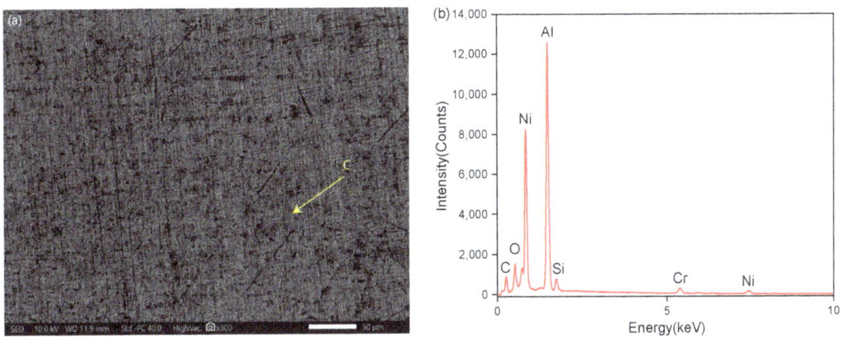

Figure 17. (**a**) The SEM micrographs of the cladding layer; (**b**) EDS analyses of point C.

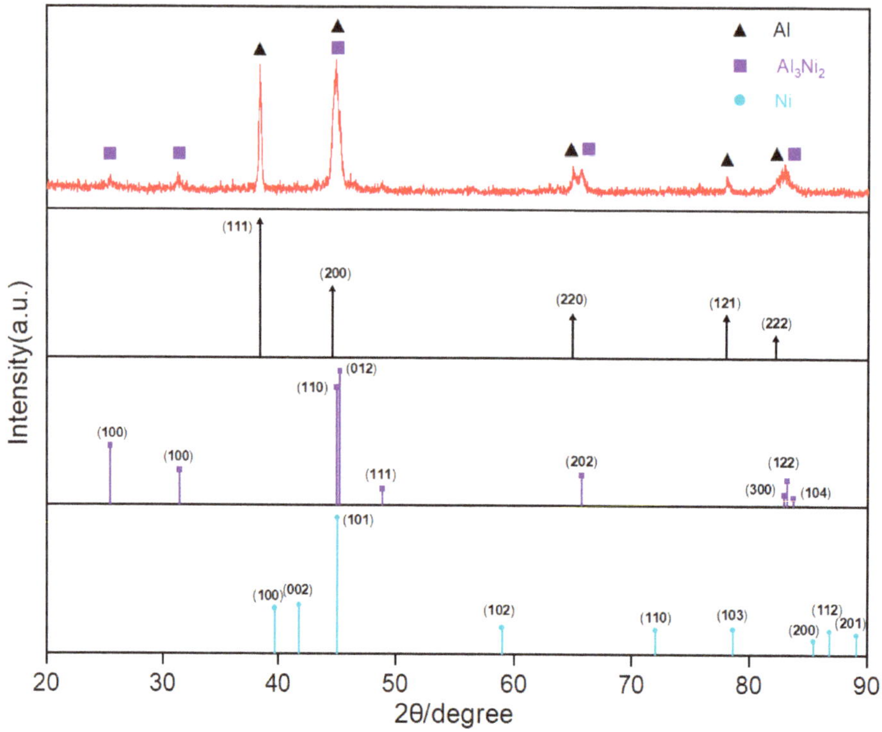

Figure 18. Comparison of XRD spectra of the cladding layer with Al_3Ni_2, Al, and Ni.

It is crucial to comprehend the thermodynamic analysis of the chemical reactions occurring within the clad layer to regulate its microstructure and properties [33]. In the laser cladding process, both the aluminum alloy substrate and Ni powder dissolve, and the chemical reactions that take place during the solidification of the melt pool are demonstrated below:

$$3Al + 2Ni \rightarrow Al_3Ni_2 \quad (5)$$

Al_3Ni_2 has high hardness and wear resistance. Figure 19 shows the microhardness data of each typical area from the surface of the cladding layer toward the substrate. As can be seen, the surface hardness of the cladding layer is 762 HV, which is 4.7 times the surface of the 7075 aluminum alloy substrate. The hardness of the coarse and fine crystal region is reduced to a certain extent compared to the 7075 aluminum alloy substrate. The hardness of the fine grain zone is 138 HV, which is 15.3% lower than that of the matrix, whereas the hardness of the coarse grain zone is 109 HV, which is 33.1% lower than that of the matrix. In reference [34], Ni was also used as the cladding layer material, and Al-Si alloy was used as the matrix material. The microhardness test results are similar to those in this paper [34]. The high hardness of Al_3Ni_2, acting as the hard phase, exhibits excellent resistance to deformation. On the other hand, the transition region and the aluminum alloy substrate, acting as the soft phase, can effectively absorb the shear stress generated during the test.

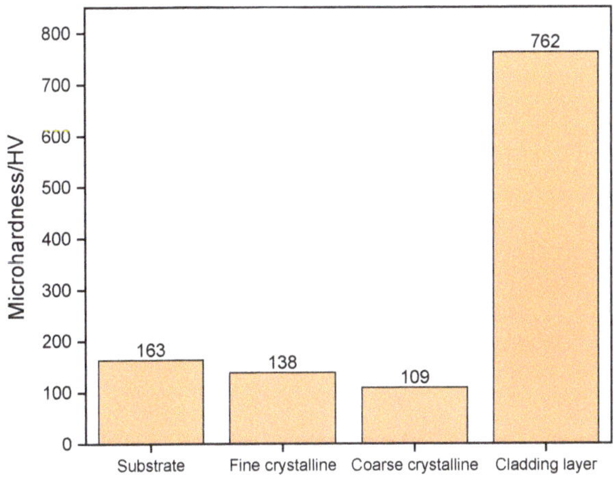

Figure 19. Microhardness test results for each region.

4.3. Data and Status of Wear Tests

Wear test results are presented in Table 5. The orange part in Figure 20 shows the comparison of the mass loss of the two structures. The comparison shows that the mass loss of the pit shape structured cladding layer is 75.4% lower compared to the normal structured cladding layer.Ignoring the surface area changes caused by cladding thickness, the green part in Figure 20 shows the comparison of the wear per area of the two structures. The result shows that the lifespan of the pit shape structure cladding layer improved by 70.0%, which is similar to the simulation results.

Table 5. Wear test results.

Model	Mass Loss (g)	Comparison of Mass Loss	Surface Area (mm^2)	Wear per Unit Area (g/mm^3)	Comparison of Life Span
Normal structure	4.10		17,203	2.38×10^{-4}	
Pit shape structure	1.01	75.4% reduction	14,012	7.21×10^{-5}	70.0% improvement

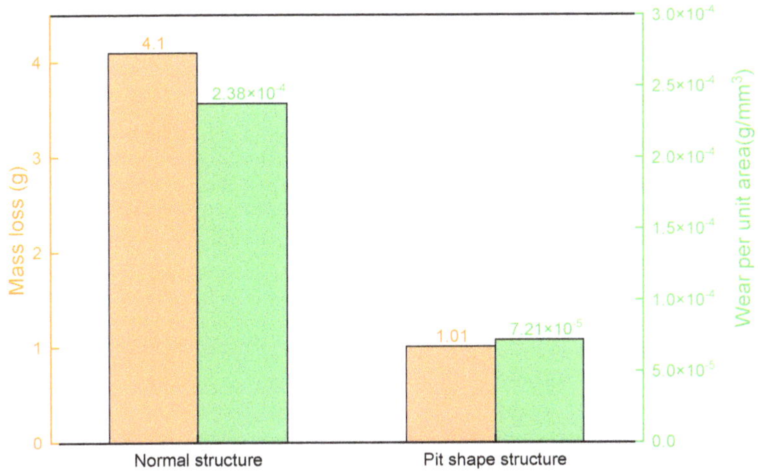

Figure 20. Comparison of wear and wear depth after test.

The surface of the two specimens after the wear test is shown in Figure 21. It shows that the area in the middle of the normal structure exhibits significant wear and noticeable shedding, whereas the edge area is relatively complete and has no obvious shedding phenomenon. The pit shape structure is severely worn around the pit unit, and the middle area is also more worn than the edge area. Two kinds of structure surface morphology after the wear test are consistent with the stress distribution in Figure 10, which verifies the accuracy of the finite element model simulation analysis.

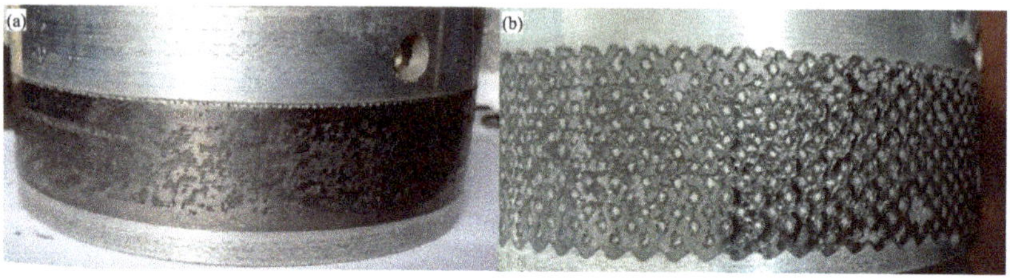

Figure 21. (**a**) Normal structural; (**b**) pit shape structural surface morphology after test.

The shedding area of the cladding layer in the normal structure exhibits obvious annular grooves, and no similar phenomenon is observed on the pit shape structure. This difference is attributed to the presence of particles that have not been washed off, which become trapped between the cladding layer and the rock. As a result, these particles undergo abrasive wear as they are cut by the rotating cladding layer. However, due to the existence of pit units, the pit shape structure can store the particles in it, effectively reducing abrasive wear, as shown in Figure 22.

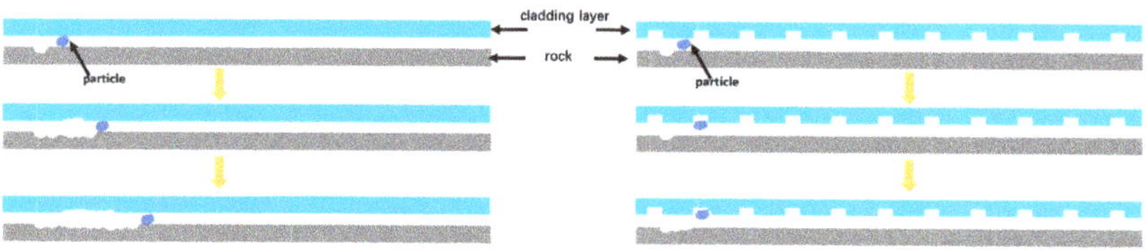

Figure 22. Wear process of two structures.

By combining wear and surface morphology, the bionic structure plays an important role in improving the stress distribution and extending the cladding life.

5. Conclusions

In this paper, the relationship between pit diameter and center distance was obtained by observing and measuring the pit shape structure on the abdomen of dung beetles. Subsequently, the stress distribution and wear conditions of the bionic structure and normal structure model during rock grinding were simulated and analyzed, and the friction and wear experiments were carried out. The microstructure, composition, and hardness behavior of the specimen were characterized using a scanning electron microscope, X-ray diffractometer, and microhardness tester. Based on the above research, the following conclusions are drawn:

(1) By studying and comparing the abdominal pit shape structure of dung beetles and measuring two kinds of parameters, pit diameter and center distance, the obtained ratio of them was around 1.98.

(2) The bionic pit shape structure model was established. Compared to the general structure, the results of finite element simulation showed that the bionic structure can significantly improve stress distribution, and the peak stress of the bionic structure was significantly reduced. A significant stress concentration phenomenon occurred near the pit unit, and the life span improved by 81.3%.

(3) The coating after laser cladding is predominantly composed of Al_3Ni_2, which exhibits high hardness. Furthermore, a transition region exists between the cladding layer and the substrate, with the composition being aluminum, which has relatively low hardness. This combination improves the ability of the drill pipe to withstand alternating loads.
(4) The surface appearance of two specimens after the wear test aligns with the stress distribution gained from ANSYS. Additionally, the lifespan of the pit shape structure increased by 70.0%. This increase closely matches the simulation results and validates the accuracy of the ANSYS analysis findings.
(5) Pit shape structure plays a crucial role in altering the stress distribution, as it effectively stores the abrasive particles and reduces the occurrence of abrasive wear. This, in turn, enhances the lifespan of the cladding layer.

Author Contributions: Data curation, C.Z.; Formal analysis, X.L. (Xiaoshu Lü); Methodology, Y.Z.; Project administration, K.G.; Software, X.X.; Visualization, H.A.; Writing, original draft, X.L. (Xu Li). All authors have read and agreed to the published version of the manuscript.

Funding: This research has been funded by the National Key R & D Program of China (Grant No. 2022YFC3005903-2), the Engineering Research Center of Geothermal Resources Development Technology and Equipment (Grant Nos. 23021 and 23022), the Graduate Innovation Fund of Jilin University (Grant No. 2023CX106), and the Ministry of Education, Jilin University, National Natural Science Foundation of China (Grant No. 41972324 and No. 42172345).

Institutional Review Board Statement: Not applicable.

Informed Consent Statement: Not applicable.

Data Availability Statement: The datasets used and/or analyzed during the current study are available from the corresponding author upon reasonable request.

Conflicts of Interest: The authors declare no conflict of interest.

References

1. Liang, J.; Yue, W.; Gu, Y.; Liu, J.; Wang, C.; Ma, H. Improving Corrosion Resistance and Corrosive Wear Resistance of Aluminum Alloy Drill Pipe by Surface Nanocrystallization and Micro-arc Oxidation. *J. Mater. Eng. Perform.* **2018**, *27*, 4462–4472. [CrossRef]
2. Kondrat'ev, S.Y.; Shvetsov, O.V. Technological and Operational Features of Drill Pipes from Aluminum Alloys 2024 and 1953. *Met. Sci. Heat. Treat.* **2018**, *60*, 32–38. [CrossRef]
3. Kondrat'ev, S.Y.; Shvetsov, O.V. Effect of High-Temperature Heating on the Structure and Properties of Aluminum Alloys in the Production of Drill Pipes. *Met. Sci. Heat. Treat.* **2013**, *55*, 191–196. [CrossRef]
4. Sun, Y.; Wang, X.; Liu, B.; Ding, D.; Meng, Q. Inverse solution to heat transfer coefficient during heat assembly of aluminum alloy drill pipes. *Adv. Mech. Eng.* **2017**, *9*, 265–276. [CrossRef]
5. Zhao, X.; Chen, X.; Wang, X. Effect of aging processes on corrosion behavior and stress corrosion sensitivity of pre-stretched 7075 aluminum alloy. *Mater. Corros.* **2018**, *69*, 850–857. [CrossRef]
6. Lourenço, M.I.; Netto, T.A.; Silva, N.d.S.d.; Ribeiro Plácido, J.C. Multiaxial Fatigue of Aluminum Drill Pipes—Experiments and Numerical Analyses. *J. Offshore Mech. Arct. Eng.* **2022**, *144*, 011802. [CrossRef]
7. Lee, I.S.; Hsu, C.J.; Chen, C.F.; Ho, N.J.; Kao, P.W. Particle-reinforced aluminum matrix composites produced from powder mixtures via friction stir processing. *Compos. Sci. Technol.* **2011**, *71*, 693–698. [CrossRef]
8. Tjong, S.C. Recent progress in the development and properties of novel metal matrix nanocomposites reinforced with carbon nanotubes and graphene nanosheets. *Mater. Sci. Eng. R Rep.* **2013**, *74*, 281–350. [CrossRef]
9. Algahtani, A.; Mahmoud, E.; Khan, S.; Tirth, V. Experimental Studies on Corrosion Behavior of Ceramic Surface Coating using Different Deposition Techniques on 6082-T6 Aluminum Alloy. *Processes* **2018**, *6*, 240. [CrossRef]
10. Chi, Y.; Gu, G.; Yu, H.; Chen, C. Laser surface alloying on aluminum and its alloys: A review. *Opt. Lasers Eng.* **2018**, *100*, 23–37. [CrossRef]
11. Wang, C.; Bai, S.; Xiong, Y. Recent advances in surface and interface engineering for electrocatalysis. *Chin. J. Catal.* **2015**, *36*, 1476–1493. [CrossRef]
12. Zhang, P.; Liu, X.; Yan, H. Phase composition, microstructure evolution and wear behavior of Ni-Mn-Si coatings on copper by laser cladding. *Surf. Coat. Technol.* **2017**, *332*, 504–510. [CrossRef]
13. Lu, X.; Liu, X.; Yu, P.; Qiao, S.; Zhai, Y.; Wang, M.; Chen, Y.; Xu, D. Synthesis and characterization of Ni60-hBN high temperature self-lubricating anti-wear composite coatings on Ti6Al4V alloy by laser cladding. *Opt. Laser Technol.* **2016**, *78*, 87–94. [CrossRef]

14. Sun, G.; Zhang, Y.; Liu, C.; Luo, K.; Tao, X.; Li, P. Microstructure and wear resistance enhancement of cast steel rolls by laser surface alloying NiCr–Cr_3C_2. *Mater. Des.* **2010**, *31*, 2737–2744. [CrossRef]
15. Wang, G.; Zhang, J.; Shu, R.; Yang, S. High temperature wear resistance and thermal fatigue behavior of Stellite-6/WC coatings produced by laser cladding with Co-coated WC powder. *Int. J. Refract. Met. Hard Mater.* **2019**, *81*, 63–70. [CrossRef]
16. Paul, C.; Sellamuthu, R. Effect of nickel content on hardness and wear behaviour of surface modified functionally graded Cu-Sn bronze alloy. *Int. J. Mater. Eng. Innov.* **2016**, *7*, 43–55. [CrossRef]
17. Xie, S.; Li, R.; Yuan, T.; Chen, C.; Zhou, K.; Song, B.; Shi, Y. Laser cladding assisted by friction stir processing for preparation of deformed crack-free Ni-Cr-Fe coating with nanostructure. *Opt. Laser Technol.* **2018**, *99*, 374–381. [CrossRef]
18. Weng, Z.; Wang, A.; Wu, X.; Wang, Y.; Yang, Z. Wear resistance of diode laser-clad Ni/WC composite coatings at different temperatures. *Surf. Coat. Technol.* **2016**, *304*, 283–292. [CrossRef]
19. Pang, Z.; Zhou, H.; Xie, G.; Cong, D.; Meng, C.; Ren, L. Effect of bionic coupling units' forms on wear resistance of gray cast iron under dry linear reciprocating sliding condition. *Opt. Laser Technol.* **2015**, *70*, 89–93. [CrossRef]
20. Jiang, H.; Liu, Y.; Zhang, Y.; Liu, Y.; Fu, X.; Han, D.; Song, Y.; Ren, L.; Sun, H. Reed Leaf-Inspired Graphene Films with Anisotropic Superhydrophobicity. *ACS Appl. Mater. Interfaces* **2018**, *10*, 18416–18425. [CrossRef]
21. Arjangpay, A.; Darvizeh, A.; Tooski, M.Y. Effects of Structural Characteristics of a Bionic Dragonfly Wing on Its Low Velocity Impact Resistance. *J. Bionic Eng.* **2018**, *15*, 859–871. [CrossRef]
22. Zhang, H.; Liu, S.; Xiao, H.; Zhang, X. Synthesis and Tribological Properties of Bio-Inspired Nacre-Like Composites. *Materials* **2018**, *11*, 1563. [CrossRef]
23. Jones, K.; Schmid, S.R. Experimental Investigation of Laser Texturing and its Effect on Friction and Lubrication. *Procedia Manuf.* **2016**, *5*, 568–577. [CrossRef]
24. Cho, M. Friction and wear of a hybrid surface texturing of polyphenylene sulfide-filled micropores. *Wear* **2016**, *346–347*, 158–167. [CrossRef]
25. Zhang, H.; Zhang, P.; Sui, Q.; Zhao, K.; Zhou, H.; Ren, L. Influence of Multiple Bionic Unit Coupling on Sliding Wear of Laser-Processed Gray Cast Iron. *J. Mater. Eng. Perform.* **2017**, *26*, 1614–1625. [CrossRef]
26. Mao, C.; Ma, Y.; Wu, S.; Wei, Y.; Li, J. Wear resistance and wet skid resistance of composite bionic tire tread compounds with pit structure. *Mater. Res. Express* **2019**, *6*, 085331. [CrossRef]
27. Wang, Y.; Liu, B.; Guo, Z. Wear resistance of machine tools' bionic linear rolling guides by laser cladding. *Opt. Laser Technol.* **2017**, *91*, 55–62. [CrossRef]
28. Qi, C.; Zhan, X.; Gao, Q.; Liu, L.; Song, Y.; Li, Y. The influence of the pre-placed powder layers on the morphology, microscopic characteristics and microhardness of Ti-6Al-4V/WC MMC coatings during laser cladding. *Opt. Laser Technol.* **2019**, *119*, 105572. [CrossRef]
29. Cheng, J.; Li, F.; Wang, Y.; Wang, Y.; Liu, X.; Zhang, J.; Wang, Z.; Li, Y.; Wang, H.; Yang, Z.; et al. Dweller and tunneler dung beetles synergistically accelerate decomposition of cattle and horse dung in a semi-arid steppe. *Agric. Ecosyst. Environ.* **2022**, *329*, 107873. [CrossRef]
30. Wang, X.; Liu, B.; Gao, K.; Meng, Q.; Sun, Y. Analysis of Thermal Deformation and Influencing Factors in Shrink-Fitting Assembly of Aluminum Alloy Drill Pipe. *Adv. Mech. Eng.* **2016**, *8*, 1687814016674099. [CrossRef]
31. Sysaykeo, D.; Linares, J.-M.; Mermoz, E. Wear Behavior of a Bio-inspired Bearing for off-center Loads. *J. Bionic Eng.* **2020**, *17*, 1251–1262. [CrossRef]
32. Rezaei, A.; Van Paepegem, W.; De Baets, P.; Ost, W.; Degrieck, J. Adaptive finite element simulation of wear evolution in radial sliding bearings. *Wear* **2012**, *296*, 660–671. [CrossRef]
33. Farayibi, P.K. Microstructural Evolution of Metal Matrix Composites Formed by Laser Deposition of Ti-6Al-4V Wire and WC-W_2C Powder. *Adv. Eng. Forum* **2018**, *26*, 22–32. [CrossRef]
34. Krishnakumar, M.; Mohnbabu, A.; Saravanan, R. Impact of Surface Alloying of Nickel on Microstructure, Hardness and Wear on Aluminium–12%Silicon Alloy. *Trans. Indian. Inst. Met.* **2019**, *72*, 2395–2402. [CrossRef]

coatings

Article

Corrosion and Micro-Abrasion Properties of an AISI 316L Austenitic Stainless Steel after Low-Temperature Plasma Nitriding

Darko Landek [1,*], Marin Kurtela [2], Ivan Stojanović [2], Jurica Jačan [1] and Suzana Jakovljević [1]

1 Department of Materials, Faculty of Mechanical Engineering and Naval Architecture, University of Zagreb, 10000 Zagreb, Croatia; jurica.jacan@fsb.hr (J.J.); suzana.jakovljevic@fsb.hr (S.J.)
2 Department of Welded Structures, Faculty of Mechanical Engineering and Naval Architecture, University of Zagreb, 10000 Zagreb, Croatia; marin.kurtela@fsb.hr (M.K.); ivan.stojanovic@fsb.hr (I.S.)
* Correspondence: darko.landek@fsb.hr; Tel.: +385-91-577-2044

Abstract: The paper investigates the effects of DC plasma nitriding on surface roughness, hardness, microstructure, micro-abrasion, and corrosion resistance of AISI 316L Austenitic Stainless (AS) steel. The nitriding has been conducted for durations ranging from 4 to 24 h at a temperature of 430 °C in a commercial vacuum furnace, Rübig PN90/70. Micro-abrasion resistance has been tested using the calotest device with a measurement diameter of craters produced on the sample surface after 10 to 60 s of wear. Corrosion resistance has been tested using the Electroimpedance Spectroscopy (EIS) method in a 3.5% NaCl water solution. The surface roughness parameters and hardness of the samples increased with longer nitriding times, attributed to the saturation of austenite and the formation of iron and chromium nitrides. Nitriding for longer than 8 h resulted in the formation of a thicker compound layer that is hard and brittle, leading to reduced wear resistance compared with shorter nitriding times. EIS measurements revealed that nitrided samples had lower corrosion resistance compared with the untreated sample. The corrosion stability was not significantly affected by nitriding time. Different nitriding times have a great influence on resistance to pitting corrosion. This study provides valuable insights into the effects of plasma nitriding on the properties of AS steel, highlighting the importance of optimizing nitriding parameters for specific applications.

Keywords: AS steel AISI 316L; plasma nitriding; corrosion resistance; micro-abrasion wear; S-phase; scanning electron spectroscopy; electrochemical DC measurements

1. Introduction

The austenitic stainless (AS) steels have a combination of good corrosion resistance, high temperature strength, and oxidation resistance, high toughness on low temperatures, good ductility, and weldability. The steel AISI 316L (EN X5CrNiMo17-12-2, 1.4401) is typical representative of the high quality AS steels and its properties are usually compared as a reference to the properties of other stainless steels. Additionally, their heat treatment after welding, cold forming, or grinding is simple and usually only annealing for reducing residual stresses and stress corrosion is recommended. This annealing can be conducted at a temperature of 850–950 °C for 2 h for a section with a wall thickness of 25.5 mm, 4 h for a section with a thickness of 50.1 mm, etc. To maintain good corrosion resistance, the weld surface of AS steels should be cleaned of residual protective pastes and oxides by acid pickling and finely ground or polished to a medium roughness of $R_a < 0.5$ μm. The AS steel AISI 316L is used in numerous structures that require corrosion resistance in marine or industrial atmospheres, clean water, heat exchangers, and preheaters, production of chemical compounds such as hydrogen peroxide, concentrated sulfuric acid, alcohol-based solvents, etc. [1,2].

Although possessing superior properties, AS steels can be attacked by pitting and stress corrosion, in the presence of hydrochloric acid or a hot caustic environment. Additionally, their hardness and wear resistance are low. This problem can be solved by replacing the AS steel with a more expensive martensitic stainless steel or by protecting the AS steel with surface modification and coating processes.

To modify the surface of AS steels, according to the multiple favorable influence on the properties, the nitriding process in plasma is highlighted over gas and salt bath nitriding. The addition of nitrogen to surface layer of AS steels will change stacking fault energy and the dislocation structure, leading to increased surface hardness, fatigue life, and enhances pitting corrosion resistance, without affecting the bulk properties [2–4]. Before plasma nitriding (PN) the workpiece is cleaned by ion bombardment (sputtering) with H_2, N_2 and Ar ions. The plasma nitriding (PN) temperature needs to be lower than 430–440 °C to avoid the precipitation of chromium nitrides at the grain boundaries and to maintain good corrosion resistance [5–7]. This low-temperature PN process enables the formation of the S-phase, which is a nitrogen-supersaturated austenite (γN) with a distorted Face Centered Cubic (FCC) crystalline lattice in a thin surface layer. Temperatures higher than 440 °C stimulate the precipitation of nitrides γ′-$(Fe,Cr)_4N$ and ε-$(Fe,Cr)_2$+xN at the grain boundaries of the S-phase and at the surface's top layer. The presence of the S-phase can be observed using X-ray diffractometry, where two characteristic peaks in a diffractogram appear at smaller angles compared to the peaks of the substrate austenite. These peaks correspond to the (111) and (200) crystallographic planes in the FCC lattice. The shift in these peaks indicates that the S-phase has a larger lattice parameter than the substrate austenite and is supersaturated with nitrogen atoms [5–8]. A homogeneous S-phase exhibits higher hardness than the austenitic phase in the substrate while preserving good corrosion resistance. However, producing it through PN requires a specific combination of parameters and has a limited range of surface layer depth. The most influential process parameters on nitride layer thickness include nitriding temperature and time, the ratio of atmosphere gases (H_2/N_2), pressure in the vacuum chamber, substrate voltage, and plasma current density.

In reference [9], a plasma nitriding (PN) process was conducted for a duration of 5 h with a mass flow ratio of H_2/N_2 = 75%/25% and a pressure of 500 Pa. Two transition temperatures were identified: T1 for producing a homogeneous S-phase and T2 for the decomposition of the S-phase and the formation of γ′-$(Fe,Cr)_4N$, Cr_2N, and CrN nitride precipitates. Plasma nitriding of AS steel AISI 316 below the temperature T1 = 460 °C results in the formation of a surface layer consisting only of the S-phase. Nitriding at temperatures above T_2 = 550 °C results in the formation of nitride precipitates and a thin γ′-$(Fe,Cr)_4N$ surface layer. The thickness of the S-phase layer obtained varied from 3 μm after PN conducted at 400 °C for 5 h to 14 μm after treatment in plasma at 450 °C for 5 h. It was also observed that the thickness of the nitride layer on a plane surface was not uniform and depended on the thickness of the plasma layer formed during the PN process. In references [9,10], the authors concluded that increasing the PN temperature above 450 °C with a nitriding time longer than 10 h causes the occurrence of nitride precipitates. Reference [11] provides a review of the literature, indicating the possibility of obtaining an S-phase layer on AS steels through conventional PN and pulsed DC discharge PN at temperatures below 420 °C. In the above mentioned low-temperature nitriding procedures, lasting up to 4 h, a surface layer composed of pure S-phase, up to 10 μm thick, and a layer with a mixture of S-phase and nitride precipitates, up to 20 μm thick, were obtained. In references [5,11], vacuum pressures between 100 and 1000 Pa, substrate voltage of 100–1000 V, and a plasma current density greater than 1 mA/cm^2, along with low temperatures below 440 °C, are recommended for obtaining a surface layer up to 10 μm thick, composed of a homogeneous S-phase on AISI 316L steel in a PN process with ionized gases N_2 or N_2–H_2.

In reference [12], the plasma nitriding (PN) process was investigated in an atmosphere of dissociated and ionized ammonia at a vacuum pressure of 300 Pa, at temperatures ranging from 350 to 500 °C, with a nitriding time of 6 h, and an anode voltage (vacuum chamber wall) of 600–800 V. The thickness of the surface layer with a compact S-phase increased from 1.5 μm after nitriding at 350 °C to 30 μm after nitriding at the upper temperature of 500 °C. However, when nitriding at temperatures above 450 °C, nitride precipitates were observed in the nitrided layer. In reference [13], a study of low-temperature PN treatment of AISI 316LM AS steel was conducted with the following parameters: a ratio of ionized gases H_2/N_2 = 15%:85%, vacuum pressure of 300 Pa, nitriding temperature at 380 °C, voltage between the anode and cathode at 490–500 V, and processing times ranging from 0.5 to 8 h. The thicknesses of the nitrided layer with S-phase and without precipitated nitrides were found to be between 1.90 and 4.31 μm.

The references [14,15] investigate the impact of ionized gas atmosphere and temperature on the microstructure and properties of the surface layer. In reference [14], AS steel AISI 316L underwent nitriding with the following parameters: a flow ratio of ionized gases H_2/N_2 = 75%:25% or H_2/N_2 = 80%:20%, a vacuum pressure of 650 Pa, a nitriding temperature of 400 °C, a voltage between the anode and cathode of 700 V, and a nitriding time of 20 h. The PN process with an N_2 content of 25% resulted in a 20 μm thick nitrided layer, while the process with a lower N_2 content created a 15 μm thick surface layer. According to the results of XRD analysis, the thinner layer consisted of the S-phase, while the thicker layer also contained nitride precipitates. In reference [15], AISI 316L steel was nitrided with the following parameters: gas flow ratio H_2/N_2 = 1:3 or H_2/N_2 = 3:1, vacuum pressure of 250 Pa, nitriding temperature of 400 °C or 550 °C, voltage between anodes and cathodes of 470 V, and a nitriding time of 12 h. Using the Rietveld method in the analysis of the X-ray diffraction diagram, it was discovered that the S-phase, along with chromium and iron nitrides, is always present in the nitrided layer. During low-temperature nitriding at 400 °C, the share of the S-phase dominates over the share of nitrides. With the ratio H_2/N_2 = 1:3, the share of the S-phase in the nitrided layer was 89.6%, while with the ratio H_2/N_2 = 3:1, it was 96.7%. In the PN process with a higher proportion of hydrogen, CrN nitride is compatible with the S-phase, while in nitriding with a higher mass flow of N2, γ′-Fe4N nitride is dominantly present. At a temperature of around 420 °C, the highest solubility of nitrogen in austenite is achieved. Nitriding at temperatures higher than 420 °C leads to the decomposition of the metastable S-phase and the release of CrN precipitates. After the PN process at 550 °C, only 7.4% of the S-phase is present in the nitrided layer.

In reference [16], AISI 316L steel underwent nitriding with the following parameters: a gas flow ratio H_2/N_2 = 80%:20%, a vacuum pressure of 100 Pa, nitriding temperatures ranging from 400 to 500 °C, and a nitriding time of 5 h. In the nitrided layer of samples processed at temperatures up to 450 °C, the S-phase dominates, while CrN nitride also appears after nitriding at higher temperatures. On samples nitrided at 470–500 °C, in addition to CrN nitride, γ′-Fe4N nitride also becomes evident. The thickness and hardness of the nitrided layer range from 4 μm/373 HK0.1 after nitriding at 400 °C to 47 μm/1605 HK0.1 after processing at 500 °C. In reference [17], AISI 316L steel was nitrided with the following parameters: a gas flow ratio H_2/N_2 = 80%:20%, a vacuum pressure of 1000 Pa, nitriding temperatures ranging from 450 to 550 °C, and nitriding times of 2, 4, and 9 h. The presence of the S-phase, along with precipitates of iron (Fe_4N) and chromium (CrN, Cr_2N) nitrides, was observed on samples nitrided at 450 °C for 2 h, resulting in a layer with a thickness of about 15 μm. After nitriding at temperatures higher than 450 °C for 9 h, a thin film composed of γ′-Fe_4N and ε-$Fe_{2-3}N$ nitrides forms on the surface above the S-phase layer. The highest thickness of the nitrided layer, 95 μm, was obtained after a PN process conducted at 500 °C for 9 h. The hardness of the S-phase depends on the amount of soluted nitrogen and can range from 330 to 1200 HV [18]. The hardness of nitride phases and CrN precipitates can reach around 2000 HV. While a nitriding temperature of around 450 °C is recommended to achieve the highest hardness, good wear resistance, and maintain good corrosion resistance, the appearance of ε-$Fe_{2-3}N$ nitride precipitates in the S-phase and the formation of a surface film of hard nitrides after nitriding at 550 °C can lead to embrittlement, cracks, and delamination of the surface layer [17].

Summarizing all the above research results for low-temperature plasma nitriding of AISI 316L AS steel, the optimal process temperature is between 420 and 450 °C can be recommended for producing a nitride layer with the maximum thickness composed from the S-phase without significant precipitation of iron or chromium nitrides along the grain boundaries. Additionally, the gas flow ratio of H_2/N_2 = 4:1 or H_2/N_2 = 3:1 is recommended. The selection of a vacuum pressure depends on the plasma furnace construction and has a significant impact on nitride layer thickness. For the PN process carried out with a gas flow ratio of H_2/N_2 = 4:1, a vacuum of 100 Pa, a temperature of 450 °C, and a time of 5 h, the thickness of the nitrided layer is approximately 13 μm [16]. In PN process with the parameters H_2/N_2 = 4:1, a vacuum of 650 Pa, a nitriding temperature of 400 °C, and time of 20 h, the nitrided layer with S-phase was 15 μm thick [14]. For the PN process carried out with the same parameters of gas flow and temperature of 450 °C in a shorter time of 4 h, with a vacuum of 1000 Pa, a thickness of the nitrided layer of 45 μm was achieved [17].

In comparison to the surface of an untreated AS steel, the nitride layer of the S-phase has an enhanced resistance to tribocorrosion [15] and to pitting corrosion in neutral chloride solutions [19]. If the plasma nitriding was conducted only with S-phase layer formation below surface, and without overheating of surface, the surface roughness obtained on workpieces after polishing or fine grinding remains practically unchanged. The Fe_4N and $(Fe,Cr)_4N$ nitride film formed over S-phase layer increase surface roughness compared with the heat-untreated AS steel [20,21], and might be associated with the lower corrosion resistance in water Na_2SO_4 and H_2SO_4 solutions [22–24]. Therefore, few proposals can be found in the literature for the selection of parameters of low-temperature plasma nitriding, which would achieve a favorable combination of increased wear and corrosion resistance. In [25] the application of the active screen plasma nitriding was proposed for producing S-phase layer with improved wear and corrosion resistance. In [26], a hybrid plasma nitriding was proposed with two processes alternately running with high and low pressures. In [27] the optimal process of plasma nitriding for an AISI 316L AS steel was proposed as a treatment carried out at 400 °C with nitriding current density between 1 and 2.5 mA/cm^2.

The surface overheating, due to the formation of cathodic arcs, may appear on sharp edges and protrusions or when nitriding a surface insufficiently cleaned of oxides. Cathodic sparking can be avoided by using a direct current (DC) pulsed plasma generator instead of a conventional DC power supply with constant voltage. Modern plasma generators enable pulsed oscillations of the cathode voltage with frequencies of 1–10 kHz. Using a generator with pulsating cathode voltage, it is possible to regulate the temperature of nitrided objects, by adjusting the duration of the voltage pulse in relation to the total duration of the pulse period [28].

The influence of the nitriding temperature on the corrosion resistance of AS steel AISI 316L in aqueous NaCl solution has been confirmed in several works. Plasma nitriding at a temperature of 400 °C increases the resistance to pitting, compared to non-nitrided steel, but the surface layer of the S-phase is too thin for efficient protection of the substrate. Nitriding at temperatures between 430 and 450 °C proved to be optimal for increasing resistance to pitting and crevice corrosion. Plasma nitriding at temperatures above 470–500 °C reduces the resistance to intergranular corrosion and localized corrosion on the surface in comparison with untreated AS steel or steel nitrided at lower-temperature PN process [16,17,29].

Additional improvement in the wear resistance of AS steel can be achieved by forming an S-phase layer covered by a thin film of $Fe_{2-4}N$, $Cr_{1-2}N$, and $(Fe,Cr)_4N$ nitrides, with a surface hardness of 800–1200 HV [5,17,29]. In reference [30], a long-duration micro-abrasion test involved a fixed rotating steel ball pressed against a test sample without abrasive slurry to assess the wear resistance of a thin nitrided and nitrocarburized layer on AISI 316L AS steel after treatment at 450 °C for 5 h and 10 h. The test was performed using a 25.4 mm ball diameter at a rotational speed of 500 rpm under dry sliding conditions, with a normal load of 2.45 N. The test durations ranged from 5 to 20 min. The results show that the S-phase layer, at the beginning of dry sliding wear, reduces the volumetric loss of material by about three times compared to the loss of material on non-nitrided samples. At the end of the test, this loss was seven times lower. Even after 20 min of testing, the nitrided or nitrocarburized layer remained unperforated. This fact highlights the advantage of micro-abrasion testing compared to macro-abrasion wear tests, such as the "sand-rubber wheel" method, as it allows for testing the wear resistance of thin surface layers on a small area without significant substrate influence, using a simple and cost-effective test system [31]. Another variant of the micro-abrasion test involves a steel ball that rotates freely between the shaft and the test specimen. A suspension containing abrasive particles is added between the ball and the sample, pressed into the sample's surface by the normal force generated from the ball's weight. As the ball rotates on the contact area with the sample, it generates a tangential force that moves abrasive particles between the ball and the sample. These particles scratch, scrape, and smooth the sample's surface, forming a wear mark in the shape of a calotte. The normal force in micro-abrasion tests typically falls in the range of 0.05–5 N. In testing with a freely rotating ball, it is usually less than 0.5 N, depending on the ball's diameter and the sample's angle of inclination to the horizontal plane. The final sliding distance should be selected in a manner that ensures a constant wear rate with increasing wear time [32,33]. In the implementation and interpretation of micro-abrasion test results, factors such as the type, shape, and size of abrasive particles, the fluid used to create the suspension, the normal force, the rotational speed at which the ball acts on the sample, the ball's diameter, the roughness of the ball and sample surfaces, and the hardness ratio between the ball and the sample must be considered. To describe abrasive wear, three basic modes of wear are proposed: two-body grooving abrasion, three-body rolling abrasion, and a mixed regime, which is a combination of the previous two. In the first mode of wear, abrasive particles adhere to the ball's surface and, as it turns, groove the sample's surface. In the second mode of wear, abrasive particles roll on the contact surface between the ball and the sample [34–36]. Due to the shallow penetration depth and low normal loads, micro-wear tests reflect the intrinsic properties of the surface layer or coating and are suitable for comparison or quality checking.

The aim of this paper is to investigate the influence of the DC plasma nitriding time on tribological and corrosion properties of surface layer on an AISI 316L AS steel with methods limited in action to the thin surface layer using micro-abrasion, cyclic polarization and electro impedance spectroscopy (EIS) testing. Based on these limited surface tests, the optimal range of nitriding temperature and time for an AISI 316L AS steel will be proposed to achieve a good combination of its wear and corrosion resistance.

2. Materials and Methods

2.1. Materials

All tests were performed on samples of a commercially available AISI 316L AS steel. The chemical composition of the steel was determined by the two-measurement apparatus: a GDS LECO 850 device (LECO Corporation Inc., St. Joseph, MO, USA) and X-ray fluorescence device Olympus Innov-X Systems, model DS 2000-C (Evident Scientific, Inc., Waltham, MA, USA). The values of chemical elements are consistent and listed in Table 1. The samples were made in the form of a thin disk with a diameter of 40 mm and a thickness of 1.5 mm. The samples surface was sanded with waterproof sandpaper with grain sizes of P120, P600, and finally P1200. Before applying the plasma nitriding processes, all samples were cleaned in an ultrasonic bath VTUSC3 (Velleman, Gavere, Belgium) in 75% ethyl alcohol for 5 min.

2.2. Plasma Nitriding Treatment

Ten cleaned and dried samples for one plasma nitriding process were put into a PC 70/90 commercial vacuum furnace (Rübig GmbH, Wels, Austria) onto a same charging pad, which was also used as a cathode of the system. The surrounding wall of cylindrical chamber was used as an anode of the system on earth potential. The furnace was equipped with a bipolar DC voltage-controlled micro pulse plasma generator. The vacuum chamber with samples was heated with an auxiliary heating system consisting of three heating zones, the temperature of which was controlled by thermocouples loaded directly into the charge of heated samples. The DC plasma nitriding consisted of a surface sputtering and plasma nitriding treatment with the parameters shown in Table 2. Four plasma nitridings were carried out at the same temperature of 430 °C ± 5 °C in duration of 4 h, 8 h, 12 h, and 24 h, with a flow ratio of gasses H_2/N_2 = 3:1 at a pressure of 200 Pa.

Table 1. Elementary composition of the AISI 316L AS steel.

Elements	wt.%	+/−
C	0.05	0.02
Si	0.94	0.09
Mn	1.53	0.16
Cr	16.59	0.30
Ni	9.46	0.42
Mo	2.00	0.04
P	0.030	0.005
S	0.018	0.005

Table 2. Parameters of sputtering pre-treatment and plasma nitriding of test samples.

Heat Treatment	Plasma Processing Conditions				
Constant parameters	Temperature	430 °C ± 5 °C			
	Discharge voltage	360 V DC			
	Pressure	2 mbar			
Sputtering phase	Sputtering duration	2 h			
	Power of DC plasma generator	1400 W			
	Flow rates of gases H_2/N_2/Ar	140 dm^3/h (H_2)/10 dm^3/h (N_2)/15 dm^3/h (Ar)			
	Duty cycle	85%			
Plasma nitriding phase	Nitriding duration	4 h	8 h	12 h	24 h
	Power of DC plasma generator	2800 W			
	Flow rates of gases H_2/N_2/Ar	120 dm^3/h (H_2)/40 dm^3/h (N_2)/10 dm^3/h (Ar)			
	Duty cycle	90%			

Figure 1 shows test samples after plasma nitriding for 8 h to 24 h. The samples were nitrided only on one side, whereas the other side was supported on a non-alloyed steel holder, which makes good electrical contact with the cathode and enables the plasma to uniformly settle around the sample. Despite that, around the edge of the sample there are characteristic ring colored light gray due to the concentration of the plasma electromagnetic field and more intense heating, to a temperature between 430 and 450 °C. The central part of the sample, in good thermal and electrical contact with cathode, was retained at a temperature of 430 °C during all nitriding process. Overheating of the edge part of AISI 316L samples at temperatures above 420 °C leads to reaching the solubility limit of nitrogen in the metastable expanded austenite and to precipitation of chromium nitrides [37].

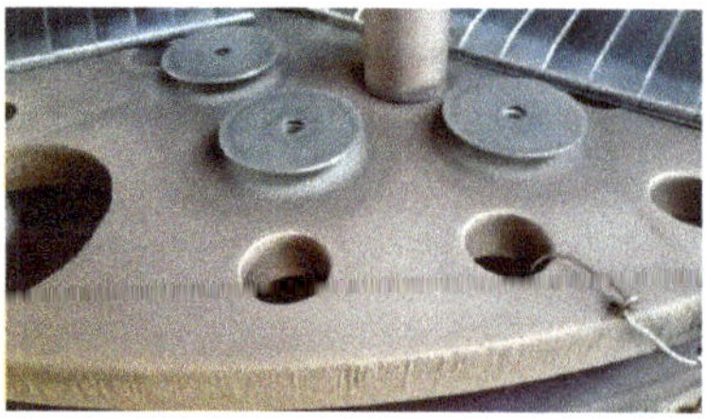

(a)

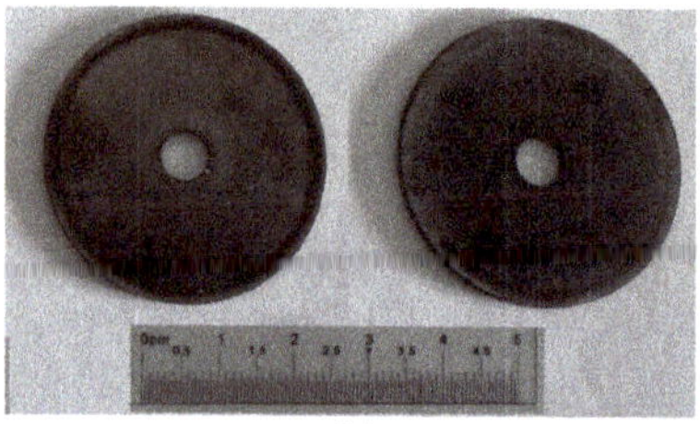

(b)

Figure 1. Appearance of test samples after nitriding in plasma at 430 °C: (**a**) samples placed on the cathode of a vacuum furnace (after nitriding for 8 h); (**b**) appearance of the samples after nitriding 12 h (left sample) and 24 h (right sample).

2.3. Mechanical Tests and Microstructural Analysis

The surface hardness of test samples was determined on a Wilson Wolpert Tukon 2100 B microhardness tester (Instron Corporation Inc., Norwood, MA, USA) using the Vickers method, with a load of 1.96 N (HV0.2) on repeated measurements at 10 randomly selected locations on a sample. The initial hardness of the samples was 228 ± 10 HV0.2 Total nitriding depth was determined on metallographically prepared samples using the Vickers method with a load of 0.49 N (HV0.05).

The surface roughness of samples was tested with 5 repetitions using an electromechanical device with a stylus, on the Surface Roughness Tester TR200 (INNOVATEST Europe BV, Maastricht, The Netherlands). The length reading for the measurement of roughness parameters was 1.75 mm and the Gauss filter was used for filtering data (λ_c = 0.8 mm). After sanding, the mean roughness of samples was R_a = 0.609 ± 0.043 μm, the average height difference was R_z = 4.768 ± 0.639 μm, and the maximum height difference was R_{max} = 8.235 ± 2.795 μm.

Microstructure analysis of nitrided AS steel samples was performed on metallographically prepared samples which, after standard preparation by grinding and polishing, were exposed to electrochemical etching in an aqueous solution with 10% oxalic acid, with DC voltage of 8 V and etching duration of 3 min. The microstructure of the electrochemically etched samples was analyzed using the scanning electron microscope Tescan Vega 5136LS(TESCAN GROUP, a.s., Brno-Kohoutovice, Czech Republic), equipped with the energy dispersive spectroscopy (EDS) device Bruker (Bruker AXS GmbH, Karlsruhe, Germany). The EDS line analysis and mapping were performed on 3 randomly selected transverse sites on the metallographic prepared section of each sample.

2.4. Micro-Abrasion Tests

The wear resistance to micro-abrasion was determined using a calotester device (TRIBOtechnic, Clichy, France) with freely rotating ball, shown in Figure 2.

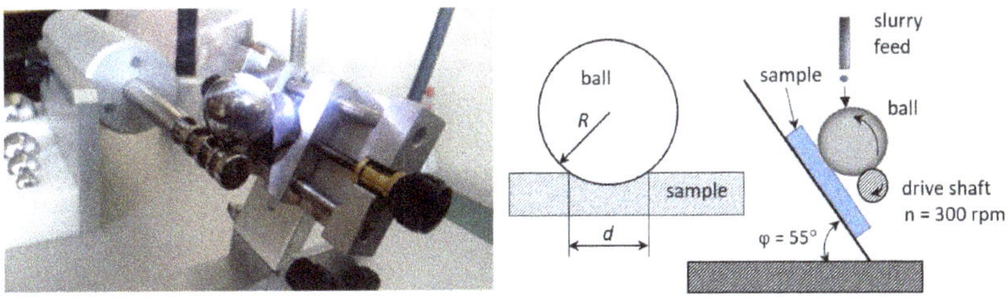

(a) (b)

Figure 2. Micro-abrasion wear by the freely rotating ball method: (**a**) a detail of testing on the TRIBOtechnic Calotester device; (**b**) the relationship between the geometric dimensions of the ball and the crater on the surface of the sample.

The ball, 2R = 20 mm in diameter, was made of EN 100Cr6 steel, with mass m = 32.6 ± 0.05 g. The abrasive slurry placed on the rotating ball consisted of 3 μm polycrystalline diamonds dispersed in the Struers DP-Suspension P (Struers, S.A.S. Champigny, France). The speed of the ball was 300 rpm. Ball rotation speed and the duration of micro-abrasion test were controlled by a microprocessor. Before tests samples were ultrasonically clean in 95% ethyl alcohol for 3 min and dry in oven at 75 °C for 10 min. The surface of all test samples was micro-abraded in durations from 10 to 60 s in steps of 10 s, with the manual addition of one drop of diamond suspension every 10 s. This feed rate ensures that the contact between the ball and sample was always wet, and that a circular ring was formed in the middle of the ball covered with the diamond suspension. At each specified wear time, a new crater was created on the surface of the sample. The angle of inclination of the sample to the horizontal plane was φ = 55°. This angle enabled continuous and uniform rotation of the ball on the sample without slipping. After micro-abrasion wear process, a spherical crater with a clearly visible diameter remains on the surface of the sample. The samples were again cleaned in 95% ethyl alcohol for 3 min and dried in an oven at 75 °C for 10 min. The spherical cap diameter was observed by an optical stereo microscope BTC STM-3C (Castell Nova Kft, Sopron, Hungary) equipped with camera MicroQ SP-51(Hangzhou ToupTek Photonic Co., Ltd., Hangzhou, China). The resolution of the camera is 5.1 Mpx, and the pixel size is 2.0 μm × 2.0 μm. Measurement of the dimensions of abraded crater was conducted in the image analysis software ToupTek ToupView, version x64, 4.11.19561.20210912 The diameter of the wear scar was measured in parallel and at a perpendicular direction to the ball rotation.

The wear scars were produced with test times of 10 s, 20 s, 30, 40 s, 50 s, and 60 s to obtain the volume loss curve. The number of parameters affect the results of the free rotating ball cratering tests. The most significant of them are [38]: the abrasive material, the abrasive size and shape, the abrasive slurry loading, the type of suspension fluid, the drive shaft rotation speed, the ball mass, its diameter, material, and surface roughness. All these parameters were kept approximately the same during all tests. Assuming that the crater is spherical in shape and that its depth and diameter are significantly smaller than the diameter of the ball, the volume loss of a monolithic material (V) removed from the crater and its depth (h) can be calculated according to the following equations [30,39]:

$$V \approx \frac{\pi \cdot d^4}{64 \cdot R}, \text{for } d << R \quad (1)$$

$$h \approx \sqrt{\frac{V}{\pi \cdot R}}, \text{for } h << R \quad (2)$$

in which d is the crater diameter (mm) and R is the ball radius (mm). The wear volume can be deduced as a function of the relative sliding distance (S), and normal load (N) on the contact area and the wear coefficient κ may be calculated from Equation (3), which is an extension of the Archard wear equation [38,40]:

$$\kappa = \frac{V}{S \cdot N} \quad (3)$$

The wear path (S) is determined based on the following equation:

$$S = 2 \cdot R \cdot \pi \cdot \frac{n}{60} \cdot t \quad (4)$$

where R is the ball radius, n is rotation speed, and t is time. The wear path obtained via ball rotation without sliding was in the range of S = 0 m to S = 14.30 m, after 60 s of rotation. The normal load N on the sample surface caused by the ball weights can be calculated from the equation:

$$N = m \cdot g \cdot \cos\varphi \quad (5)$$

where m is the ball mass, g = 9.81 ms^{-2}, and φ is the inclination angle of the sample to the horizontal plane. For the test conditions shown in Figure 2, the normal load was N = 0.1834 N.

2.5. DC Electrochemical Tests

For measuring the electrochemical corrosion resistance, samples were researched by direct current DC and alternating current AC techniques in 3.5 wt. % NaCl solution (pH = 6.6) at room temperature (20 ± 2) °C. For electrochemical measurement, the surface of samples was cleaned with ethyl alcohol. Electrochemical DC measurements were obtained using the standard three-electrode system. The counter electrode was platinum, and the reference electrode was a saturated calomel electrode (SCE). Stainless steel specimens (type 1.4401) with different surface nitriding time (0 h, 8 h, 12 h and 24 h) and exposed test area of 78 mm^2 were used as the working electrode. Measurements were performed using

Potentiostat/Galvanostat VersaSTAT 3 (AMETEK, Princeton Applied Research, Berwyn, PA, USA) with use of the software package VersaStudio v2.44 (AMETEK Scientific Instruments, Berwyn, PA, USA). The volume capacity of the used electrochemical cell (K0235 AMETEK Scientific Instruments, Berwyn, PA, USA) was 250 mL. Changes in the corrosion potential of a metal substrate with coating were tested using the open circuit potential. The reading of the open circuit potentials (E_{ocp} or E_{corr}) was performed after stabilizing the system over a time interval of 1000 s. Besides the E_{corr} measurement, in order to define susceptibility to pitting corrosion, the pitting potential E_{pitt} and nucleation resistance ΔE were measured. The span of ΔE ($E_{pitt} - E_{corr}$), as small difference between E_{pitt} and E_{corr}, indicates higher susceptibility to localized corrosion. The initial potential for cyclic polarization was −0.1 V vs. E_{corr} and the vertex potential was 1.2 V vs. E_{corr}.

2.6. AC Electrochemical Tests

Electrochemical behavior at the interface of metal/coating/electrolyte phases has a key role in understanding metal behavior (corrosion resistance) in an electrolyte. For that reason, electrochemical impedance spectroscopy (EIS) as a useful and nondestructive technique was performed using VersaSTAT 3 Potentiostat/Galvanostat (AMETEK Scientific Instruments, Princeton applied research, Berwyn, PA, USA), with the application of the appropriate ZsimpWin 3.60 software package. Changes in the corrosion properties were tested in accordance with EN ISO 17475:2008, in 3.5 wt. % NaCl solution (pH = 6.6) at room temperature (20 ± 2) °C. The application of EIS enabled the quantification of the surface layer resistance and the definition of the mechanism of corrosion progression without any degradation of the test surface. It is also highly sensitive and can detect changes on the metal surface over time, allowing for the monitoring of the corrosion process at the interface in real-time. Generally, interface was described with an equivalent electrical circuit (EEC) that contains a specific combination of resistors, inductors, and capacitors. A comparison of experimental results which were obtained by mathematical aligning the measurement data with the theoretical function of the equivalent electrical circuit, represents the basis for interpreting corrosion processes. Measurements were performed after 0 h, 4 h, 8 h, 12 h, and 24 h and are given with Nyquist's and Bode's graphs over a wide range of applied potential frequencies (0.1 Hz to over 100 kHz), recording 10 points per decade. The amplitude of the sinusoidal voltage signal was 10 mV. To checks the repeatability of the data; each measurement was executed in three replications.

3. Results and Discussion

3.1. Surface Roughness

Measured results of the surface roughness before and after the plasma nitriding are shown in Table 3 using several amplitude parameters [41]: the arithmetic mean of the magnitude of the deviation of the profile from the mean line (R_a), the root-mean-square value (R_q),the ISO 10-point height parameter (R_z), the maximum height of the profile above the mean line within the sampling length (R_p), the maximum depth of the profile below the mean line (R_v), the maximum peak to valley height of the profile (R_t), and the maximum peak to valley height of the profile in a sampling length (R_{max}).

Table 3 and Figure 3 show a decrease in all roughness amplitude parameters after plasma nitriding for 4 h. This phenomenon is a consequence of the sputtering process, in which the surface of the samples is cleaned from the remains of organic impurities, the oxide film is removed, and the surface is smoothed by the action of DC plasma composed of ionized hydrogen, argon, and nitrogen gases. Bombarding the surface of stainless steel with argon and nitrogen ions removes the peaks of micro-unevennesses, sputter iron ions and alloying elements, while hydrogen ions break down the oxide film. After such preparation, in addition to obtaining a physically clean surface, there is also a reduction in surface roughness. In the continuation of the nitriding process, FeN nitrides are deposited on the cleaned surface from the plasma, and nitrogen diffusion into the surface layer and the formation of metastable expanded austenite occurs. By nitriding stainless steel for more than 12 h, a compound layer composed of iron nitride Fe_xN_y and chromium nitride Cr_xN_y with different stoichiometric ratios of nitrogen and metal elements begins to form on the surface. The creation and increase in the thickness of the compound layer causes an increase in the value of almost all the amplitude parameters of the surface roughness. After 24 h of nitriding, the surface roughness is practically similar to the roughness in the initial state.

Table 3. Surface roughness parameters of the plasma-nitrided AISI 316L AS steel.

	R_z, µm	R_a, µm	R_q, µm	R_t, µm	R_p, µm	R_v, µm	R_{max}, µm
Heat-untreated	4.768 ±0.320	0.609 ±0.021	0.843 ±0.038	8.232 ±1.397	1.606 ±0.088	3.162 ±0.232	8.235 ±1.398
PN 430 °C/4 h	2.960 ±0.029	0.448 ±0.008	0.558 ±0.003	3.625 ±0.094	1.390 ±0.040	1.571 ±0.015	3.470 ±0.362
PN 430 °C/8 h	3.474 ±0.041	0.459 ±0.006	0.626 ±0.002	5.186 ±0.101	1.249 ±0.050	2.226 ±0.010	4.860 ±0.240
PN 430 °C/12 h	4.003 ±0.028	0.570 ±0.003	0.741 ±0.002	5.841 ±0.327	1.652 ±0.050	2.352 ±0.023	5.635 ±0.428
PN 430 °C/24 h	6.213 ±0.604	0.761 ±0.039	1.048 ±0.065	10.070 ±2.719	2.492 ±0.599	3.721 ±0.005	9.995 ±2.795

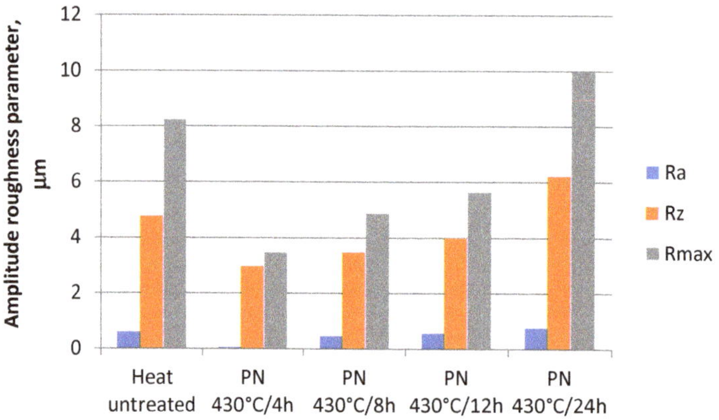

Figure 3. Changes in the main surface roughness parameters after plasma nitriding of an AISI 316L steel at 430 °C with a process duration of 4 h, 8 h, 12 h, and 24 h compared to the heat-untreated state.

3.2. Hardness and Nitriding Depth

The mean values of 10 surface hardness measurements per individual sample, after plasma nitriding AS steel AISI 316L at 430 °C for 4 h, 8 h, 12 h, and 24 h, are shown in Figure 4. The deviation of the measured hardness values from the mean value is between 8% and 10%. By comparing the hardness of the nitrided sample with the hardness of the sample in the initial state, an increase in hardness is observed with a longer nitriding time. After 4 h and 8 h of nitriding, the hardness of the surface increased very slightly compared to the untreated state. This 60–80% increase in hardness is the result of the formation of a diffusion zone under the surface of the sample, the dissolution of nitrogen in the austenite and the formation of metastable expanded austenite. Nitriding for 12 h increased the surface hardness by 100% compared to the untreated state, whereas after 24 h of nitriding the surface hardness increased by 162% compared to the non-nitrided surface. This increase in hardness indicates the formation of an iron nitride zone on the sample surface and the precipitation of chromium nitride in the compound layer.

The distribution of hardness according to the depth of the nitrided layer, shown in Figure 5, was tested by the Vickers method with a load of 0.491 N by measuring two sets of hardness with distances between impressions of 50 μm. The hardness of the core was 190–200 HV0.05.

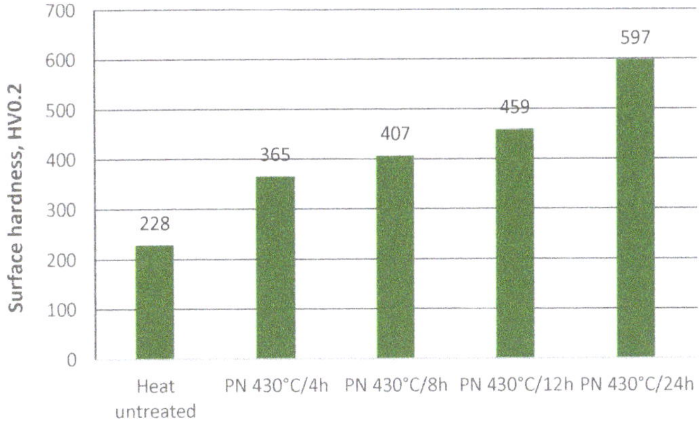

Figure 4. Surface hardness of the AS steel AISI 316L after plasma nitriding at 430 °C with a process duration of 4 h, 8 h, 12 h, and 24 h compared to a heat-untreated state.

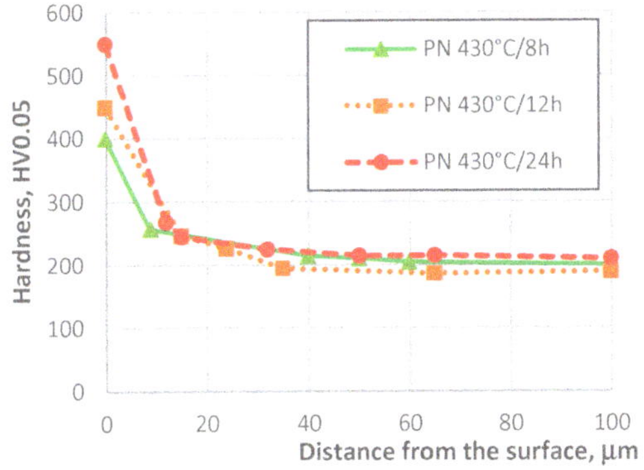

Figure 5. Distribution of hardness according to the depth of the nitrided AS steel AISI 316L after plasma nitriding at 430 °C with a process duration of 8 h, 12 h, and 24 h.

The effective depth of nitriding, determined as the distance from the surface where the measured hardness is greater than the hardness of the core by 50 HV0.05, is 6–8 μm for the sample nitrided for 8 h, 10–12 μm for the sample nitrided for 12 h, and 18–20 μm for the sample nitrided for 24 h. From the distribution of hardness by layer depth, an increase in nitriding depth and surface hardness can be observed with longer nitriding time.

3.3. EDS and Microstructure Analysis

Figure 6 shows the microstructure of the nitrided layer on an AISI 316L steel after nitriding at 430 °C for 4 h and 8 h observed with light microscopy (LM) on a Metallurgical microscope KERN OKO 178 (KERN & SOHN GmbH, Balingen, Germany), and after 12 h and 24 h observed with the scanning electron microscopy (SEM) on a device Tescan Vega 5136LS. The LM measured thickness of a nitride layer after PN treatment for 4 h and 8 h was 3–4 μm and 6–9 μm, respectively. A nitride layer achieved after PN treatment for the duration of 12 h has a thickness between 11 and 14 μm and, as can be observed in Figure 6c, it consists of the expanded austenite with a very thin top layer of brittle nitrides. In the expanded austenite sub layer, grains were of equal size and similar shape, without visible nitride precipitates. Figure 6d shows a significantly thicker nitride layer, compared with the previously described one, with precipitates of iron and chromium nitrides on the grain boundaries. The thickness of this layer, which was formed after 24 h of plasma nitriding, had values between 16 and 20 μm. The results of the SEM analysis of the nitrided layer coincide with the results of the microhardness test.

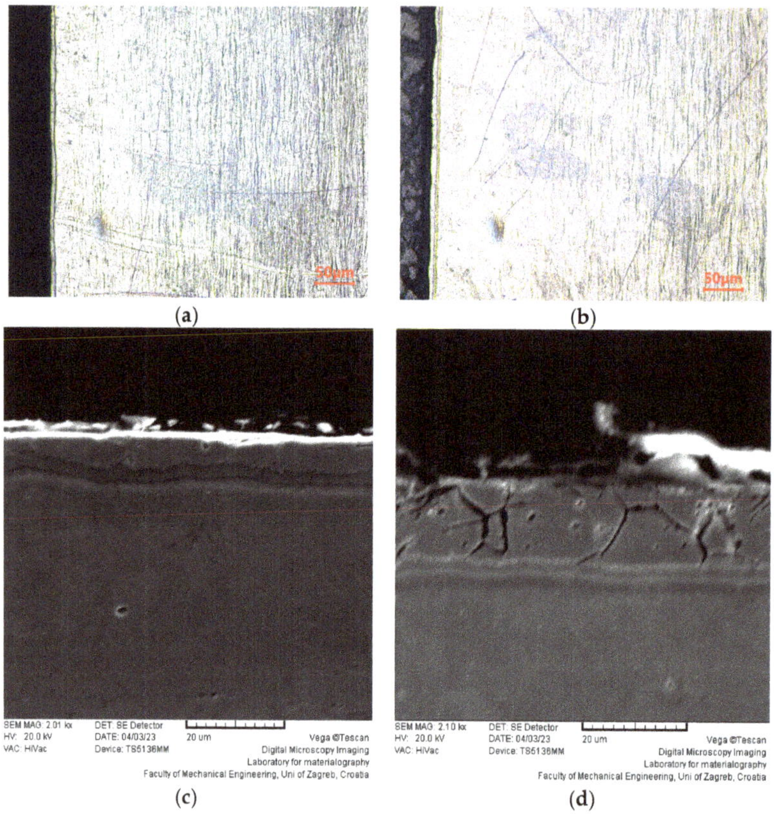

Figure 6. Microstructure of the surface layer on an AISI 316L AS steel after plasma nitriding at 430 °C with a process duration of: (**a**) 4 h (LM); (**b**) 8 h (LM); (**c**) 12 h (SEM); (**d**) 24 h (SEM).

The elemental composition of the nitrided layers on an AISI 316L steel was determined via line EDS analysis with device Bruker. The distribution of elements in the surface layer after nitriding at 430 °C/12 h is shown in Figure 7. The formation of iron and chromium nitrides is confirmed by the increased presence of elements iron (Fe), chromium (Cr), and nitrogen (N) in the compound layer. Below, there is a thin layer of the expanded austenite (S-phase) with decreased content of Fe and Cr. Above the S-phase toward to the surface, an increased presence of Fe, Cr, and N can be observed at distances from 22 to 25 μm in Figure 7. This is caused by formation of compound layer formed from metal elements (Fe, Cr) diffused from steel core and nitrogen (N) diffused from the ionized H_2-N_2 plasma.

The distribution of elements in the surface layer after nitriding of an AS steel AISI 316L at 430 °C/24 h is shown in Figure 8. In this sample, there is an even greater amount of chromium (Cr) in the compound layer than in the sample nitrated for 12 h, while the amount of iron (Fe) is smaller. This relationship between iron and chromium is a consequence of the local concentration of chromium at the grain boundaries in the form of CrN precipitates.

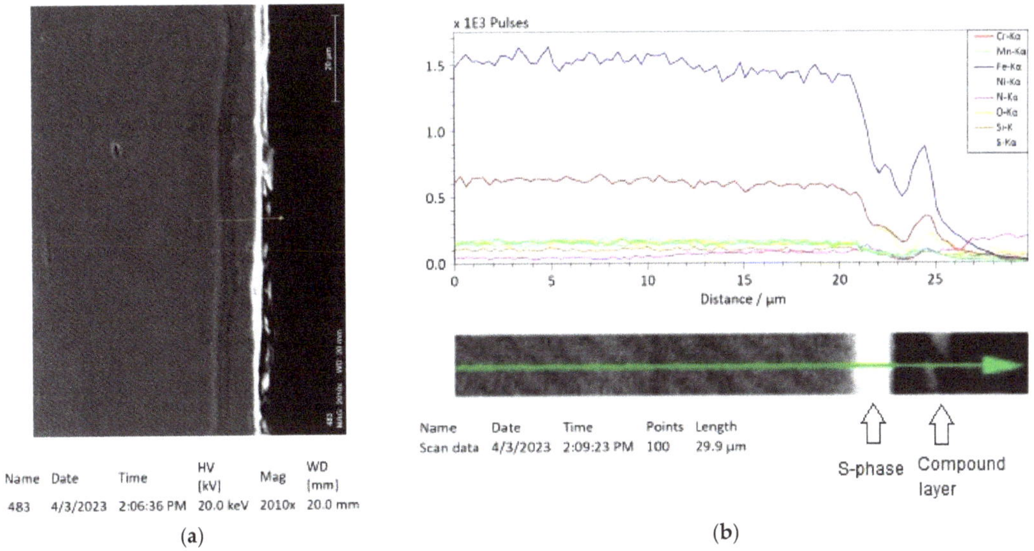

Figure 7. AISI316L AS steel after plasma nitriding at 430 °C in duration of 12 h: (**a**) microstructure; (**b**) chemical composition of the surface layer.

Microstructure analysis and EDS distribution of chemical elements by depth of nitrided layers on an AISI 316L steel samples confirmed the assumptions about the formation of a hard and brittle zone of increased hardness and increased values of roughness amplitude parameters.

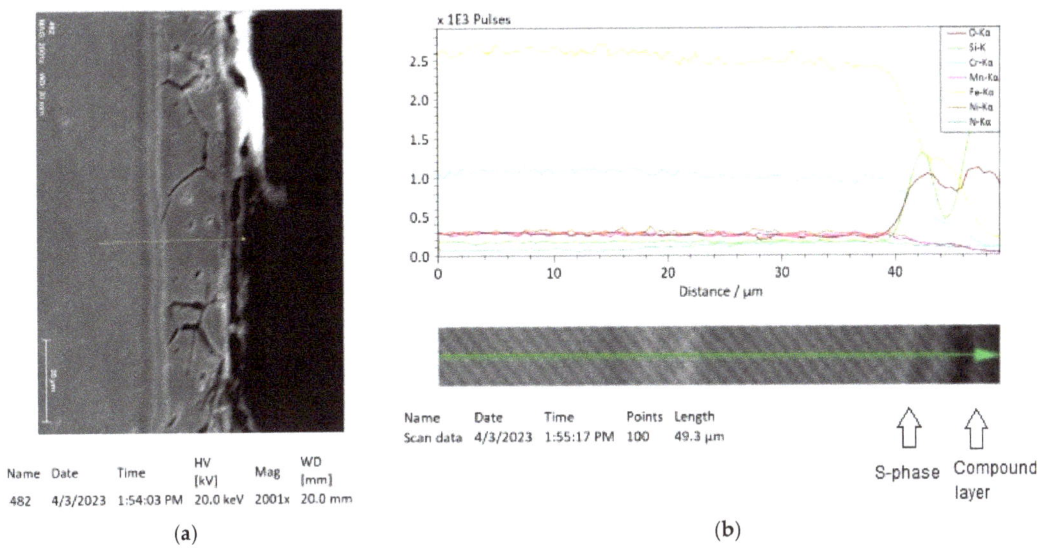

Figure 8. AISI 316L AS steel after plasma nitriding at 430 °C in duration of 24 h: (**a**) microstructure; (**b**) chemical composition of the surface layer.

3.4. Micro-Abrasion Wear Resistance

The micro-abrasive wear results of nitrided samples of an AISI 316L AS steel are shown in Figure 9. Using Equation (1), the volume of the sample material (V) removed from the crater after wear for 10–60 s was determined. Using Equations (4) and (5), the product $S{\cdot}N$ was determined, consisting of the wear path (S) after the given wear time and the normal force at the point of contact between the ball and the sample (N). The dependence of the wear volume V on the wear path, i.e., the SN product, was determined using linear regression analysis. The coefficient of micro-abrasive wear κ, which physically represents the wear rate, is calculated according to Equation (3) for wear time between 10 and 60 s. From the change in the coefficient of micro-abrasion wear depending on the SN product, shown in Figure 8b, unstable wear can be observed during 10–40 s, then the values of the coefficient are set at the characteristic values for each nitrated sample. The amounts of the coefficient of micro abrasive wear after 60 s of wear are shown in Table 4. The depth of the crater (h) after 60 s of micro-abrasion wear was calculated by Equation (2) and shown in Figure 9c.

From the comparison of the coefficients of micro-abrasion wear, a significantly higher rate of wear can be observed in samples nitrated for 12 h and 24 h compared to samples nitrated for 4 h and 8 h and the non-heat-treated sample. These results derive from the microstructure and properties of the nitrided layer and the test conditions. During micro-abrasive wear of the soft surface of non-nitrided 316L steel, larger abrasive particles are pressed into the surface of the sample and the abrasive suspension accumulates around the crater, while a smaller number of abrasive particles remain stuck to the ball of the calotester. Therefore, with the rotation of the calotester balls, the abrasive particles between the ball and the soft surface of the sample simultaneously rotate and move. The result of such micro-abrasive wear with the simultaneous rotation of the ball and microparticles is a relatively small diameter and depth of craters in the surface of non-nitrided steel, as well as a small wear coefficient $\kappa = 1.2336 \times 10^{-4}$ mm^3/Nm. Additionally, a clearly visible sign of wear remains on the calotester ball after wear. Micro abrasion wear of the nitrided layer with ductile expanded austenite formed after 4 h and 8 h of nitriding shows wear characteristics similar to non-nitrided austenitic steel. At the same time, the amounts of wear coefficients are slightly higher than the coefficient of non-nitrided steel.

The amount of the wear coefficient of samples nitrated for 12 h and 24 h are significantly higher than the coefficients of samples nitrated for 4 h and 8 h. This phenomenon can be explained by the appearance of a different mechanism of micro abrasive wear and the effect of precipitated nitrides on the grain boundaries of expanded austenite. In the case of these, longer nitrided samples, the surface hardness is higher than in non-nitrided austenite steel, and the abrasive particles are not pressed into the nitrided layer but remain mostly stuck to the ball of the calotester. Therefore, in this case, pure abrasive wear occurs with grooves and removal of particles from the nitrided surface with minimal rolling of abrasive particles between the sample and the calotester ball. On the other hand, in the nitrided layer, hard and brittle nitrides of iron and chromium are secreted along the grain boundaries, which additionally promote abrasive wear, after they break off and mix with the diamond grains in the abrasive suspension. During the wear of the nitrided sample for 12 h and 24 h, significantly less wear of the calotester ball was observed than when the wear of the soft samples of non-nitrided steel was observed. The shown action of micro abrasives on the soft or hard surface of the sample and the results of the micro abrasion test are described in refs. [34–36]. From the comparison of the coefficients of micro-abrasive wear and the calculated depth of the crater, it can be concluded that short-term plasma nitriding of AS steel, lasting 4–8 h, is more favorable than long-term plasma nitriding of 12–24 h from the aspect of better resistance to abrasive wear.

The appearance of the crater after 60 s of micro abrasion wear of the nitrided sample with a ball with a diameter of 20 mm at 300 rpm is shown in Figure 10 for nitriding times of 4 h, 8 h, 12 h, and 24 h.

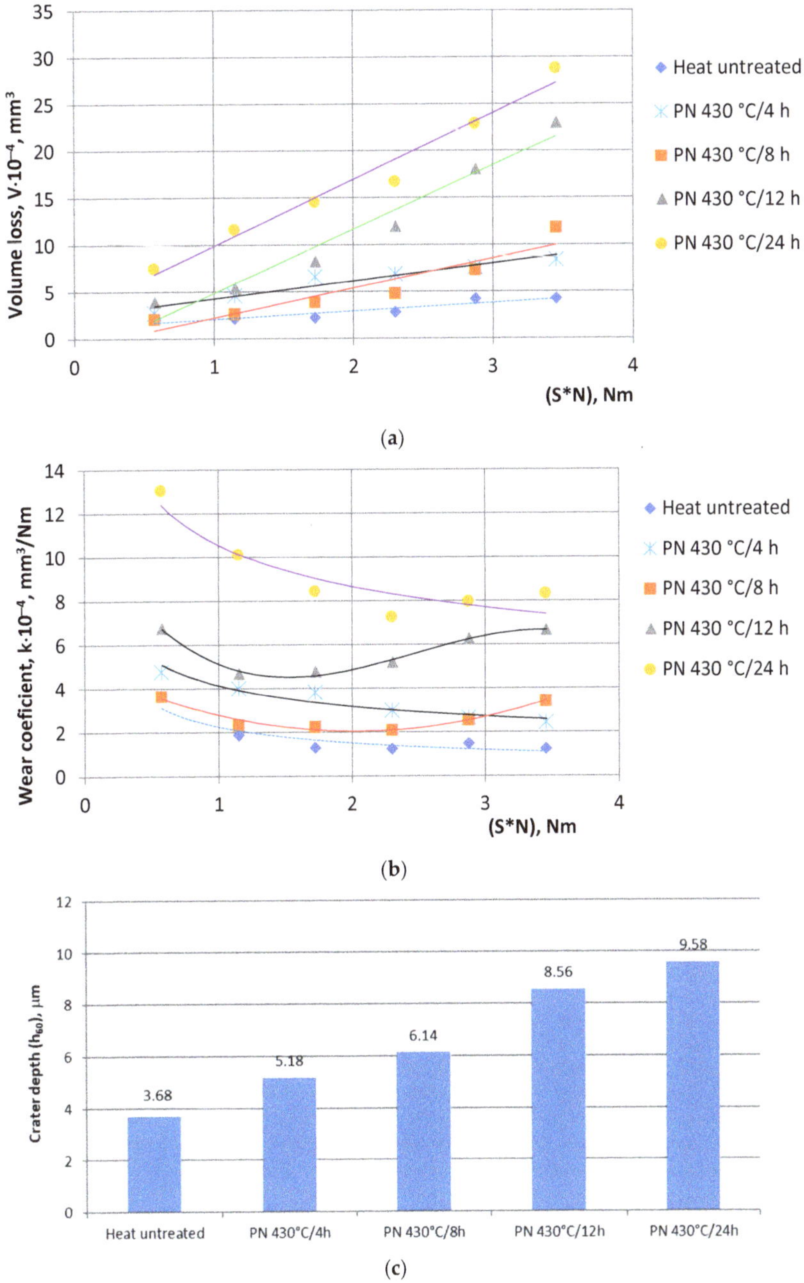

Figure 9. Results of micro-abrasive wear of a nitrided AS steel AISI 316L: (**a**) loss of volume from the calotte crater against SN product; (**b**) wear coefficient κ; (**c**) crater depth after 60 s wear.

Table 4. The mean crater diameter (D_c) and coefficient of micro-abrasive wear κ after 60 s testing of AISI 316L steel samples.

Sample Treatment	D_c, μm	$\kappa \cdot 10^{-4}$, mm³/Nm
Heat-untreated	542 ± 28	1.2336
Plasma-nitrided 430 °C/4 h	486 ± 52	2.4370
Plasma-nitrided 430 °C/8 h	553 ± 25	3.4282
Plasma-nitrided 430 °C/12 h	964 ± 55	6.6560
Plasma-nitrided 430 °C/24 h	868 ± 40	8.3489

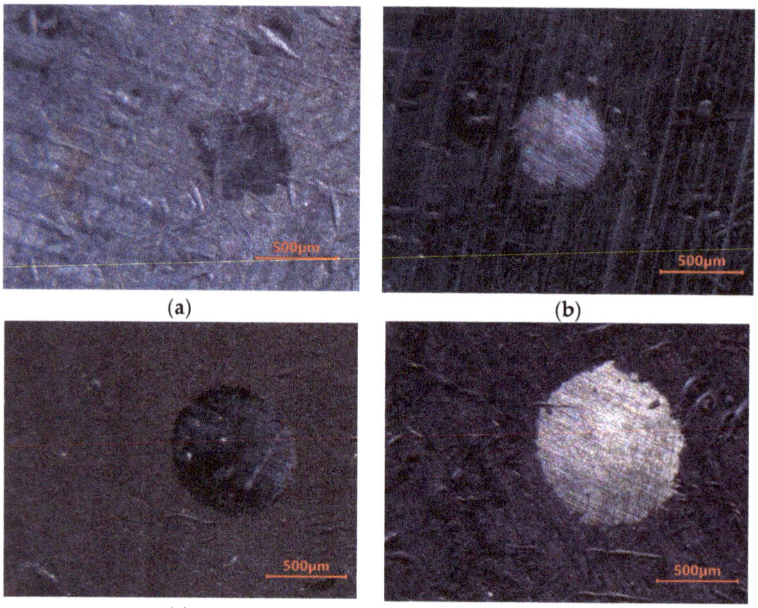

(a) (b) (c) (d)

Figure 10. Crater on the surface of AISI 316L steel samples nitrided in plasma at 430 °C after 60 s of micro-abrasive wear with a 20 mm diameter ball coated with a diamond suspension. Nitriding time: (**a**) 4 h; (**b**) 8 h; (**c**) 12 h; (**d**) 24 h.

Photographs of the worn surface were taken under the same conditions on a stereo microscope BTC STM-3C equipped with camera MicroQ SP-51 at a magnification of 40:1. The appearance of the worn surface in the crater is characteristic for pure abrasion wear. In the crater, one can clearly see parallel grooves and scratches created during removal of sample material by action of diamond grains from the suspension enveloped around a steel ball. The grooves are orientated in the ball rolling direction. The craters which are formed on the samples after 60 s of testing are circular in shape with a mean diameter shown in Table 4. By comparing the diameter of the craters, a significant increase in craters dimension can be observed on samples nitrated for 12 h and 24 h compared to samples nitrated for 4 h and 8 h. In the interior of the crater in Figure 10b, on the sample that was nitrided for 24 h, micro porosity and microcracks can be observed, which were created by the formation of a brittle and partially porous nitride layer on the surface and the release of nitride needles along the grain boundaries. In the process of abrasive wear, such weakened places contribute to a greater loss of mass and an increase in the diameter of the crater.

From the results of micro-abrasion wear, shown in Figure 9 and in Table 4, it can be seen that nitriding of an AS steel AISI 316L in plasma at a temperature of 430 °C for a duration of more than 8 h does not achieve an increase in wear resistance, but an unfavorable decrease. This unfavorable influence of the nitriding time at a temperature of 430 °C on the wear resistance of an AS steel AISI 316L agrees with the results and conclusions of several papers [17,18,42]. Increased wear resistance can be expected when nitriding at temperatures between 350 °C and 400 °C, with the formation of a surface layer of expanded austenite (S-phase), whereas the formation of a layer with a mixture of hard iron and chromium nitrides on the surface or their precipitation along grain boundaries in the edge layer reduces the micro-abrasion wear resistance of AS steel. As a rule, it can be observed that a thicker nitride layer and a larger amount of nitride precipitates cause a greater reduction in the wear resistance. In [42], it was concluded that plasma nitriding temperatures between 350 and 400 °C led to the best wear properties of AS steel due to the maximum amount of the S-phase and no chromium nitrides. At this temperature interval, increasing the duration of nitriding has no critical impact on wear resistance, but only on the increasing nitriding depth and microhardness of the S-phase. When nitriding at a temperature of 430 °C and higher, a longer nitriding duration of 8 h causes the formation of a thicker layer of iron and chromium nitrides, which is hard and brittle and wears out faster than the softer, but tougher, layer of expanded austenite.

3.5. Electrochemical DC Measurements

In the initial phase of the polarization measurement, the system was stabilized in such a way that after immersion in in 3.5 wt. % NaCl water solution at room temperature (20 ± 2) °C, the electrical circuit between the working and counter electrodes was left open, and the potential difference between the reference and working electrodes was recorded as a function of time [43,44]. The measured numerical values of open circuit potentials for heat-untreated and nitrided AS steel AISI 316L samples (4 h, 8 h, 12 h and 24 h) were recorded after stabilization in 3.5 wt. % NaCl water solution at room temperature (20 ± 2) °C and are shown in Table 5. Graphical values of corrosion potentials are given in Figure 10.

Table 5. The results of electrochemical DC techniques applied on heat-untreated and nitrided AS steel AISI 316L.

Sample	T_s [°C]	E_{corr} vs. SCE [mV]	E_{pitt} vs. SCE [mV]	ΔE [mV]
Heat-untreated	20 ± 2	−175	376	551
PN 430 °C/4 h	20 ± 2	−246	141	387
PN 430 °C/8 h	20 ± 2	−305	192	497
PN 430 °C/12 h	20 ± 2	−369	198	567
PN 430 °C/24 h	20 ± 2	−344	470	814

From the measured numerical values of the corrosion potential presented in Table 5 and graphically interpreted in Figure 11, information on the corrosion stability of the samples in the test electrolyte was obtained. After closely established stationary state (t = 1000 s), when the test samples are in equilibrium with the environment (anodic and cathodic currents are equal in magnitude, but in opposite directions), it is visible and expected that heat-untreated (blank) sample has a more positive E_{corr} than samples with an inhomogeneous nitrided surface. This fact indicates that the oxide film on a AS steel AISI 316L is more resistant to corrosion then surface films on the nitrided samples, under the same electrolyte corrosion load. According to curve *E-t* for the heat-untreated sample, an oscillation of the curve can be observed. Such potential oscillation can be correlated with the increased corrosion activity caused by surface instability and material dissolution.

From the numerical and graphical values of E_{corr} (Table 5 and Figure 10), it can be concluded that nitriding time has no significant influence on the corrosion stability of the samples of nitrided AISI 316L steel in the test medium. It is also possible to notice that the slope of the curve is much milder in the case of the heat-untreated sample, and that all four curves were shifted in a negative direction, which indicates the corrosion instability of the working electrodes in the chloride medium. All samples showed the same trend and good repeatability of the results [45,46].

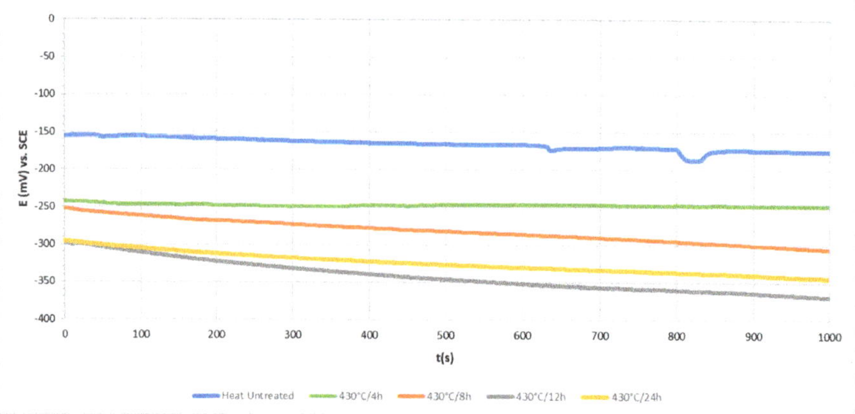

Figure 11. Open circuit potential (E_{corr}) vs. time curves of tested samples.

From the numerical and graphical values of E_{corr} (Table 5 and Figure 11), it can be concluded that nitriding time has no significant influence on the corrosion stability of the samples of nitrided AISI 316L steel in the test medium. It is also possible to notice that the slope of the curve is much milder in the case of the heat-untreated sample, and that all four curves were shifted in a negative direction which indicates the corrosion instability of the working electrodes in the chloride medium. All samples showed the same trend and good repeatability of the results [45,46].

To define susceptibility to pitting corrosion, the pitting potential (E_{pitt}) was measured. The initial potential for cyclic polarisation was −0.1 V vs. E_{corr} and the vertex potential was 1.2 V vs. E_{corr}. The cyclic polarisation curves obtained in 3.5 wt. % NaCl solution at room temperature (20 ± 2) °C for heat-untreated and plasma-nitrided samples are shown in Figure 12. Table 5 presents the quantitative electrochemical parameters, E_{pitt} and ΔE, collected from the polarization curves in Figure 12.

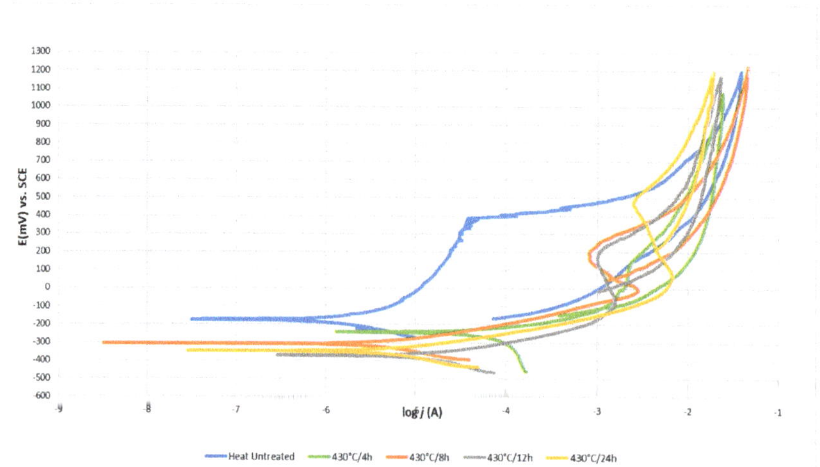

Figure 12. Cyclic polarization curves of heat-untreated and nitrided samples of AS steel AISI 316L in 3.5% NaCl water solution.

In the case of the heat-untreated AS steel AISI 316L, it can be seen that the heat-untreated sample does not show an evident passive area like samples with a nitrided surface do, and that the current density increases sharply when the potential is higher than about 376 mV vs. SCE (Table 5) [47]. It is well known that this sudden increase in current density is caused by the break-through of the passive layer on stainless steel and present the potential E_{pitt} at which pitting corrosion initiates in the chloride ion containing environment [46,48]. On the other hand, samples with a nitrided surface generally showed a tendency for passivation, which indicates that the nitrided layer can protect the substrate, providing a good barrier against localized attack at a lower potential range (from −78 to 200 mV vs. SCE), except for the sample where nitriding lasted longer than 12 h. In that case, a sample with 24 h surface nitriding showed better resistance against localized attacks at higher potential.

The value of E_{pitt} for the sample PN 430 °C/24 h was 470 mV vs. SCE, which is 94 mV more positive than the E_{pitt} value of the blank sample (Table 5). However, after E_{pitt}, the anodic current density rapidly raised with the potential which indicates that the nitrided layer had become porous or too thin to withstand the corrosion loading caused by chloride ions [46,48]. Specific behaviours of different nitrided samples to local corrosion loads can be reasonably explained considering two reasons: the nitrogen surface concentration and the thickness of the modified layers. According to data from the literature, it can be concluded that the different amounts of nitrogen in AS steels have a great influence on their resistance to pitting corrosion [49,50]. According to Table 5 and Figure 12, due to the high chloride concentration, pitting branch starts at relatively low anodic potential values (about +200 mV) [51]. At the highest polarization potentials, the anodic current values of the nitrided samples are sensibly reduced in respect of the untreated steel sample and are limited only by concentration polarization phenomena [50]. When it comes to susceptibility to localized corrosion, it can be concluded from Table 5 that the smallest value of pitting nucleation resistance (ΔE) or the highest resistance to localized corrosion has a sample PN 430 °C/24 h (ΔE = 814 mV) with the with the thickest compound layer. ΔE values for other three samples are approximately equal which strongly demonstrate the important role of the nitriding time on the adsorption mechanism and the pit initiation [52–54]. The anodic potentiodynamic polarization curves obtained in the present investigation were similar to those previously published for the nitrided AS steel in chloride-containing solutions [55–58].

3.6. Electrochemical AC Measurements

Electrochemical behaviour at borders of metal/electrolyte phases is a key factor in understanding metal behaviour in an electrolyte [43]. For that reason, electrochemical impedance spectroscopy (EIS) was used to evaluate the resistance and protection of the untreated and nitrided surface. Since the application of the AC measuring technique, during the implementation of EIS, enables the quantification of the resistance of surface layers and the definition of the mechanism of corrosion progression without any degradation of the test surface, the numerical and graphical data obtained by this method are extremely important in determining the corrosion resistance of metal surfaces. For this reason, the dominant data needed to determine the resistance of the base substrate and barrier layers is manifested through the variation of resistance and capacity values; thus, information about the degradation of protective layers and substrates is obtained by indicating the oscillation of the resistance and capacity values, which ultimately results in a change in the impedance value [43,44].

Constant phase element (CPE), used in equivalent circuit, is applied in the case when the inhomogeneity (coating) of the system is affected by surface roughness and porosity and when the porosity of the newly created nitrided film needs to be compensated. Its impedance is defined by the following equations:

$$Z_{CPE} = Q^{-1}(j\omega)^{-n} \quad (6)$$

$$C = Q(\omega_{max})^{n-1} \tag{7}$$

In which Q denotes the proportionality constant (CPE), n the empirical exponent or phase shift ($-1 \leq n \leq 1$), ω the angular frequency, root of -1 (imaginary unit), C the capacity, and $\omega_{\max}$ the angular frequency in which the imaginary component impedance Z'' reaches the maximum of the time constant. In this way, the impedance spectrum of a distributed system is explained, which cannot be interpreted by a finite number of ideal electrical elements. In the case when $0.6 < n \leq 1$, CPE represents a capacitor of capacity Q; for $n = 1$, CPE represents an ideal coil $Q = L$; for $n = 0.5$, CPE represents Warburg impedance $Q = W$; for values of $n = 0$, CPE represents an ideal resistor $Q = R$ [43,44].

In order to further study the electrochemical behavior of the of AS steel 316L samples with heat-untreated and plasma-nitrided surfaces, equivalent circuits were used to simulate the EIS data. The layer shown in Figure 12 is porous (different thickness) with electrochemical reactions occurring only on the exposed electrode surface at the end of the pore. The equivalent circuit with two constant models corresponding to the scheme of the coated electrode is presented in Figure 13.

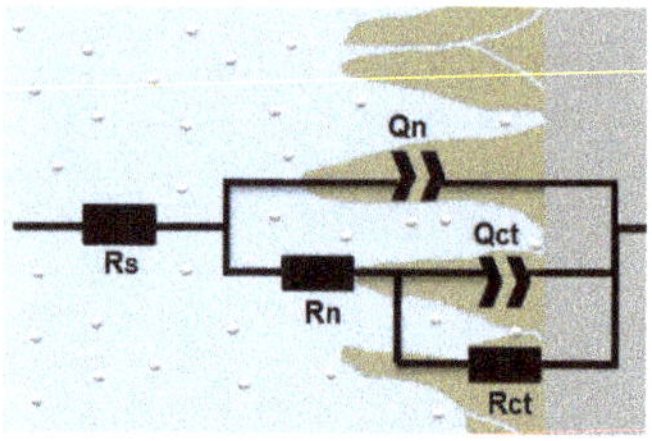

Figure 13. The equivalent circuit model R(Q(R(QR))) that was used to analyze the EIS plot.

On the equivalent circuit meaning of the labels are as follows: R_s is the solution resistance. R_n and Q_n are the pore resistance and constant phase element, i.e., an imperfect capacitor of oxide/nitride layer. Furthermore, R_{ct} represents the charge transfer resistance; the double electric layer constant phase element is defined with size Q_{ct} and the chi-square value (χ^2) method was used to assess the goodness of fit between measured and simulated data.

The EIS measurements were carried out in 3.5 wt. % NaCl solution at room temperature (20 ± 2) °C. The impedance spectrum was recorded at the open circuit potential. The amplitude of the sinusoidal voltage signal was 10 mV, while the detection of the current response was performed in the frequency range of 100 kHz–100 mHz [54]. The results of the EIS measurements and corresponding fitting results are shown in Table 6.

Table 6. The results of electrochemical impedance spectroscopy (AC) technique applied on the heat-untreated and plasma-nitrided samples of the AS steel AISI 316L.

Sample	T_s [°C]	R_s [Ωcm^2]	Q_n [Fcm^{-2}]	n_1	R_n [Ωcm^2]	Q_{ct} [Fcm^{-2}]	n_1	R_{ct} [Ωcm^2]	χ^2
Heat-untreated	20 ± 2	2.109×10^1	5.434×10^{-5}	0.780	1.311×10^2	2.761×10^{-5}	0.842	9.947×10^3	1.874×10^{-4}
PN 430 °C/4 h	20 ± 2	2.268×10^1	2.774×10^{-5}	0.746	3.300×10^2	2.824×10^{-4}	0.603	2.794×10^3	4.075×10^{-4}
PN 430 °C/8 h	20 ± 2	2.772×10^1	3.216×10^{-4}	0.749	6.860×10^1	2.952×10^{-4}	0.817	1.697×10^3	2.008×10^{-4}
PN 430 °C/12 h	20 ± 2	2.067×10^1	4.797×10^{-4}	0.798	3.695×10^1	3.848×10^{-4}	0.781	1.890×10^3	1.406×10^{-4}
PN 430 °C/24 h	20 ± 2	2.263×10^1	1.715×10^{-4}	0.820	1.621×10^2	3.291×10^{-4}	0.718	2.201×10^3	6.132×10^{-5}

The Bode and Nyquist diagrams of the heat-untreated and nitrided AS steel samples with the corresponding equivalent circuit model applied for the simulation of EIS results are presented in Figure 14a,b. The response of the EIS spectrum at higher, medium, and lower frequencies is shown in Figure 15.

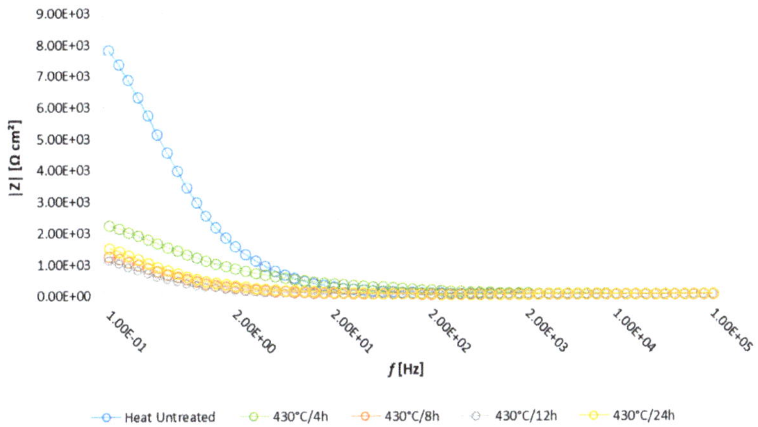

(a)

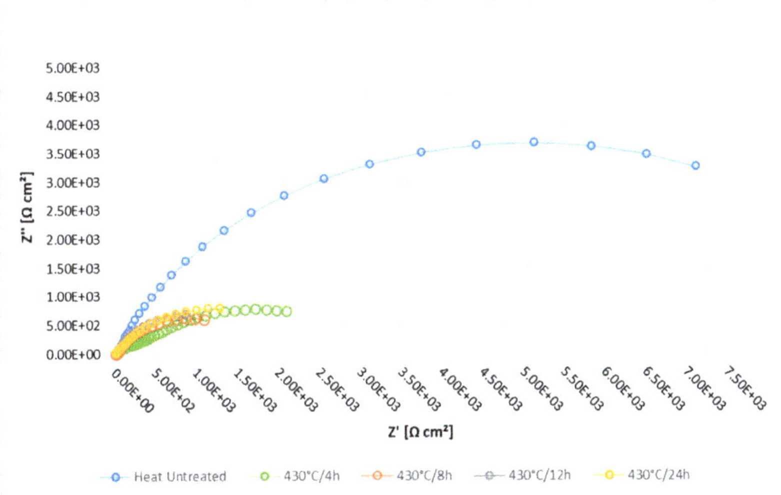

(b)

Figure 14. *Cont.*

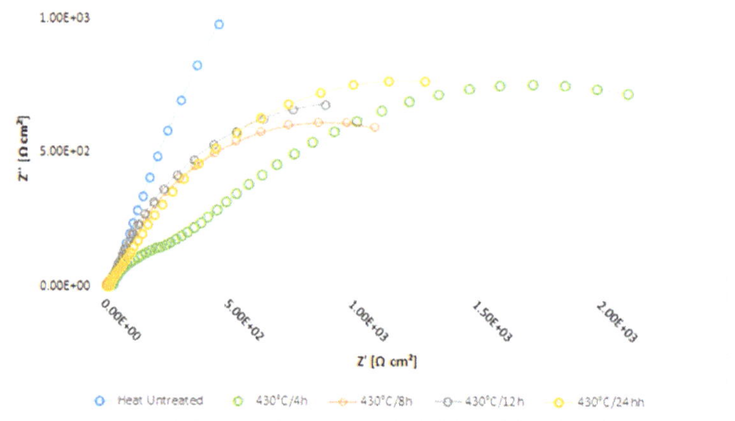

(c)

Figure 14. The results of EIS tests conducted on heat-untreated and nitrided AS steel AISI 316L: (**a**) Bode plot; (**b**) Nyquist plot; (**c**) detail of the Nyquist plot for low values of Z′.

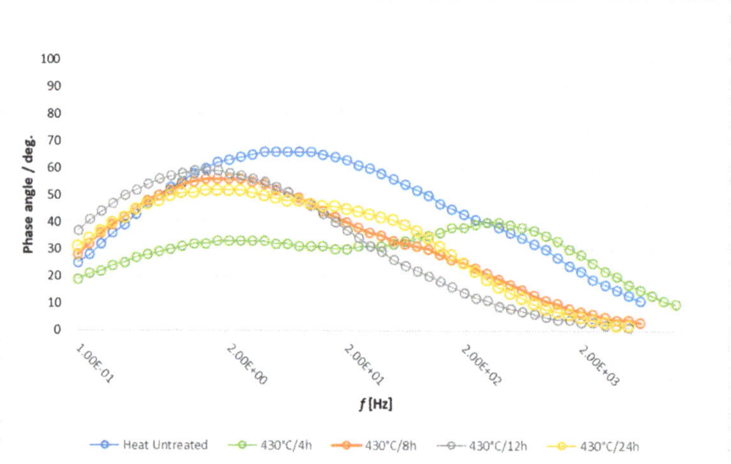

Figure 15. Phase angle as a function of the frequency of heat-untreated and nitrided AS steel AISI 316L.

The results of EIS tests shown in Table 6, confirm that heat-untreated sample has the highest corrosion resistance, which is evident from the increased value of the charge transfer resistance (R_{ct} = 9.947 × 10^3 Ωcm^2) at the borders between substrate and 3.5 wt. % NaCl solution. In contrast to untreated samples, nitrided samples have smaller and approximately equal corrosion resistance values, PN 430 °C/4 h (R_{ct} = 2.794 × 10^3 Ωcm^2), PN 430 °C/8 h (R_{ct} = 1.697 × 10^3 Ωcm^2), PN 430 °C/12 h (R_{ct} = 1.890 × 10^3 Ωcm^2), PN 430 °C/24 h. Additionally, resistance values are confirmed by the radius *r*, which represents capacitive semicircles in the Nyquist diagrams, shown in Figure 14b. The radius of the capacitive loop for blank (untreated) sample is more extensive than nitrided samples, where a decreased diameter is associated with the lower corrosion resistance of the porose-nitrided film. Lower values of the nitrided surface resistance and higher values of their capacity are a common indicator of the electrolyte absorption through the pores, i.e., worse preservation and durability of the surface layer (Table 6), which can also be a partial consequence of too-thin layers. The Bode plot also confirms that the blank untreated sample has the higher value of the resistance and lower value of capacity which is evident from the curve in Figure 14a. It is also evident from Bode's diagram that the curves of the nitrided samples have approximately equal values and that their slope remains constant. According to Figure 15, all phase angles (except PN 430 °C/4 h) of nitrided samples are between 50° and 70°, with a shift of the approximately constant height peak to the left (towards lower frequencies), which proves that there was an increase in the double layer capacity. Untreated sample phase angle is around 65° and the maximum is detected at medium frequency values (15 Hz) [43,54,59].

4. Conclusions

DC plasma nitriding leads to improved tribological and corrosion properties of the surface layer on AS steel AISI 316L. From previous research conducted with a low-temperature plasma nitriding of 316L austenitic stainless steel, the optimal process temperature/time window was found between 420 and 450 °C/4 and 20 h. Such a combination of parameters might result in a nitrided layer composed of the expanded austenite (S-phase) without significant precipitation of iron or chromium nitrides on the grain boundaries or in the top of the surface. However, the desire to realize the processing process with lower energy consumption sets the requirements for shortening the processing time while achieving good resistance to corrosion and wear. In plasma nitriding experiments of AS steel AISI 316L at a temperature of 430 °C, the effect of nitriding time 4–24 h on surface roughness, and mechanical, tribological, and corrosion properties was investigated.

The all-surface roughness amplitude parameters were decreased after PN (except for PN process lasting 24 h) compared to the untreated state, because of the sputtering process applied before plasma nitriding treatment. With prolongation of nitriding time at 4–12 h, all surface roughness parameters were increased due to the formation of the nitrogen-supersaturated S-phase and Fe and Cr nitrides in the surface layer.

The surface hardness after nitriding for 4–12 h is in the value range of 365–459 HV0.2, while after nitriding for 24 h the hardness increases significantly to 597 HV0.2. The increase in surface hardness and roughness after nitriding at 430 °C/24 h parameters is an indication of the formation of a thin nitride layer on the S-phase layer. However, the increase in surface hardness achieved by nitriding longer than 8 h did not result in an increase in resistance to micro abrasion or to corrosion in salt solution. The nitrided layer with hard and brittle Fe and Cr nitrides precipitated after 430 °C/12–24 h nitriding wears out faster than the softer but tougher S-phase created during 430 °C/4–8 h PN treatment.

Heat-untreated sample has a more positive E_{corr} than samples with inhomogeneous nitrided surfaces, which indicates that the oxide film on stainless steel is more resistant to corrosion then surface films on the nitrided samples, under the same electrolyte corrosion load. It can also be concluded that nitriding time has no significant influence on the corrosion stability of nitrided AS steel AISI 316L in the test medium.

Heat-untreated AS steel AISI 316L does not show an evident passive area as samples with a nitrided surface and that the current density increases sharply at pitting potential E_{pitt} = 376 mV vs. SCE. Samples with a nitrided surface generally showed tendency for passivation, which indicates that the nitrided layer provides a good barrier against localized attacks at a lower potential range (from −78 to 200 mV vs. SCE). Samples with 24 h surface nitriding showed the best resistance against pitting corrosion at higher potential. The value of E_{pitt} for the sample PN 430 °C/24 h was 470 mV vs. SCE. It can be concluded that the different amount of nitrogen in AS steels and nitriding time has a great influence on their resistance to pitting corrosion due to nitrogen surface concentration and the thickness of the modified porous layer.

The results of EIS shown in Table 6, confirm that heat-untreated sample has the highest corrosion resistance. Opposite to untreated sample, nitrided samples have smaller and approximately equal corrosion resistance values. Lower values of the nitrided surface resistance and higher values of their capacity are a common indicator of the electrolyte absorption through the pores, i.e., worse preservation and durability of the surface layer.

The conducted tests of plasma-nitrided AS steel AISI 316L provided an insight into the dependence of mechanical, tribological and corrosion properties on the nitriding time, with minimal influence of the substrate on the tests results and measured properties of the thin nitrided layer. The applied test methods (surface roughness measurement, microhardness test, metallographic examination, micro abrasion testing, EIS, and cyclic polarization testing) complement each other and are confirmed by the obtained results in a way that they can be recommended as a reliable procedure for the basic characterization of a thin nitrided layer's properties on stainless steels.

Author Contributions: Conceptualization, D.L. and M.K.; methodology, D.L. and M.K.; software, J.J. and S.J.; validation, D.L., I.S. and M.K.; formal analysis, D.L. and M.K.; investigation, I.S. and J.J.; resources, S.J.; data curation, S.J., I.S. and J.J.; writing—original draft preparation, D.L. and M.K.; writing—review and editing, D.L., M.K. and I.S.; visualization, D.L., M.K., I.S., J.J. and S.J.; supervision, D.L., M.K. and I.S. All authors have read and agreed to the published version of the manuscript.

Funding: This research was funded by the Ministry of Science and Education (MSE) of Republic Croatia together with the Central Agency for Financing and Contracting Programs and Projects of the European Union (SAFU), Croatia through the funding grant KK.01.1.1.07.0045—Development of an anti-corrosion protection system for the multi-purpose use of pipes.

Institutional Review Board Statement: Not applicable.

Informed Consent Statement: Not applicable.

Data Availability Statement: Not applicable.

Acknowledgments: The authors wish to acknowledge the projects KK.01.1.1.07.0045 "Development of an anti-corrosion protection system for the multi-purpose use of pipes", funded from the Ministry of Science and Education of Republic Croatian and The Central Agency for Financing and Contracting Programs and Projects of the European Union (SAFU) for their financial support.

Conflicts of Interest: The authors declare no conflict of interest.

References

1. Dillon, C.P. *Corrosion Resistance of Stainless Steels*, 1st ed.; Marcel Dekker, Inc.: New York, NY, USA, 1995.
2. Lo, K.H.; Shek, C.H.; Lai, J.K.L. Recent developments in stainless steels. *Mater. Sci. Eng. R Rep.* **2009**, *65*, 39–104. [CrossRef]
3. Kim, D.W.; Chang, J.H.; Ryu, W.S. Evaluation of the creep–fatigue damage mechanism of Type 316L and Type 316LN stainless steel. *Int. J. Press. Vessel. Pip.* **2008**, *85*, 378–384. [CrossRef]
4. Rokhlin, S.I.; Kim, J.-Y.; Nagy, H.; Zoofan, B. Effect of pitting corrosion on fatigue crack initiation and fatigue life. *Eng. Facture Mech.* **1999**, *62*, 425–444. [CrossRef]
5. Sharma, M.K.; Saikia, B.K.; Phukan, A.; Ganguli, B. Plasma nitriding of AS steel in N_2 and N_2–H_2 dc pulsed discharge. *Surf. Coat. Technol.* **2006**, *201*, 2407–2413. [CrossRef]
6. Egawa, M.; Ueda, N.; Nakata, K.; Tsujikawa, M.; Tanaka, M. Effect of additive alloying element on plasma nitriding and carburizing behavior for AS steels. *Surf. Coat. Technol.* **2010**, *205*, S246–S251. [CrossRef]
7. Fewell, M.P.; Mitchell, D.R.G.; Priest, J.M.; Short, K.T.; Collins, G.A. The nature of expanded austenite. *Surf. Coat. Technol.* **2000**, *131*, 300–306. [CrossRef]
8. Bell, T. Surface engineering of AS steel. *Surf. Eng.* **2002**, *18*, 415–422. [CrossRef]
9. Sun, Y.; Bell, T.; Kolosvary, Z.; Flis, J. The Response of AS Steels to Low-temperature Plasma Nitriding. *Heat Treat. Met.* **1999**, *1*, 9–16.
10. Thaiwatthana, S.; Li, X.Y.; Dong, H.; Bell, T. Mechanical and chemical properties of low temperature plasma surface alloyed 316 AS steel. *Surf. Eng.* **2002**, *18*, 140–141. [CrossRef]
11. Czerwiec, T.; Renevier, N.; Michel, H. Low-temperature plasma-assisted nitriding. *Surf. Coat. Technol.* **2000**, *131*, 267–277. [CrossRef]
12. Li, G.Y.; Lei, M.K. Microstructure and Properties of Plasma Source Nitrided AISI 316 AS Steel. *J. Mater. Eng. Perform.* **2017**, *26*, 418–423. [CrossRef]
13. Keddama, M.; Thiriet, T.; Marcos, G.; Czerwiec, T. Characterization of the expanded austenite developed on AISI 316 LM steel by plasma nitriding. *J. Min. Metall. Sect. B Metall.* **2016**, *53*, 47–52. [CrossRef]

14. De Las Heras, E.; Ybarra, G.; Lamas, D.; Cabo, A.; Dalibon, E.L.; Brühl, S.P. Plasma nitriding of 316L stainless steel in two different N_2-H_2 atmospheres—Influence on microstructure and corrosion resistance. *Surf. Coat. Technol.* **2017**, *313*, 47–54. [CrossRef]
15. Mingolo, N.; Tschiptschin, A.P.; Pinedo, C.E. On the formation of expanded austenite during plasma nitriding of an AISI 316L AS steel. *Surf. Coat. Technol.* **2006**, *201*, 4215–4218. [CrossRef]
16. Fossati, A.; Borgioli, F.; Galvanetto, E.; Bacci, T. Corrosion resistance properties of glow-discharge nitrided AISI 316L AS steel in NaCl solutions. *Corros. Sci.* **2006**, *48*, 1513–1527. [CrossRef]
17. Gokcekaya, O.; Ergun, C.; Gulmez, T.; Nakano, T.; Yilmaz, S. Structural Characterization of Ion Nitrided 316L AS Steel: Influence of Treatment Temperature and Time. *Metals* **2022**, *12*, 306. [CrossRef]
18. Aizawa, T.; Yoshino, T.; Morikawa, K.; Yoshihara, S.-I. Microstructure of plasma nitrided AISI420 martensitic stainless steel at 673 K. *Crystals* **2019**, *9*, 60. [CrossRef]
19. Larisch, B.; Brusky, U.; Spies, H.J. Plasma nitriding of stainless steels at low temperatures. *Surf. Coat. Technol.* **1999**, *119*, 205–211. [CrossRef]
20. Singh, G.P.; Alphonsa, J.; Barhai, P.K.; Rayjada, P.A.; Raole, P.M.; Mukherjee, S. Effect of surface roughness on the properties of the layer formed on AISI 304 stainless steel after plasma nitriding. *Surf. Coat. Technol.* **2006**, *200*, 5807–5811. [CrossRef]
21. Cisquini, P.; Ramos, S.V.; Viana, P.R.P.; Lins, V.D.F.C.; Franco, A.R., Jr.; Vieira, E.A. Effect of the roughness produced by plasma nitrocarburizing on corrosion resistance of AISI 304 AS steel. *J. Mater. Res. Technol.* **2019**, *8*, 1897–1906. [CrossRef]
22. Flis, J.; Kuczynska, M. Impedance of Cr18Ni10 stainless steel in sulphate solutions after a low-temperature plasma nitriding. *Mater. Corros.* **2003**, *54*, 953–956. [CrossRef]
23. Flis, J.; Kuczynska, M. Effect of Low-Temperature Plasma Nitriding on Corrosion of 304L Stainless Steel in Sulfate and Chloride Solutions. *J. Electrochem. Soc.* **2004**, *151*, B573–B580. [CrossRef]
24. Nakajima, M.; Nakamura, Y.; Suzuki, K.; Bai, Y.; Uematsu, Y. Effect of solution treatment after nitriding on fatigue properties in type 304 stainless steel. *Int. J. Fatigue* **2014**, *68*, 103–110. [CrossRef]
25. Zhao, C.; Li, C.X.; Dong, H.; Bell, T. Study on the active screen plasma nitriding and its nitriding mechanism. *Surf. Coat. Technol.* **2006**, *201*, 2320–2325. [CrossRef]
26. Oliveira, R.M.; Ueda, M.; Silva, L.L.G. Characteristics of AS Steel Nitrided in a Hybrid Glow Discharge Plasma. *Braz. J. Phys.* **2009**, *39*, 554–558. [CrossRef]
27. Díaz-Guillén, J.C.; Granda-Gutiérrez, E.E.; Vargas-Gutiérrez, G.; Díaz-Guillén, M.R.; Aguilar-Martínez, J.A.; Álvarez-Contreras, L. Effect of Nitriding Current Density on the Surface Properties and Crystallite Size of Pulsed Plasma-Nitrided AISI 316L. *J. Mater. Sci. Chem. Eng.* **2015**, *3*, 45–51. [CrossRef]
28. Alves, C., Jr.; Rodrigues, J.A.; Martinelli, A.E. The effect of pulse width on the microstructure of dc-plasma-nitrided layers. *Surf. Coat. Technol.* **1999**, *122*, 112–117. [CrossRef]
29. Wang, J.; Xiong, J.; Peng, Q.; Fan, H.; Wang, Y.; Li, G.; Shen, B. Effects of DC plasma nitriding parameters on microstructure and properties of 304L stainless steel. *Mater. Charact.* **2009**, *60*, 197–203. [CrossRef]
30. Fernandes, F.A.P.; Heck, S.C.; Pereira, R.G.; Lombardi-Neto, A.; Totten, G.E.; Casteletti, L.C. Wear of plasma nitrided and nitrocarburized AISI 316L AS steel. *J. Achiev. Mater. Manuf. Eng.* **2010**, *40*, 175–179.
31. Staia, M.H.; Enriquez, E.; Puchi, E.S.; Lewis, D.B.; Jeandin, M. Application of Ball Cratering Method to Study Abrasive Wear. *Surf. Eng.* **1998**, *14*, 49–54. [CrossRef]
32. de Mello, J.D.B.; Binder, C.; Binder, R.; Klein, A.N. Effect of nature of nitride phases on microabrasion of plasma nitrided sintered iron. *Tribol. Mater. Surf. Interfaces* **2010**, *4*, 191–196. [CrossRef]
33. Stack, M.M.; Mathew, M.T. Mapping the micro-abrasion resistance of WC/Co based coatings in aqueous conditions. *Surf. Coat. Technol.* **2004**, *183*, 337–346. [CrossRef]
34. Axén, N.; Jacobson, S.; Hogmark, S. Influence of hardness of the counterbody in three-body abrasive wear—An overlooked hardness effect. *Tribol. Int.* **1994**, *27*, 233–241. [CrossRef]
35. Camerini, R.V.; de Souza, R.B.; de Carli, F.; Pereira, A.S.; Balzaretti, N.M. Ball cratering test on ductile materials. *Wear* **2011**, *271*, 770–774. [CrossRef]
36. Resendiz-Calderon, C.D.; Farfan-Cabrera, L.I.; Oseguera-Pen, J.E.; Cazares-Ramırez, I.; Gallardo-Hernandez, E.A. Friction and Wear of Metals under Micro-abrasion, Wet and Dry Sliding Conditions. *J. Mater. Eng. Perform.* **2020**, *220*, 6228–6238. [CrossRef]
37. Olzon-Dionysio, M.; Campos, M.; Kapp, M.; de Souza, S. Influences of plasma nitriding edge effect on properties of 316L stainless steel. *Surf. Coat. Technol.* **2010**, *204*, 3623–3628. [CrossRef]
38. Shipway, P.H. The role of test conditions on the microabrasive wear behaviour of soda-lime glass. *Wear* **1999**, *233–235*, 191–199. [CrossRef]
39. Rutherford, K.L.; Hutchings, I.M. Micro-scale abrasive wear testing of PVD coatings on curved substrates. *Tribol. Lett.* **1996**, *2*, 1–11. [CrossRef]
40. Allsopp, D.N.; Hutchings, I.M. Micro-scale abrasion and scratch response of PVD coatings at elevated temperatures. *Wear* **2001**, *251*, 1308–1314. [CrossRef]
41. Whitehouse, D. *Surfaces and Their Measurement*; Hermes Penton Science: London, UK, 2002.
42. Li, G.J.; Peng, Q.; Li, C.; Wang, Y.; Gao, J.; Chen, S.Y.; Wang, J.; Shen, B.L. Effect of DC plasma nitriding temperature on microstructure and dry-sliding wear properties of 316L stainless steel. *Surf. Coat. Technol.* **2008**, *202*, 2749–2754. [CrossRef]

43. Kurtela, M.; Šimunović, V.; Stojanović, I.; Alar, V. Effect of the Cerium (III) Chloride Heptahydrate on the Corrosion Inhibition of Aluminium Alloy. *Mater. Corros.* **2019**, *71*, 125–147. [CrossRef]
44. Barsoukov, E.; Macdonald, J.R. *Impedance Spectroscopy Theory, Experiment, and Applications*, 2nd ed.; John Wiley & Sons Inc.: Hoboken, NJ, USA, 2005. [CrossRef]
45. Aleksandrov Fabijanić, T.; Šnajdar, M.; Kurtela, M.; Šimunović, V.; Marciuš, M.; Klaić, M. Corrosion Resistance of Nanostructured Cemented Carbides with Alternative FeNi and FeNiCo Binders. *Nanomaterials* **2023**, *13*, 1407. [CrossRef]
46. Fathy, M.H.; Wafaa, A.G. Effect of nitrogen on the corrosion behavior of AS steel in chloride solutions. *Mater. Lett.* **2005**, *59*, 3311–3314. [CrossRef]
47. Aleksandrov Fabijanić, T.; Kurtela, M.; Škrinjarić, I.; Pötschke, J.; Mayer, M. Electrochemical Corrosion Resistance of Ni and Co Bonded Near-Nano and Nanostructured Cemented Carbides. *Metals* **2020**, *10*, 224. [CrossRef]
48. Yang, L.; Yongyong, H.; Jianxun, Q.; Jun, Z.; Qianwen, Y.; Yijie, Z.; Junyuan, M. Enhancement of Pitting Corrosion Resistance of AS Steel Through Deposition of Amorphous/Nanocrystalline Oxy-nitrided Phases by Active Screen Plasma Treatment. *Mater. Res.* **2018**, *21*, 1–10. [CrossRef]
49. Baba, H.; Kodama, T.; Katada, Y. Role of nitrogen on the corrosion behavior of AS steels. *Corros. Sci.* **2002**, *44*, 2393–2407. [CrossRef]
50. Yang, L.; Zhuo, W.; Liang, W. Surface properties of nitrided layer on AISI 316L AS steel produced by high temperature plasma nitriding in short time. *Appl. Surf. Sci.* **2014**, *298*, 243–250. [CrossRef]
51. Borgioli, F.; Fossati, A.; Galvanetto, E.; Bacci, T. Glow-discharge nitriding of AISI 316L AS steel: Influence of treatment time. *Surf. Coat. Technol.* **2006**, *200*, 3511–3517. [CrossRef]
52. Olzon-Dionysio, M.; Souza, S.D.; Basso, R.L.O. Application of Mössbauer spectroscopy to the study of corrosion resistance in NaCl solution of plasma nitrided AISI 316L stainless steel. *Surf. Coat. Technol.* **2008**, *202*, 3607–3614. [CrossRef]
53. Borgioli, F.; Fossati, A.; Galvanetto, E.; Bacci, T. Glow-discharge nitriding of AISI 316L AS steel: Influence of treatment temperature. *Surf. Coat. Technol.* **2005**, *200*, 2474–2480. [CrossRef]
54. Yang, L.; Shangzhou, Z.; Yongyong, H.; Lei, Z. Characteristics of the nitrided layer formed on AISI 304 AS steel by high temperature nitriding assisted hollow cathode discharge. *Mater. Des.* **2014**, *64*, 527–534. [CrossRef]
55. Singh, V.; Marchev, K.; Cooper, K.V.; Meletis, E.I. Intensified plasma-assisted nitriding of AISI 316L stainless steel. *Surf. Coat. Technol.* **2002**, *160*, 249–258. [CrossRef]
56. Lei, M.K.; Zhu, X.M. Plasma-based low-energy ion implantation of AS steel for improvement in wear and corrosion resistance. *Surf. Coat. Technol.* **2005**, *193*, 22–28. [CrossRef]
57. Baranowska, J.; Arnold, B. Corrosion resistance of nitrided layers on austenitic steel. *Surf. Coat. Technol.* **2006**, *200*, 6623–6628. [CrossRef]
58. Yang, W.J.; Zhang, M.; Zhao, Y.H.; Shen, M.L.; Lei, H.; Xu, L.; Xiao, J.Q.; Gong, J.; Yu, B.H.; Sun, C. Enhancement of mechanical property and corrosion resistance of 316L stainless steels by low temperature arc plasma nitriding. *Surf. Coat. Technol.* **2016**, *298*, 64–72. [CrossRef]
59. Tian, S.; Liu, Z.; Shen, L.; Pu, J.; Liu, W.; Sun, X.; Li, Z. Performance evaluation of mercapto functional hybrid silica sol–gel coating and its synergistic effect with f-GNs for corrosion protection of copper surface. *RSC Adv.* **2018**, *8*, 7438–7449. [CrossRef]

coatings

MDPI

Article

Physical and Chemical Properties of High-Temperature Silicone-Based Polymer Coatings Applied on Different Surface Roughnesses

Ivan Stojanović *, Borna Škrlec, Marin Kurtela, Vesna Alar and Martina Odeljan

Faculty of Mechanical Engineering and Naval Architecture, University of Zagreb, 10000 Zagreb, Croatia; borna.skrlec@fsb.unizg.hr (B.Š.); marin.kurtela@fsb.unizg.hr (M.K.); vesna.alar@fsb.unizg.hr (V.A.); martina.odeljan@fsb.unizg.hr (M.O.)

* Correspondence: ivan.stojanovic@fsb.unizg.hr

Abstract: High-temperature coatings play a crucial role in protecting surfaces exposed to extreme temperatures, corrosion, and other harsh environments. This paper focuses on the physical and chemical properties of solvent-borne and water-borne high-temperature silicone-based polymer coatings applied on two types of surface roughness of carbon steel plates. The corrosion protection performance of the coatings was characterized using a salt spray chamber, humidity chamber, electrochemical impedance spectroscopy (EIS) measurements, and differential scanning calorimetry (DSC). The physical properties of high-temperature coatings were determined using the Shore D hardness method and cross-cut adhesion tests. This study investigates the effects of different surface preparation methods on coating adhesion, which is considered to be a crucial property of organic coatings for corrosion protection durability. The thermal stability of the coating was tested using furnace cyclic testing. The results show that high-temperature coatings in general exhibit excellent thermal stability, high adhesion strength, and good resistance to warm and humid environments, except in the conditions of a salty atmosphere. This study reveals that coating performance is affected by the composition and surface preparation method. This study can be useful for coating manufacturers and researchers interested in understanding the physical and chemical properties of high-temperature coatings and their applications in various environments.

Keywords: corrosion; high-temperature coatings; physical properties; chemical properties

1. Introduction

Corrosion protection significantly affects the functionality of a structure, its durability, and operational safety, thereby playing a substantial role in construction and maintenance. Coating protection is the most widely used method for safeguarding various steel structures against corrosion. In the technology of coating protection, it is crucial to correctly execute all technological operations involving surface preparation and coating application. Alongside quality, protective processes are increasingly needed to meet environmental and human non-harmfulness requirements.

High temperature is a critical factor influencing the occurrence and development of corrosion, greatly reducing the product or machinery component's lifespan. High-temperature corrosion is a type of corrosion that emerges where temperatures are well above 100 °C, such as in chimney flues, automobile exhaust systems, diesel engines, power plants, gas turbines, and other machinery in contact with hot gases containing certain impurities [1,2]. For higher temperatures up to 900 °C, the most commonly used materials are aluminum oxide and chromium (III) oxide, while silicon dioxide can withstand temperatures up to 1800 °C [3]. These oxide films represent a special type of coating that creates an oxide protective layer between the high-temperature oxidative environment and the material itself. For this oxide layer to fulfill its material protection purpose, it needs to remain stable at high temperatures, adhere well to the material's substrate, possess a certain density, and be resistant to the thermal shock caused by rapid temperature changes.

Among corrosion protection methods, organic coating protection stands out due to its effectiveness, which depends on various factors like environmental corrosiveness, the type of protected surface, coating system durability, surface preparation, and the coating application technology. Materials employed in high-temperature environments must meet specific criteria like mechanical or thermodynamic fatigue strength, while the coating itself must provide suitable environmental resistance and be chemically and mechanically compatible with the substrate [4]. High-temperature paint must withstand high temperatures without decomposing, cracking, or delaminating [5]. The purpose of these high-temperature protective coatings and paints is to maintain external appearance and heat resistance over extended periods at high temperatures [6]. The thermal stability limit of commonly available coating systems is typically around 60 °C [7]. Epoxy systems using conventional bisphenol-A-based resin are generally limited to a service temperature of around 80–90 °C [8]. At temperatures ranging from 200 °C to 300 °C, most organic-based coatings undergo carbonization, and in some cases, this carbonization is accompanied by the development of aggressive gases such as hydrogen chloride from PVC [9]. Coatings for wood-burning fireplaces must endure temperatures up to 600 °C. For such applications, coatings based on silicone resins are widely used due to their excellent high-temperature resistance [3]. Compared with homologous carbon-based polymers, silicone resins show superior performance in thermal stability due to the physicochemical properties of the siloxane (-Si-O-Si) bond [10]. In high-temperature applications, solvent- and water-based inorganic zinc-rich silicate coatings are also used [11].

A coating protection study is a complex and comprehensive issue that requires a lot of experimental work to obtain the proper physical and chemical properties of the coatings. Authors [12,13] used electrochemical impedance spectroscopy (EIS) as a method to evaluate the condition of the coating surface and its barrier properties, which are key to achieving satisfactory corrosion resistance of the coating. Chen et al. [14] used a neutral salt spray test to accelerate the indoor aging of cold-spray aluminum zinc, hot-spray zinc, and hot-dip zinc coatings. After the salt spray test adhesion was determined by the method of a scratch test. Tsai et al. [15] evaluated high-performance polyurethane/graphene composite coatings using various corrosion and mechanical tests, namely electrochemical impedance spectroscopy, salt spray tests, cross-cut tape tests, and dynamic mechanical analysis. The cross-cut test is a standard test method for the adhesion of organic coatings. The grade of adhesion quality in the cross-cut test is rated from 0 to 5; 0 and 5 represent the best and poorest adhesion performances, respectively [16]. The humidity test for paint helps assess coating quality in 100% relative humid conditions and elevated temperatures. Aračić et al. [17] performed research on the new generation of paint systems for the anticorrosion protection of steel structures using a humidity chamber. Panels were placed inside a humidity chamber and exposed continually for a specified period according to ISO 6270 [18], then evaluated for degradation according to ISO 4628 [19]. The standard ISO 4628 refers to the designation of quantity and size of defects and of intensity of uniform changes in appearance, namely blistering, rusting, cracking, flaking, chalking, delamination, and degree of filiform corrosion. The testing in the water vapor atmosphere served to determine the behavior of the material exposed to the moisture-saturated warm air in the presence of condensed water.

Previous studies on silicone-based coatings included the type of silicone resin for the long-term heat resistance of the coating [3], adhesion properties, variety of precursors and polymer-to-ceramic conversion of silicone-based coatings [20], chemical properties of silane-based coatings [21], flake ZnAl alloy as an effective pigment in silicate coatings for the corrosion protection of steel [22], and the interface properties between water-based inorganic zinc silicate coating modified by organosilicon and iron substrate [11].

To our knowledge, a detailed comparison of physical and chemical properties between the water-borne and solvent-borne silicone-based polymer coatings applied on different surface roughnesses has not been extensively reported. The evaluation of their resilience against varied environmental conditions and thermal stress in this work provides a different approach to a critical aspect that has remained insufficiently unexplored in the existing research literature.

This study aims to investigate the physical and chemical properties of two solvent-borne compared to two water-borne high-temperature coatings used for the protection of wood-burning fireplaces using a salt spray chamber, humidity chamber, cyclic low–high temperature exposure, electrochemical impedance testing (EIS), and differential scanning calorimetry (DSC). Special attention is given to the examination of the influence of surface roughness and profile on the adhesion of coatings. The corrosion properties and thermal stability of commercially available water-based high-temperature paints were evaluated to assess their suitability for prospective applications and to provide researchers with reference data compared to traditionally used solvent-based paints. Based on these tests, the properties and appearance of coatings that align with the real-use application conditions for wood-burning stoves will be defined.

2. Materials and Methods

2.1. Materials

The metallic substrate used for this study was EN 10111 DD13 low-carbon hot-rolled steel with the chemical composition 0.04 C, 0.20 Mn, 0.008 P, 0.05 S, and Fe in balance (wt.%). The material is used for stove manufacturing. The dimension of the sample was 100 × 150 × 3 mm, and the surface of the samples was prepared by abrasive blasting in two ways: half of the samples were sandblasted (letter P in the sample designation), and the other half of the samples were shot-blasted (letter S in the sample designation). The surface roughness was measured using a TMR200 surface roughness gauge (TMTeck Instrument Co., LTD., Beijing, China).

2.2. Preparation of the Coatings

The study was conducted on the four types of high-temperature silicone-based polymer coatings, two of which are solvent-based (Solvalitt Black, Resist 78), and the other two coatings (Senotherm UHT 2K-Hydro 3590, Thermodur 600 Aqua) are new water-based high-temperature formulations. Their characteristics are presented in Table 1. The described coatings were prepared, applied, and dried in room condition according to the manufacturer's instructions.

Table 1. Properties of the tested coatings.

Coating	Label	Base	Temperature Resistance, °C	Density kg/L	Wet Film Thickness, μm	Dry Film Thickness, μm	Volume Sold, %	Theoretical Spreading Rate
Solvalitt Black	SO	Silicone acrylic coating	600	1.30	50–70	20–30	43 ± 2	21–14 m^2/L
Thermodur 600 Aqua	TH	Water-based modified silicone resin	600	1.29	53–107	15–30	28	11 m^2/kg
Senotherm UHT 2K-Hydro 3590	SE	Water-based	600	-	47	25	53	9 m^2/kg
Resist 78	RE	Inorganic zinc ethyl silicate	540	2.50	70–125	50–90	72 ± 2	14.4–8 m^2/L

The preparation of samples is visually represented in the schematic diagram in Figure 1.

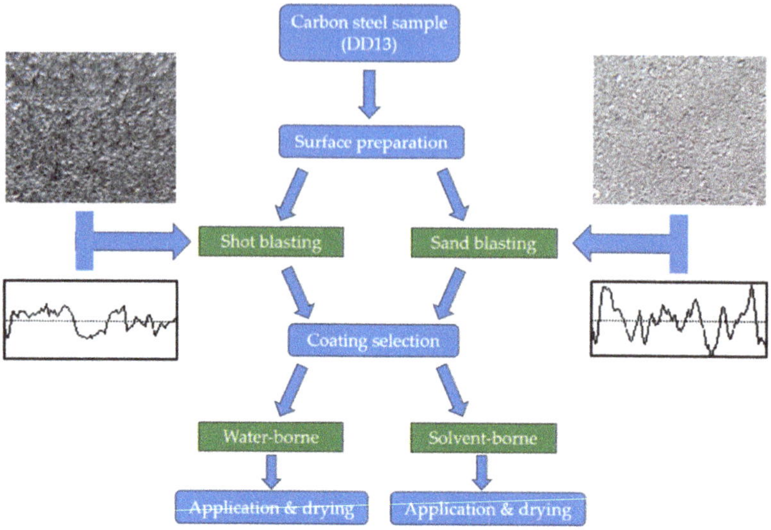

Figure 1. Flow diagram of the procedures for the preparation of the samples used in this research.

2.3. *Dry Film Thickness, Adhesion, and Gloss*

After the coating application, the coating thickness was determined using an Elcometer 456 device (Elcometer Limited, EdgeLane, Manchester, UK), according to ISO 2808 [23]. Measurements were made at 10 different places per sample with instrument accuracy of ±2.5 μm, and the mean value was calculated. Adhesion was tested using the "Cross-cut" method and a Zehntner ZCT 2160.123 G device (Zehntner GmbH Testing Instruments, Sissach, Switzerland), according to ISO 2409 [16]. Also, hardness testing was conducted using the Shore D method and a PosiTector SHD durometer (DeFelsko Corporation, NY, USA) with an accuracy of ±1.

2.4. *Accelerated Testing in Chambers*

One set of samples was tested in a salt spray chamber, the second in a humidity chamber, and the third was tested for temperature cycle endurance, with a maximum achieved temperature of 450 °C. After the accelerated testing in chambers, the coatings were evaluated according to ISO 4628.

Regarding the salt spray chamber, the samples were tested in the Ascott S450 salt spray chamber (Ascott Analytical Equipment Limited, Staffordshire, UK), according to ISO 9227 [24], with the parameters provided in Table 2. The samples were subjected to corrosion conditions in a salt atmosphere for 96 h, i.e., four days.

Table 2. Testing parameters in the salt chamber.

Testing Parameters	Standard	Testing Conditions
Testing Duration, hours	As agreed	96
Temperature, °C	35 ± 2	35 ± 0.1
Testing Chamber Volume, liters	min 400	450
Air Humidifier Temperature, °C	45–50	47
Compressed Air Pressure, bar	0.7–1.4	0.98
Used Solution	NaCl	NaCl
Solution Concentration, %	5	5
Collected Condensate Amount, mL/80 cm^2/h	1.5 ± 0.5	2.0
Condensate pH Value at 25 ± 2 °C	6.5–7.2	7.1
Conductivity of Distilled Water, μS/cm at 25 ± 2 °C	max 20	<10

On the samples that were in the salt spray chamber, a scribe was made to assess the corrosion under the coating. According to ISO 12944-6 [25], corrosion around the scribe should not exceed 1.5 mm, calculated as:

$$M = (C - W)/2 \quad (1)$$

where C is the maximum width of corrosion across the scratch, and W is the original width of the scribe in millimeters.

As for the humidity chamber, it simulates conditions of a warm and humid atmosphere with water condensation. The testing in the humidity chamber was conducted in the Humidity Cabinet Model AB6 (C&W Specialist Equipment, Belrose, NSW, Australia), following the ISO 6270-2 standard [18]. The testing temperature was 40 °C, and the relative humidity in the chamber was approximately 100%, resulting in the samples experiencing condensation.

The furnace cyclic testing was carried out as part of the coating adhesion and appearance validation after cyclic exposure to low/high temperature environments.

Views of the samples at the beginning of testing in the salt spray chamber and humidity chamber are provided in Figure 2.

Figure 2. Samples ready for testing in the salt spray chamber (**left**) and humidity chamber (**right**).

2.5. Electrochemical Characterization

Electrochemical characterizations of the coatings were performed by measuring the electrochemical impedance spectroscopy (EIS). Electrochemical impedance spectroscopy has a key role in understanding electrochemical behavior at the metal/coating interface and determining the resistance properties of a coating [12]. The application of EIS enabled the quantification of the surface layer resistance and the definition of the mechanism of corrosion progression without any degradation of the test surface. EIS was conducted using a VersaSTAT 3 Potentiostat/Galvanostat (AMETEK Scientific 131 Instruments, Princeton applied research, Berwyn, PA, USA), with the application of ZsimpWin 3.60 software. The tested electrochemical cell consisted of a metal substrate with a coating as a working electrode, a saturated calomel electrode as a reference electrode, and two graphite rods as counter electrodes. The coating properties were determined in a 3.5 wt.% NaCl solution, at a frequency range of 0.1 to 10^5 Hz and at room temperature (23 ± 2 °C).

2.6. Differential Scanning Calorimetry (DSC)

A Mettler Toledo differential scanning calorimeter DSC 3+ (Mettler Toledo, Greifensee, Switzerland) was used to carry out the differential thermal analysis and determine the glass transition temperature (T_g). Samples of 10 mg were analyzed in a stream of nitrogen (40 mL/min) with a heating and cooling rate of 10 °C/min in a temperature range from −100 °C to 150 °C by a double heating/cooling cycle. Samples were first heated from 25 °C to 150 °C at a rate of 10 °C/min and held at that temperature for 5 min to erase the thermal history of the sample during the preparation process. Samples were then cooled from 150 °C to −100 °C at a cooling rate of 10 °C/min, held at −100 °C for 5 min, and reheated from −100 °C to 150 °C at a rate of 10 °C/min [13]. Liquid nitrogen was used to cool the samples to low temperatures. From the second heating cycle, values of the glass transition temperature T_g were obtained. The glass transition temperature represents the temperature at which an amorphous polymer changes from a rigid, glassy state to a more flexible, rubbery state. Knowing the T_g of coatings is crucial, as it provides valuable insights into the material's mechanical and thermal properties and helps to define the appropriate temperature range for the application and utilization of the coating.

Methods and procedures for the evaluation of coating stability are visually represented in the diagram in Figure 3.

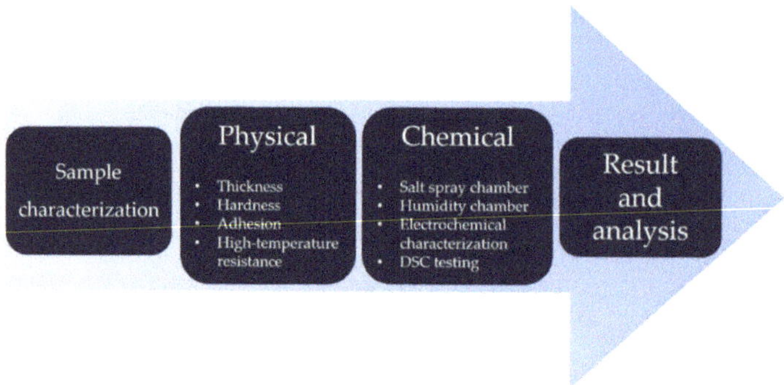

Figure 3. Flow diagram of the methods for the evaluation of coating stability.

3. Results and Discussion

3.1. Surface Roughness, Coating Thickness, and Hardness Testing

The average surface roughness value on sandblasted samples was 23.58 µm with irregular sharp surfaces, while on shot-blasted ones, it was 18.65 µm with rounded profiles. The substrates on both sample groups were prepared to the required cleanliness of Sa 2.5, according to ISO 8501-1 [26].

After coating the steel plates, the coating thickness was measured. Thickness was measured at 10 different points on each sample, and the average value is recorded in Tables 3 and 4. The lowest coating thickness was observed on water-based modified silicone resin TH: from 10.3 µm to 11.3 µm. The measurement of coating thickness revealed that inorganic zinc ethyl silicate RE has the greatest thickness, from 27.6 µm to 33.9 µm.

Table 3. Coating thickness, hardness, and adhesion on sandblasted samples.

Sandblasted Samples				
Sample	**$\bar{x}$ (µm)**	**Shore D Hardness**	**Cross-Cut**	**Test Type**
SO-P1	21.2	93.6		Salt chamber
SO-P2	22.0	92.6		Humidity chamber
SO-P3	15.9	93.2	0	Cyclic temperature
SE-P1	14.4	93.8		Salt chamber
SE-P2	15.3	94.6		Humidity chamber
SE-P3	17.1	94.2	0	Cyclic temperature
TH-P1	10.9	91.6		Salt chamber
TH-P2	11.1	93.6		Humidity chamber
TH-P3	11.3	92.4	0	Cyclic temperature
RE-P1	33.9	93.8		Salt chamber
RE-P2	31.4	93.6		Humidity chamber
RE-P3	31.2	93.1	0	Cyclic temperature

Table 4. Coating thickness, hardness, and adhesion on shot-blasted samples.

Shot-Blasted Samples				
Sample	$\bar{x}$ (μm)	**Shore D Hardness**	**Cross-Cut**	**Test Type**
SO-S1	26.7	93.6		Salt chamber
SO-S2	24.0	92.4		Humidity chamber
SO-S3	20.3	93.3	1	Cyclic temperature
SE-S1	17.1	93.6		Salt chamber
SE-S2	12.0	94.6		Humidity chamber
SE-S3	13.3	93.2	1	Cyclic temperature
TH-S1	10.3	92.0		Salt chamber
TH-S2	11.2	93.8		Humidity chamber
TH-S3	10.3	92.1	0	Cyclic temperature
RE-S1	27.6	93.8		Salt chamber
RE-S2	33.8	93.8		Humidity chamber
RE-S3	27.8	93.6	1	Cyclic temperature

The coating hardness testing was conducted by the Shore D method. Ten measurements were taken on each sample, and the average value is recorded in Tables 3 and 4. It is evident that the results are similar, indicating that hardness is independent of the type of tested coating but mostly influenced by the steel substrate. Coating hardness testing was carried out to assess resistance to abrasion, which is a crucial property indicating the integrity of the coating during use. A low coating hardness makes it susceptible to damage [27].

3.2. Coating Adhesion Testing

Tables 5 and 6 present the results of adhesion testing by the cross-cut method, according to the ISO 2409 standard, on samples after accelerated testing in the salt spray chamber and humidity chamber. The adhesion test was conducted at room temperature according to ISO 2409.

Table 5. The results of adhesion on coatings after 96 h of testing in the salt spray chamber.

Surface Preparation	**Shot-Blasted (S)**		**Sandblasted (P)**	
Coating	**Sample Appearance**	**Grade**	**Sample Appearance**	**Grade**
SO		5	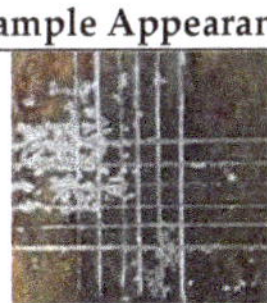	3
SE		5	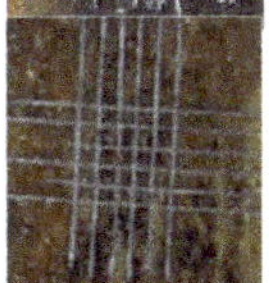	1
TH		3		1
RE		2		0

Table 6. The results of adhesion on coatings after 96 h of testing in the humidity chamber.

Surface Preparation	Shot-Blasted (S)		Sandblasted (P)	
Coating	**Sample Appearance**	**Grade**	**Sample Appearance**	**Grade**
SO	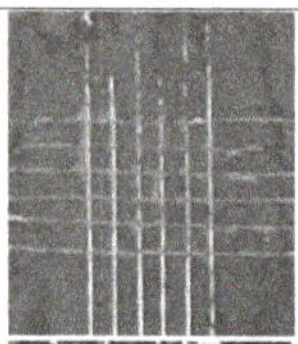	1	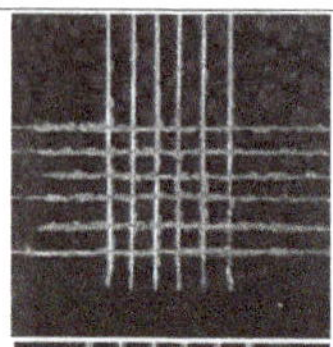	1
SE		5	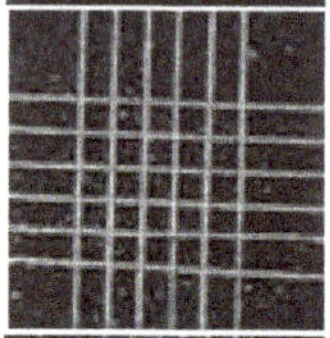	0
TH	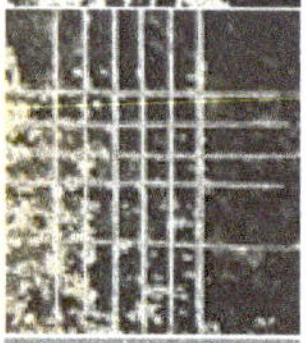	3	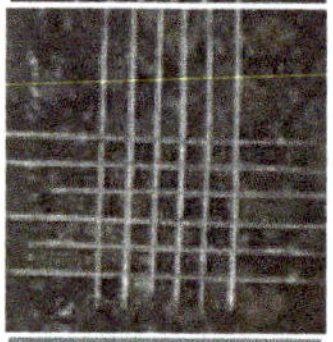	0
RE	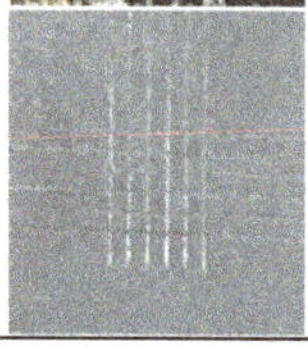	1	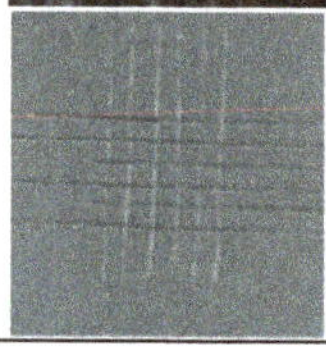	1

Six cuts were performed using a scalpel, spaced 1 mm apart in both the horizontal and vertical directions, forming a grid. According to the classification of test results according to ISO 2409, the samples were rated on a scale from 0 to 5. Classification 0 represents the best adhesion result; that is, the edges of the cuts are completely smooth and none of the squares of the lattice are detached [16]. Tables 6 and 7 show the results of coating adhesion testing after 96 h of exposure to accelerated testing in the salt spray and humidity chambers.

Table 7. Evaluation of degradation of coatings after 96 h of testing in the salt spray chamber.

Sample	ISO 4628-2	ISO 4628-3	ISO 4628-4	ISO 4628-8
	Blistering	**Rusting**	**Cracking**	**Corrosion around the Scribe, mm**
SO-P1	0(S0)	Ri 2	0(S0)	0.134
SO-S1	2(S3)	Ri 1	0(S0)	0.100
SE-P1	3(S2)	Ri 2	0(S0)	0.192
SE-S1	3(S3)	Ri 4	0(S0)	0.165
TH-P1	3(S2)	Ri 5	0(S0)	0.150
TH-S1	3(S2)	Ri 5	0(S0)	0.175
RE-P1	0(S0)	Ri 0	0(S0)	0
RE-S1	0(S0)	Ri 0	0(S0)	0

When examining the results, it is clear that the sandblasted samples exhibited significantly better coating adhesion after testing in the chambers compared to the shot-blasted samples. The surface had rounded small peaks after shot blasting or sharp notches after grit blasting [28]. The influence of surface preparation and roughness profile on coating adhesion was demonstrated through these tests. Inorganic zinc ethyl silicate RE coating showed the best adhesion after testing in both chambers. Both waterborne coatings SE and TH showed poor adhesion after testing in the humidity chamber and salt spray chamber on shot-blasted samples, while SO coatings exhibited the worst adhesion after testing in the salt spray chamber on the shot-blasted samples. Generally, the samples tested in the salt spray chamber yielded the worst results. Adhesion performance is one of the most important properties of protective coatings [29]. The greater the adhesion between the coating and the substrate, the better the coating protection performance, which improves the corrosion resistance of the coating and prolongs the service life of the coating [14]. The results showed that sandblasted samples have better adhesion.

For solid coatings, adhesion strength was determined by mechanical and chemical interactions. The mechanical interactions were associated with the surface roughness of the substrate, which may lead to a mechanical interlocking between coating and substrate after coating solidification. In addition to the mechanical adhesion, chemical interactions are generally responsible for the adhesion of silicon polymers [20]. The chemical bonding between Si-polymer and substrates takes place mostly by hydrolysis and condensation reactions [20,21].

3.3. *Assessment after Accelerated Testing in Salt Spray Chamber and Humidity Chamber*

After 96 h of accelerated corrosion testing in the salt spray chamber (ISO 9227), all coated samples showed signs of rusting and blistering, except for the Zn-silicate coating RE, which was rust-free, according to ISO 4628. Cracking was not noticed. Corrosion around the scribe was under the maximum-allowed 1.5 mm for all samples, as according to ISO 12944-6, while the RE samples did not show any corrosion creep. The water-based modified silicone coating TH showed the worst corrosion protection in a salty environment, possibly due to low dry film thickness, from 10.3 µm to 11.3 µm. A significant extent of rust was also noticed on the SE samples, with lower dry film thickness.

The results after 96 h exposure to the salt spray chamber are shown in Figure 4 and Table 7.

(**a**)

(**b**)

Figure 4. Coating assessment on samples RE, TH, SE, and SO (from left to right) on (**a**) sandblasted surface P and (**b**) shot-blasted surface S after 96 h exposure in the salt spray chamber, as according to ISO 9227.

After 96 h of accelerated corrosion testing in the humidity chamber (ISO 6270-2), high-temperature water-borne coatings SE and TH showed signs of rusting and blistering, whereas solvent-borne coatings SO and RE were rust-free, according to ISO 4628. Rusting and blistering were dominant on the TH-coated samples, which were protected in the lowest dry film thickness. The cracking of the tested coatings was not noticed. The edges of the samples were not considered.

The results after 96 h exposure to the humidity chamber are shown in Figure 5 and Table 8.

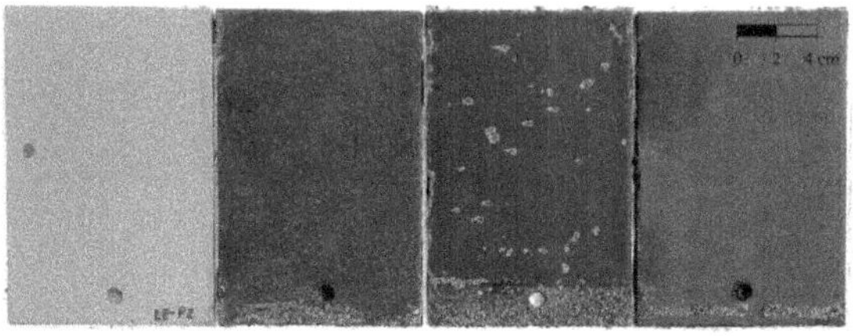

(a)

(b)

Figure 5. Coating assessment on samples RE, TH, SE, and SO (from left to right) on (**a**) sandblasted surface P and (**b**) shot-blasted surface S after 96 h exposure in the humidity chamber, as according to ISO 6270.

Table 8. Evaluation of degradation of coatings after 96 h of testing in the humidity chamber.

Uzorak	ISO 4628-2	ISO 4628-3	ISO 4628-4
	Blistering	**Rusting**	**Cracking**
SO-P2	0(S0)	Ri 0	0(S0)
SO-S2	0(S0)	Ri 0	0(S0)
SE-P2	0(S0)	Ri 1	0(S0)
SE-S2	3(S2)	Ri 1	0(S0)
TH-P2	4(S2)	Ri 1	0(S0)
TH-S2	3(S2)	Ri 1	0(S0)
RE-P2	0(S0)	Ri 0	0(S0)
RE-S2	0(S0)	Ri 0	0(S0)

3.4. High-Temperature Testing

To simulate the environment of a wood-burning fireplace or stove, the samples were subjected to a cyclic temperature test. They were first heated to a temperature of 450 °C for one hour, then held at that temperature for 5 h, and finally cooled to room temperature for 3 h (Figure 6). The total duration of the test was 120 h.

After the test and simulation of high-temperature cycles, the adhesion of the coatings was tested using the "Cross-cut" method, and the results obtained are shown in Table 9.

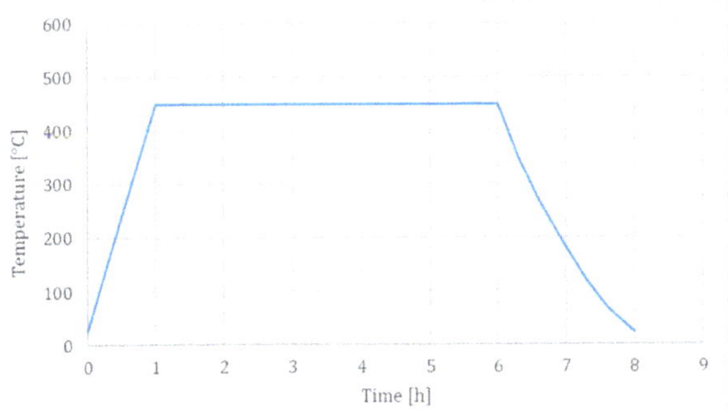

Figure 6. Test cycle time and temperature.

Table 9. The results of adhesion on coatings after high-temperature testing.

Surface Preparation	Shot-Blasted		Sandblasted	
Coating	**Sample Appearance**	**Grade**	**Sample Appearance**	**Grade**
Solvalit ATM (SO-S3, SO-P3)	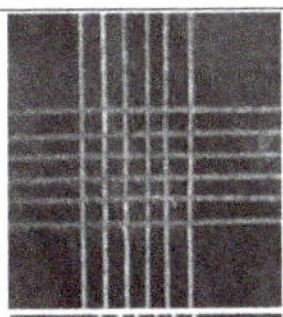	0	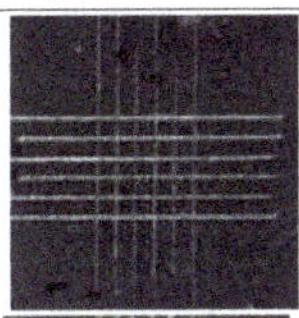	0
Senoterm UHT 2K Hydro ATM (SE-S3, SE-P3)	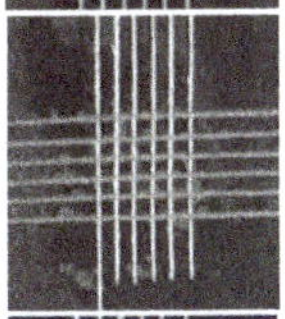	2	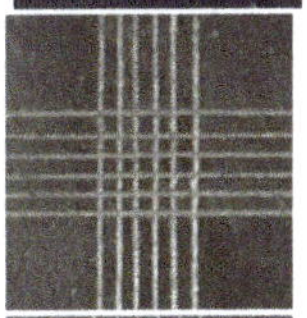	0
Thermodur 600 Aqua ATM (TH-S3, TH-P3)	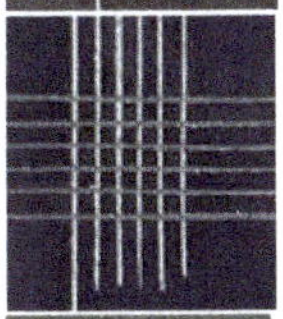	0	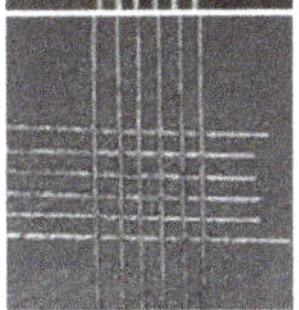	0
Resist 78 (RE-S3, RE-P3)	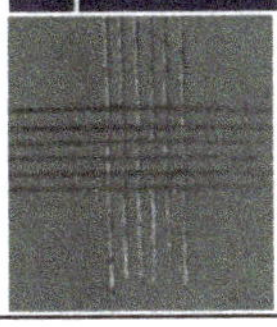	1	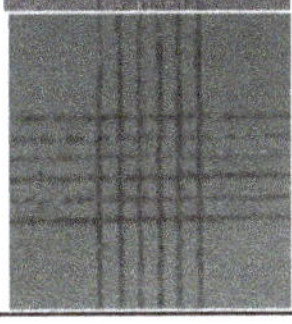	0

As is evident, nearly all samples exhibited excellent adhesion properties after exposure to high-temperature cycles. Also, the samples showed no defects or changes in appearance, which is extremely important as the testing environment closely resembled real conditions in a wood-burning fireplace or stove.

This property of good thermal resistance can be attributed to the chemical composition of the coating. According to [30], the major chemical compositions of the silicone-based coatings constitute Si, O, and alkali groups attached to the Si–O backbone structure. The flexible bond angle between Si and O (100–180 degrees) renders the coatings flexible, and the higher average bond energy of the Si–O bond (452 kJ/mol) makes the polymer durable at high temperatures, which is a very desirable coating property in high-temperature applications.

3.5. Electrochemical Characterization

To assess the protective performance of the coatings, the impedance modulus at the low frequency was observed. High values of absolute impedance and high coating resistance directly manifest better barrier properties [31].

Equivalent electrical circuits used to describe EIS results are shown in Figure 7a,b. Circuits consist of electrolyte resistance (R_S), coating resistance (R_c), coating capacitance (C_c), charge transfer resistance (R_{ct}), and double-layer capacitance (C_{dl}). The nonideal capacitance behaviors of the coating and double layer were described with constant phase elements Q_c and Q_{dl}, respectively [31]. Diffusion of the electrolyte to the metal substrate occurred on sample SE, so the equivalent circuit in Figure 7b was used. This circuit contained an additional element W, i.e., Warburg impedance, which describes the phenomenon of diffusion [12,13,32].

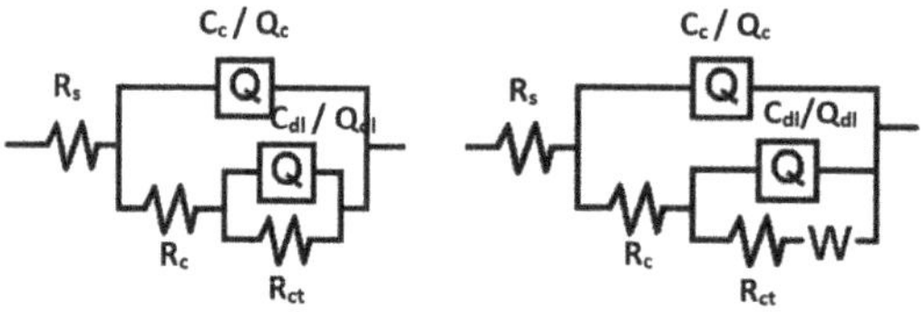

(**a**) (**b**)

Figure 7. Equivalent electric circuit models: (**a**) electric circuit with three resistors and two capacitors, (**b**) electric circuit with three resistors, two capacitors, and Warburg impedance element W [12,13].

Nyquist and Bode plots obtained by mathematical models, according to equivalent electric circuits, together with the values of corresponding modeling errors (Chsq), are shown in Figure 8. The chi-square value (Chsq) was used to assess the goodness of fit between measured and simulated data. Better-fitting results were attained with lower chi-square values. Fitted values of equivalent circuit elements (Rs, Qc, Rc, Cdl/Qdl, and Rct) after 24 h in 3.5% NaCl are presented in Table 10.

Table 10. Fitted values of equivalent circuit elements (R_S, Q_c, R_c, C_{dl}/Q_{dl}, and R_{ct}) after 24 h in 3.5% NaCl.

Sample	R_S, (10^2 Ωcm^2)	n_c	CPE_C, (10^{-6} Fcm^2)	R_C, (10^2 Ωcm^2)	n_{dl}	C_{dl}/CPE_{dl}, (10^{-7} Fcm^2)	R_{ct}, (10^2 Ωcm^2)	W (10^{-4} $\Omega cm^2 s^{1/2}$)	DFT (μm)
SO	71.49	0.478	2.763	9.380	-	0.169	398.0	-	28.0
SE	18.07	-	0.044	56.65	0.70	151.5	69.36	3.098	29.8
TH	0.136	0.409	0.059	0.872	0.78	1239	3.328	-	30.2
RE	1.133	0.800	84.03	11.39	-	8419	13.43	-	39.1

The solvent-based coating SO exhibited the highest resistance (3.979×10^4 Ωcm^2) after exposure to the NaCl electrolyte. All coatings were applied in a similar thickness range, except for the inorganic zinc ethyl silicate RE, the thickness of which was somewhat higher. RE coating showed lower immersion resistance than expected, possibly due to porosity in the film containing a high amount of zinc dust. The protective properties of the TH water-borne coating deteriorated the most; that is, it exhibited the lowest coating resistance (3.328×10^2 Ωcm^2). This test proved that the type of coating has the greatest influence on the effectiveness of protection. Generally, coated samples with polarization resistance higher than 10^6 Ωcm^2 showed good barrier properties [31]. All tested coatings in this study showed lower resistance, probably due to their chemical composition and purpose for indoor protection when applied in low dry film thickness. Due to the flexible Si–O bond angle, the disadvantage of using silicone coatings is related to the high permeation rate for atmospheric constituents such as moisture, oxygen, and other gaseous elements [30], which may result in weaker corrosion protection properties.

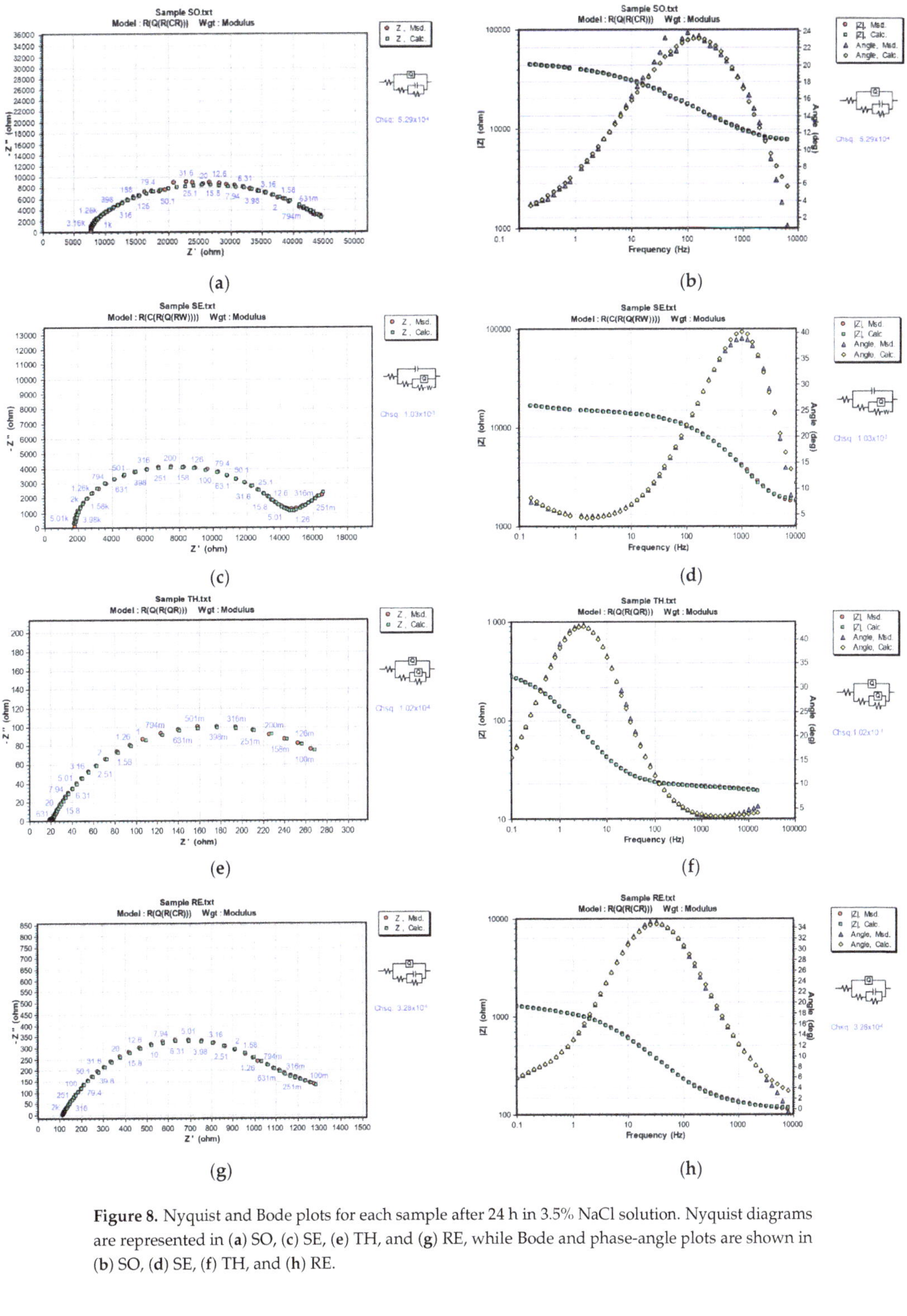

Figure 8. Nyquist and Bode plots for each sample after 24 h in 3.5% NaCl solution. Nyquist diagrams are represented in (**a**) SO, (**c**) SE, (**e**) TH, and (**g**) RE, while Bode and phase-angle plots are shown in (**b**) SO, (**d**) SE, (**f**) TH, and (**h**) RE.

3.6. *Differential Scanning Calorimetry (DSC)*

To determine the phase transitions of silicone-based coatings, a differential scanning calorimetry was used. DSC thermograms for the second heating cycle are shown in Figure 9. Sample RE achieves the highest glass transition temperature among the samples, reaching 100.86 °C, potentially indicating superior thermal stability compared to other samples. Sample SE closely approaches the highest glass transition temperature, measuring 88.86 °C. The lowest temperatures are observed in samples SO and TH, presented in Figure 9a,b, at 48.59 °C and 26.72 °C, respectively. These temperatures closely correspond to the glass transition values for silicone acrylic coatings as detailed in [33].

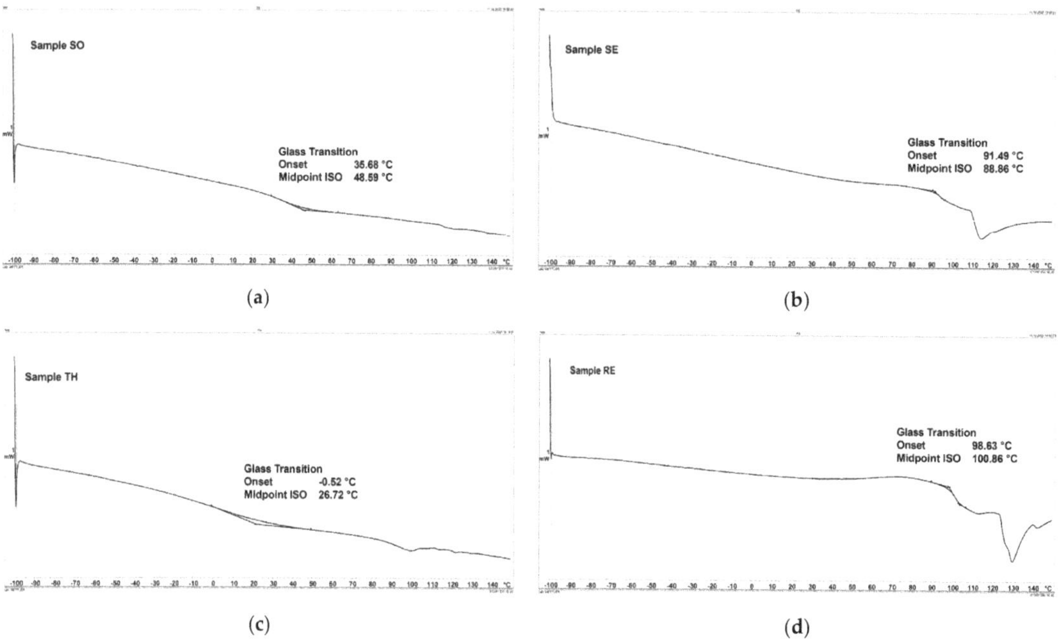

Figure 9. DSC thermograms for high-temperature coatings (**a**) SO, (**b**) SE, (**c**) TH, and (**d**) RE.

4. Conclusions

High-temperature coatings are widely used for protecting metal objects exposed to high temperatures, such as furnaces and stoves. In addition to requiring good corrosion and cyclic temperature resistance, they need to have a visually appealing surface, which can be challenging to maintain if the coatings are exposed to a salty atmosphere, as shown by this research.

The following conclusions can be drawn from this study:

- The sandblasted samples exhibited significantly better coating adhesion after testing in the salt spray and humidity chambers compared to the shot-blasted samples. After the salt spray chamber test, with the SO and SE samples, there was a complete detachment (classification 5 according to ISO 2409) of the coatings applied on the shot blasted samples.
- The best-performing coating in the accelerated chamber testing was inorganic zinc ethyl silicate, which not only exhibited excellent properties but also maintained a decent external appearance from an esthetic perspective.
- Coating hardness was equal for all tested coatings, probably due to the great influence of the steel substrate.

- The EIS results show that solvent-based coatings have better barrier properties than water-based coatings. The solvent-based coating SO exhibited the highest resistance, which is 10^1 Ωcm^2 orders higher than water-based coating SE and 10^2 Ωcm^2 orders higher than water-based coating TH.
- The DSC results show that sample RF achieves the highest glass transition temperature, measuring 100.86 °C, while sample TH exhibits the lowest temperature, measuring 26.72 °C.
- The properties and appearances of the coatings are more favorable in humid conditions and high-temperature environments, which align with the real-world application conditions for coating protection of wood-burning stoves.

Author Contributions: B.Š. and I.S. performed the experiments and wrote the paper. I.S. designed the experiments and analyzed the data. V.A. and M.K. revised and edited the paper. M.O. performed and analyzed DSC. All authors have read and agreed to the published version of the manuscript.

Funding: This research was funded by the European Regional Development Fund, grant number KK 01.2.1.02.0226.

Institutional Review Board Statement: Not applicable.

Informed Consent Statement: Not applicable.

Data Availability Statement: The data presented in this study are available on request from the corresponding author.

Acknowledgments: This study is based on the project "Development of a Highly Energy Efficient Solid Fuel Heating System from Renewable Energy Sources Using Innovative Technological Processes", which is co-financed within the Operational Programme Competitiveness and Cohesion from the European Regional Development Fund under reference number KK. 01.2.1.02.0226.

Conflicts of Interest: The authors declare no conflict of interest.

References

1. Bose, S. *High Temperature Coatings*; Butterworth-Heinemann: Oxford, UK, 2017.
2. Lai, G.Y. High-Temperature Corrosion and Materials Applications. In *High-Temperature Corrosion and Materials Applications*; ASM International: Detroit, MI, USA, 2007; Volume 2.
3. Zielecka, M.; Rabajczyk, A.; Cygańczuk, K.; Pastuszka, Ł.; Jurecki, L. Silicone Resin-Based Intumescent Paints. *Materials* **2020**, *13*, 4785. [CrossRef] [PubMed]
4. Mévrel, R. State of the art on high-temperature corrosion-resistant coatings. *Mater. Sci. Eng. A* **1989**, *120*, 13–24. [CrossRef]
5. Bellamy, G.T.; DiDomizio, M.J.; Patel, M.K.; McKinnon, M.B. Characterization of high-temperature paints for infrared thermography in fire research. *Fire Saf. J.* **2023**, *137*, 103775. [CrossRef]
6. Chew, K.W.; Yahaya, A.H.; Arof, A.K. Studies on the properties of silicone resin films on mild steel. *Pigment. Resin Technol.* **2000**, *29*, 364–368. [CrossRef]
7. Mathivanan, L.; Selvaraj, M.; Azim, S.S.; Balakrishnan, K. Evaluation of heat resistant properties of silicone-based coatings by SEM and a.c. Impedance techniques. *Prog. Org. Coat.* **1996**, *28*, 113–116. [CrossRef]
8. Ramakrishnan, M. Epoxy High Temperature Coatings and Considerations. In Proceedings of the Corrosion 2017, New Orleans, LO, USA, 26–30 March 2017; NACE International: Houston, TX, USA, 2017; p. 9430.
9. Mathivanan, L.; Arof, A.K. The degradation of silicone coatings in high-temperature atmospheres. *Anti-Corros. Methods Mater.* **1998**, *45*, 403–412. [CrossRef]
10. Xu, J.; Long, J.; Zhang, R.; Du, J.; Guan, S.; Wang, Y.; Huang, L.; Wei, H.; Liu, L.; Huang, Y. Greatly improving thermal stability of silicone resins by modification with POSS. *Polym. Degrad. Stab.* **2020**, *174*, 109082. [CrossRef]
11. Gao, H.; Xiong, Y.; Zhang, K.; Cao, S.; Hu, M.; Li, Y.; Zhang, P.; Liu, X. Molecular Dynamics Simulation of Interface Properties between Water-Based Inorganic Zinc Silicate Coating Modified by Organosilicone and Iron Substrate. *J. Renew. Mater.* **2023**, *11*, 1715–1729. [CrossRef]
12. Alsamuraee, A.M.A.; Jaafer, I.H.; Ameen, A.H.; Abdullah, Q.A. Electrochemical impedance spectroscopic evaluation of cor-rosion protection properties of polyurethane/polyvinyl chloride blend coating on steel. *Am. J. Sci. Res.* **2011**, *2*, 761–768. [CrossRef]
13. Stojanović, I.; Logar, M.; Fatović, I.; Alar, V.; Rakela-Ristevski, D. Experimental Study of Atmospherically and Infrared-Dried Industrial Topcoats. *Coatings* **2023**, *13*, 1343. [CrossRef]
14. Chen, J.; Liu, J.; Wang, H.; Li, B.; Hu, Q.; Shao, T.; Yang, R.; Wang, B.; Wan, Q.; Li, Z.; et al. Experimental Study on Neutral Salt Spray Accelerated Corrosion of Metal Protective Coatings for Power-Transmission and Transformation Equipment. *Coatings* **2023**, *13*, 480. [CrossRef]

15. Tsai, P.-Y.; Chen, T.-E.; Lee, Y.-L. Development and Characterization of Anticorrosion and Antifriction Properties for High Performance Polyurethane/Graphene Composite Coatings. *Coatings* **2018**, *8*, 250. [CrossRef]
16. *ISO 2409*; Paints and Varnishes—Cross-Cut Test. International Organization for Standardization: Geneva, Switzerland, 2020.
17. Aračić, S.; Krumes, S.; Matičević, T. Research into the possibility of using the new generation paint systems for the anticorrosion protection of steel structures. *Tech. Gaz.* **2009**, *16*, 75–79.
18. *ISO 6270*; Paints and Varnishes—Determination of Resistance to Humidity—Part 1: Condensation (Single-Sided Exposure). International Organization for Standardization: Geneva, Switzerland, 2017.
19. *ISO 4628*; Paints and Varnishes—Evaluation of Degradation of Coatings—Designation of Quantity and Size of Defects, and of Intensity of Uniform Changes in Appearance—Part 1: General Introduction and Designation System. International Organization for Standardization: Geneva, Switzerland, 2016.
20. Barroso, G.; Li, Q.; Bordia, R.K.; Motz, G. Polymeric and ceramic silicon-based coatings—A review. *J. Mater. Chem. A* **2019**, *7*, 1936–1963. [CrossRef]
21. Indumathy, B.; Sathiyanathan, P.; Prasad, G.; Reza, M.S.; Prabu, A.A.; Kim, H. A Comprehensive Review on Processing, Development and Applications of Organofunctional Silanes and Silane-Based Hyperbranched Polymers. *Polymers* **2023**, *15*, 2517. [CrossRef]
22. Hoang, N.; Khoa, T.A.; Nhung, L.T.; Phuong, P.M.; Binh, T.D.; Hang, T.T.X.; Chi, N.V.; Nguyen, T.-D. Flake ZnAl Alloy as an Effective Pigment in Silicate Coatings for the Corrosion Protection of Steel. *Coatings* **2022**, *12*, 1046. [CrossRef]
23. *ISO 2808*; Paints and Varnishes—Determination of Film Thickness. International Organization for Standardization: Geneva, Switzerland, 2019.
24. *ISO 9227*; Corrosion Tests in Artificial Atmospheres—Salt Spray Tests. International Organization for Standardization: Geneva, Switzerland, 2022.
25. *ISO 12944-6*; Paints and Varnishes—Corrosion Protection of Steel Structures by Protective Coating Systems—Part 6: Laboratory Performance Test Methods. International Organization for Standardization: Geneva, Switzerland, 2018.
26. *ISO 8501-1*; Preparation of Steel Substrates before Application of Paints and Related Products—Visual Assessment of Surface Cleanliness—Part 1: Rust Grades and Preparation Grades of Uncoated Steel Substrates and of Steel Substrates after Overall Removal of Previous Coatings. International Organization for Standardization: Geneva, Switzerland, 2007.
27. Kralj, M.; Pavković, K.; Stojanović, I.; Anđal, J. Adhesion and anticorrosive properties of DTM coating as related to primer coating. *Građevinar* **2018**, *71*, 401–408. [CrossRef]
28. Draganovská, D.; Brezinová, J.; Guzanová, A. Surface Characterization after Blasting. In *Tribology of Machine Elements*; Pintaude, G., Cousseau, T., Rudawska, A., Eds.; IntechOpen: London, UK, 2022. [CrossRef]
29. Lyon, S.B.; Bingham, R.; Mills, D.J. Advances in corrosion protection by organic coatings: What we know and what we would like to know. *Prog. Org. Coat.* **2017**, *102*, 2–7. [CrossRef]
30. Ambat, R.; Piotrowska, K. Preventive measures for corrosion in electronics: Intrinsic and extrinsic strategies. In *Humidity and Electronics*; Ambat, R., Piotrowska, K., Eds.; Woodhead Publishing: Sawston, UK, 2022; pp. 285–337. [CrossRef]
31. Stojanović, I.; Cindrić, I.; Turkalj, L.; Kurtela, M.; Rakela-Ristevski, D. Durability and Corrosion Properties of Waterborne Coating Systems on Mild Steel Dried under Atmospheric Conditions and by Infrared Radiation. *Materials* **2022**, *15*, 8001. [CrossRef]
32. Šoić, I.; Martinez, S.; Dubravić, M. Gel-Electrolyte EIS setup used for probing of IR Dried/Cured industrial coatings. *Prog. Org. Coat.* **2019**, *137*, 105331. [CrossRef]
33. Park, H.S.; Kim, S.R.; Park, H.J.; Kwak, C.; Hahm, H.S.; Kim, S.K. Preparation and Characterization Of Weather Resistant Silicone/Acrylic Resin Coatings. *J. Coat. Technol.* **2003**, *75*, 55–64. [CrossRef]

coatings

Review

A Review of Corrosion-Resistant PEO Coating on Mg Alloy

Chao Yang [1], Pinghu Chen [2,*], Wenxing Wu [2], Liyuan Sheng [3], Yufeng Zheng [3] and Paul K. Chu [4,5,6]

[1] National Engineering Research Center of Light Alloy Net Forming, School of Materials Science and Engineering, Shanghai Jiao Tong University, Shanghai 200240, China; chaoyang0315@163.com
[2] Key Laboratory of Hunan Province of Equipment Safety Service Technology under Extreme Environment, College of Mechanical Engineering, University of South China, Hengyang 421001, China; wuzhiaifrank@hotmail.com
[3] Shenzhen Institute, Peking University, Shenzhen 518057, China; lysheng@yeah.net (L.S.); yfzheng@pku.edu.cn (Y.Z.)
[4] Department of Physics, City University of Hong Kong, Tat Chee Avenue, Kowloon, Hong Kong 999077, China; paul.chu@cityu.edu.hk
[5] Department of Materials Science & Engineering, City University of Hong Kong, Tat Chee Avenue, Kowloon, Hong Kong 999077, China
[6] Department of Biomedical Engineering, City University of Hong Kong, Tat Chee Avenue, Kowloon, Hong Kong 999077, China
* Correspondence: chenpinghu1986@163.com; Tel.:+86-13786161244

Abstract: The corrosion problem of Mg alloy limits its application in many engineering fields. Plasma electrolytic oxidation (PEO) is an economical and eco-friendly technology that can create a dense oxide layer on Mg alloy, offering a solution to the corrosion issue. This research summarizes the use of PEO technology in developing corrosion-resistant coatings for Mg alloys and examines the growth mode and corrosion process of PEO coatings. It is concluded that current efforts to enhance the corrosion resistance of PEO coatings on Mg alloys can be categorized into two approaches: improving the internal structure of the coating and enhancing the phase composition. This includes optimizing coating thickness, roughness, and density; repairing micropores and cracks; and introducing corrosion-resistant compounds by doping. Micropores and cracks are identified as vulnerable points for corrosion, and sealing is an effective strategy to address this. By modifying the phase composition of the coating, corrosion occurrence can be minimized, significantly boosting the corrosion resistance of Mg alloys. Finally, future challenges and potential advancements in corrosion-resistant PEO coatings for Mg alloys are discussed.

Keywords: Mg alloy; PEO; coating; corrosion resistance

1. Introduction

Mg alloy is the lightest metal material in engineering applications, with a density of only 2/3 that of Al and 1/4 that of Fe. It possesses excellent properties such as low density, high strength, good processability, and recyclability, resulting in minimal environmental pollution [1–3]. The use of Mg alloy has significantly reduced the weight of vehicles, enhanced the aerodynamic performance of transportation vehicles, and contributed to energy conservation, emission reduction, impact resistance, and shock absorption. As a result, it is extensively utilized in the aerospace and transportation manufacturing industries. However, due to the high chemical activity of the Mg alloy, it exhibits poor corrosion resistance, particularly in marine environments. Surface treatment technology serves as an effective and common method to enhance the corrosion resistance of Mg alloys. Common techniques for corrosion protection include chemical electrodeposition, electroplating, anodizing, vapor deposition, and organic coating [4–8]. In recent years, plasma electrolytic oxidation technology (PEO), based on anodizing, has gained recognition and popularity among scholars [9,10]. This technology, also known as micro-arc oxidation or anodic spark deposition technology, offers advantages such as a simple process, ease of operation, and no limitation by the shape of the workpiece. It has been extensively studied and applied to enhance the surface hardness and corrosion resistance of materials. This study presents a summary of recent research on corrosion-resistant PEO coatings of Mg alloy, starting with the generation process and corrosion process of PEO coatings on Mg alloy. It further elaborates on the primary strategies to enhance corrosion resistance, which include improving the internal structure of coatings and regulating the phase composition of coatings. Lastly, the challenges of PEO technology on Mg alloy are discussed, and the design of corrosion-resistant PEO coatings on Mg alloy is projected for the future.

2. Growth and Corrosion Process of PEO Coating on Mg Alloy

2.1. Growth of PEO Coating

The growth process of PEO coatings on the Mg alloy is influenced by the type of electrolyte and the electrical parameters. Based on the plasma discharge behavior of the coating, its growth process can be divided into four stages (Figure 1) [11]. Initially, a rapid formation of an oxide film of about 1 μm occurs on the surface without any arc light generation. As the voltage increases, the dissolution rate of the oxide film increases, leading to the first stage—the traditional anodizing stage. When the actual voltage reaches the oxide breakdown voltage, the oxide film formed by anodizing is punctured, and bright low-density arcs appear on the surface of the Mg alloy, marking the beginning of the second stage—the spark discharge stage (as shown in Figure 2). As the voltage continues to rise, the arcs gradually grow larger and brighter, causing the surface temperature of the Mg alloy to sharply rise. This leads to rapid melting and cooling of the oxide, forming a dense PEO coating, and transitioning the reaction into the third stage—the micro-arc oxidation stage, which is the primary stage for ceramic film formation. With a prolonged reaction time, the process enters the fourth stage. With an increase in coating thickness, breakdown discharge requires more energy, resulting in two trends in discharge: either the arc light disappears due to insufficient energy density on the sample surface or intense arc discharge occurs in several regions, causing intense and prolonged arc light that results in coating burn damage [12].

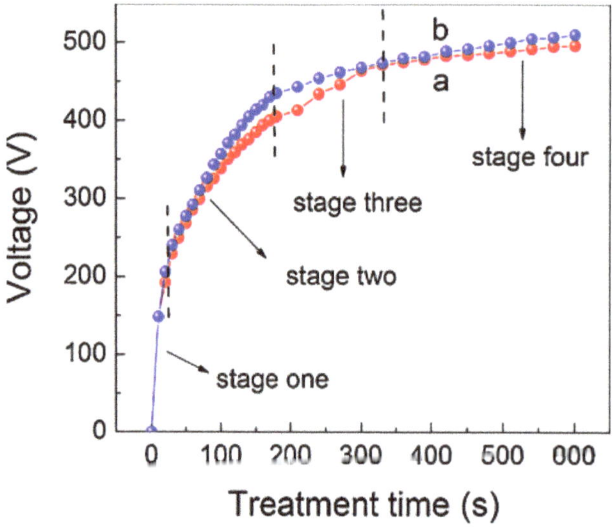

Figure 1. V-T diagram of the PEO process on Mg alloy: (**a**) silicate (red dotted line) and (**b**) silicate and phosphate mixing system (blue dotted line) [11] (Reprinted/adapted with permission from Elsevier, 2019).

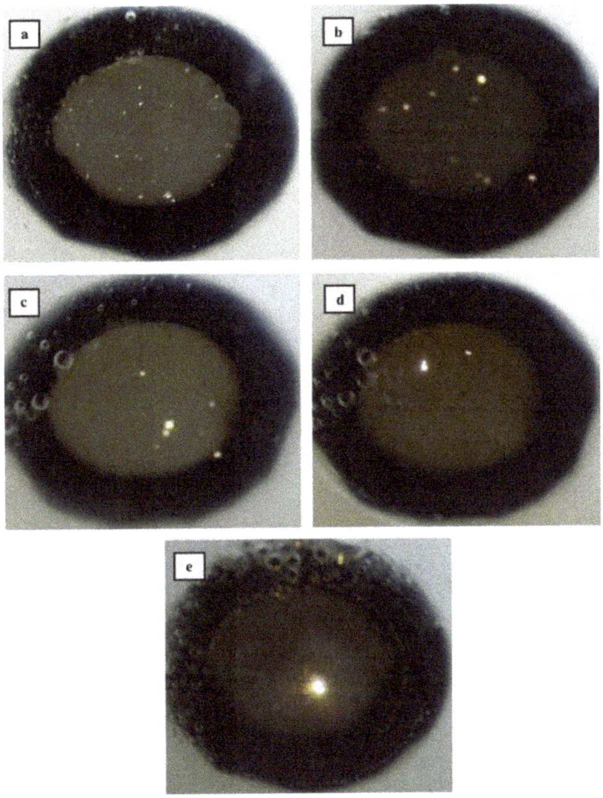

Figure 2. Surface arc discharge images of the PEO process on Mg alloy: (**a**) 10 s, (**b**) 60 s, (**c**) 150 s, (**d**) 300 s, and (**e**) 500 s [12] (Reprinted/adapted with permission from Elsevier, 2005).

The composition of the coatings prepared from different electrolytes varies, with the most common systems being silicate and phosphate electrolytes. The following reactions take place on the Mg alloy during the PEO process:

$$Mg \rightarrow Mg^{2+} + 2e^{-}. \quad (1)$$

As the reaction progresses, a large amount of MgO is generated in the coating. The reaction is as follows:

$$Mg^{2+} + OH^{-} \rightarrow Mg(OH)_2, \quad (2)$$

$$Mg(OH)_2 \rightarrow MgO + H_2O. \quad (3)$$

In the silicate electrolyte system, the phase composition of its PEO coating is depicted in Figure 3. The electrolyte is typically alkaline, and as it diffuses into the coating, the following reactions take place:

$$2Mg^{2+} + SiO_3{}^{2-} + 2OH^{-} \rightarrow Mg_2SiO_4 + 2H_2O, \quad (4)$$

$$Mg^{2+} + SiO_3{}^{2-} \rightarrow MgSiO_3. \quad (5)$$

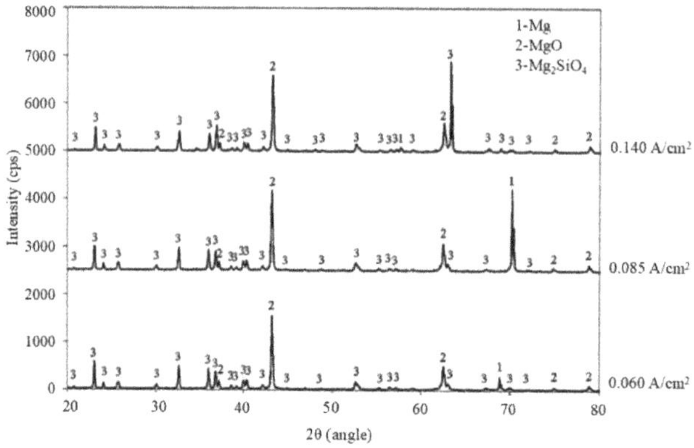

Figure 3. XRD of PEO coating prepared by silicate electrolyte [13] (Reprinted/adapted with permission from Elsevier, 2011).

Mg^{2+} ions from the matrix reach the interface of the coating/electrolyte and react with the silicates in the solution. Meanwhile, the MgO and Mg_2SiO_4 in the coating undergo a transformation into the stable $MgSiO_3$ [13]. For the phosphate electrolyte, the composition of its PEO coating is depicted in Figure 4. The reaction is as follows:

$$Mg^{2+} + PO_4{}^{3-} \rightarrow Mg_3(PO_4)_2, \quad (6)$$

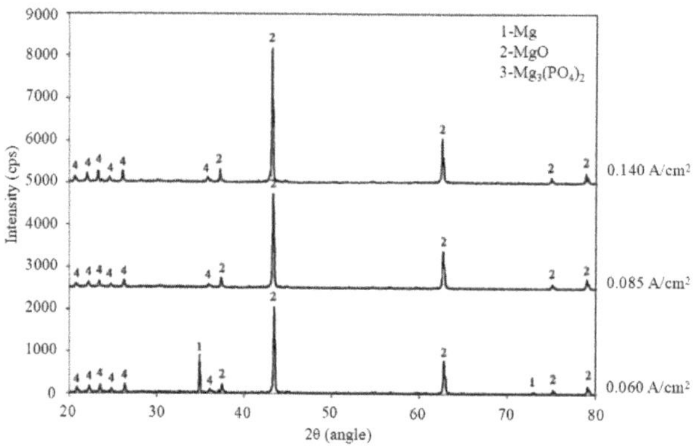

Figure 4. XRD of PEO coating prepared by phosphate electrolyte [13] (Reprinted/adapted with permission from Elsevier, 2011).

Mg^{2+} ions from the matrix reach the interface of the coating/electrolyte and react with $PO_4{}^{3-}$ to form $Mg_3(PO_4)_2$. Magnesium phosphate itself is a sparingly soluble stable phase, making it difficult to further react with MgO. In this process, $Mg_3(PO_4)_2$ mainly exists in an amorphous phase [13].

2.2. Corrosion of PEO Coating

Although the chemical composition of coatings prepared with different electrolytes varies, the overall structure of PEO coatings remains essentially the same, consisting of an internal dense layer and an external porous layer (Figure 5). Due to differences in reaction stages and electrolyte characteristics, there are often significant variations in the distribution and structure of elements within the inner and outer layers [14]. The outer layer typically contains micropores and cracks, with a distinct boundary separating it from the inner layer, serving as an initial barrier against corrosion media. The inner layer, on the other hand, is characterized by higher density and resistance levels approximately 8–10 times greater than those of the outer layer, effectively slowing down the rate of corrosion electrochemical reactions [15]. Ultimately, the inner layer structure acts as the primary barrier for corrosion protection in PEO coatings.

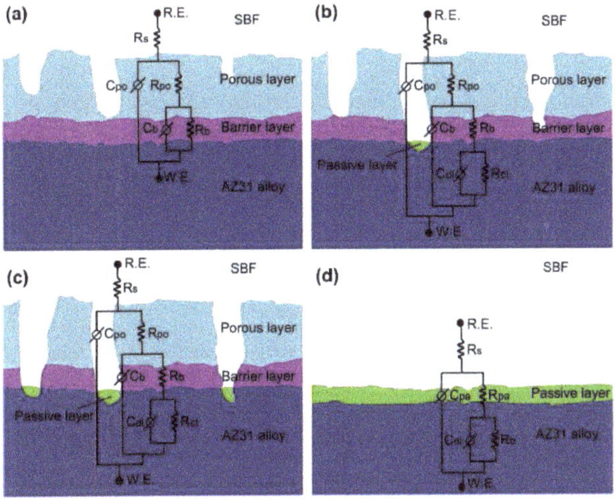

Figure 5. A physical model of the corrosion process of PEO-coated AZ31 alloy immersed in the SBF with equivalent circuits for fitting EIS data: (**a**) R(Q(R(QR))), (**b**) R(Q(R(Q(R(QR))))), (**c**) R(Q(R(Q(R(QR))))), and (**d**) R(Q(R(QR))) [16] (Reprinted/adapted with permission from Elsevier, 2012).

However, as the contact time with the corrosive medium is prolonged, the coating will eventually be destroyed by SBF [16]. Taking the typical NaCl solution as an example, the corrosion process is divided into two stages. Before corrosion occurs, the outer layer resistance R_{po} obtained by EIS fitting is about 1000–2000 Ω, while the inner layer resistance R_b represents about 7000–10,000 Ω [15]. In the first stage, when the corrosive liquid NaCl solution contacts the outer surface, it preferentially enters holes and cracks, causing MgO to begin to dissolve, reacting to generate $Mg(OH)_2$ attached to the surface. Additionally, due to the influence of alloy elements, the second phase, and β-Mg of the Mg alloy, the coating has structural defects such as the AlMn phase, causing localized stress during the coating growth process, making the coating more susceptible to corrosion [17]. The generation of corrosion products, to some extent, blocks the entry of the NaCl solution, slowing down the ion exchange process. As Cl gradually replaces O, forming $MgCl_2$, the corrosion enters the second stage. In this stage, the dissolution rate of the coating accelerates and is greater than the rate of $Mg(OH)_2$ generation, causing the corrosion products to gradually dissolve in the NaCl solution, forming pit corrosion, eventually penetrating through the entire coating [18]. At the same time, the directional corrosion at the surface defects connects to other pits, allowing more NaCl solution to enter and react with the Mg matrix, increasing the amount of corrosion products and gas [10], increasing stress, and ultimately leading to the failure and detachment of the entire coating [9], as shown in Figure 6. For coatings prepared with silicate and phosphate, there are some differences in the corrosion process. In the early stage of the phosphate film formation, the passivation film generated is more stable, making it easier for corrosion products to accumulate on the surface once corroded to the base, while the silicate overall coating is more uniform, allowing corrosion products to directly overflow along the holes, resulting in a greater surface difference after corrosion between the two, as shown in Figure 7 [19]. At this point, the outer layer resistance R_{po}

obtained by EIS fitting decreases to 400 Ω, while the inner layer resistance R_b can still remain far higher than that of the outer layer, maintained at 3000 Ω.

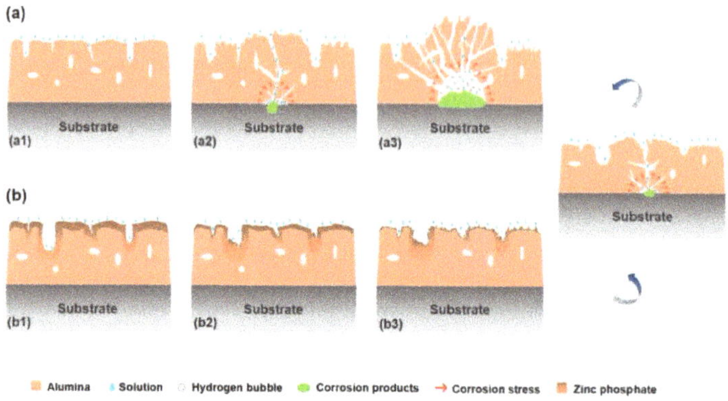

Figure 6. Schematic diagrams of the corrosion process of (**a**) PEO coating corrosion failure and (**b**) Zn-doped PEO coating without corrosion failure [9] (Reprinted/adapted with permission from Elsevier, 2020).

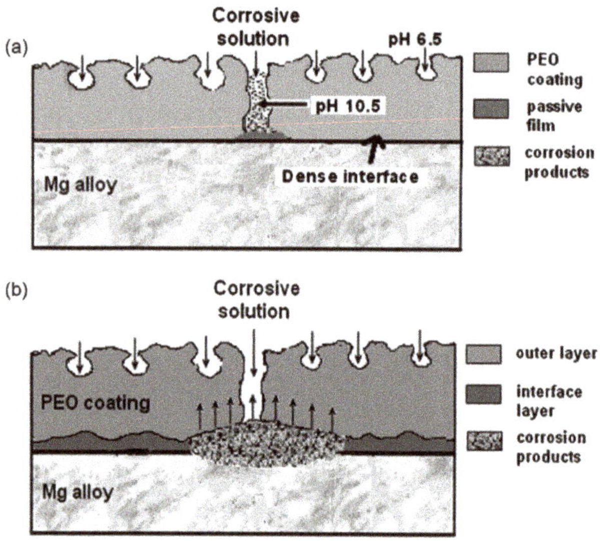

Figure 7. Schematic representation of the corrosion process of PEO-coated specimens upon immersion for 50 h in a 0.1 M NaCl solution: (**a**) Si-PEO coating and (**b**) P-PEO coating [19] (Reprinted/adapted with permission from Elsevier, 2009).

3. Improvement of the Coating Structure

In the preparation process, structural defects in PEO coatings are unavoidable, which limits their corrosion resistance improvement. The formation of PEO coatings involves various disciplines, such as chemistry, electrochemistry, thermodynamics, and plasma chemistry. However, the growth of coatings is primarily controlled by reaction energy. The energy of the coating is reflected on a macroscopic level in the arc discharge behavior, which, in turn, affects the internal structure of the coating. The micro-pores and cracks that form during coating growth are susceptible to corrosion liquid infiltration, which is essential for the coating's corrosion resistance. Additionally, the roughness, uniformity, and density of the coating all impact the infiltration of corrosion liquid and the occurrence of corrosion reactions.

3.1. The Influence of Discharge Behavior on the Coating Structure

The growth of the coating is determined by the reaction energy, which is mainly reflected in the discharge behavior. Increasing the energy of PEO coatings can be most directly achieved by changing the electrical parameters. This involves increasing the input energy to the coating, but simply increasing the input energy is often unsatisfactory because the PEO process is complex and energy control does not follow Faraday's law for discharge. Discharge is generally divided into solid breakdown and gas breakdown, with each method resulting in different types of discharge [11].

Currently, there are three different discharge modes on the PEO coating, and the discharge behavior is closely related to the coating structure [20]. As shown in Figure 8, the discharge modes can be distinguished by the different element contents near the micro-holes. The first type (type A) is the breakdown of the oxide film in a strong electric field, which often occurs in small holes on the surface at the interface between the electrolyte and the coating, with more electrolyte solution elements. The corresponding fine hole on the surface corresponds to a discharge with fewer defects due to the weak discharge intensity. The second type (type B) is gas discharge through initial dielectric breakdown in micro-holes, occurring on the surface of the metal-oxide film. The surface morphology formed is crater-like, also known as a "thin pancake", inevitably carrying molten material ejected from the discharge channel during surface contact with the electrolyte. The high content of matrix elements near the micro-holes, originating from the matrix, leads to a higher temperature, and excess energy enhances the transformation from the amorphous phase to the crystalline phase. However, as the central hole penetrates deep into the coating, the corrosion resistance of the coating will decrease. The third type (type C) is arc discharge between free electrons in the gas medium on the coating surface and the electrolyte, occurring in deeper surface apertures with higher matrix elements compared to the first type of discharge behavior. It occurs at high discharge intensity and causes coating damage, overflowing with a large amount of molten material, forming a spherical surface. Dehnavi et al. [21] mainly used 20 g/L NaOH, 80 g/L $Na_2SiO_3{\cdot}5H_2O$, and 80 g/L $Na_2B_4O_7{\cdot}10H_2O$ solutions to prepare coatings with type A and type C discharges as the main types, significantly improving the resistance value of the coatings and enhancing their corrosion resistance (Figure 9).

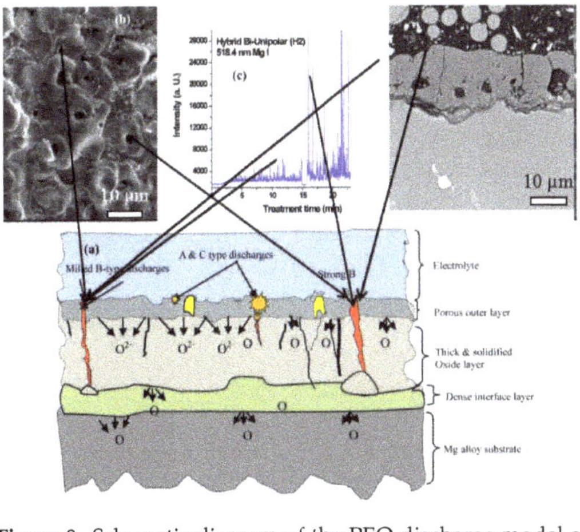

Figure 8. Schematic diagram of the PEO discharge model on the Mg alloy: (**a**) types A,B, and C discharges; (**b**) cross-sectional SEM image of the coating by type B discharge; and (**c**) XRD of the coating by type B discharge [20].

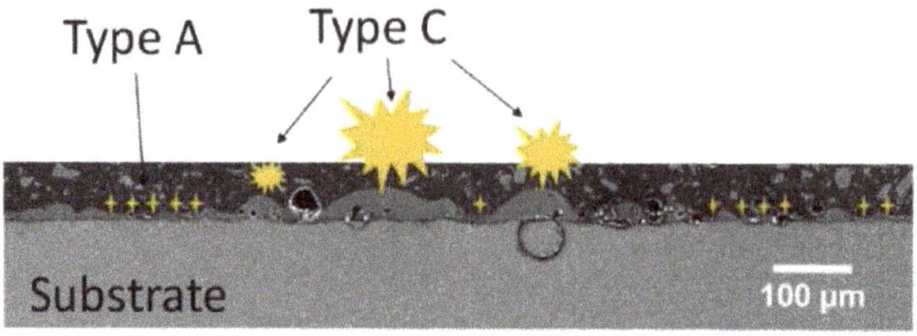

Figure 9. The corresponding pore positions of discharge in types A and C [21].

3.2. The Influence of Coating Roughness on the Corrosion

The roughness of the coating refers to the size of the spacing between two peaks or valleys, and splashing molten oxides generated during the coating growth process will result in a rough surface with certain geometric errors. Reducing the relative surface roughness can improve the contact area between the coating surface and the corrosive liquid, thus reducing the probability of corrosion [22]. The rough surface of the coating easily allows gas or liquid to penetrate into the inner layer through the valleys on the surface, thereby affecting the corrosion resistance of the coating. Furthermore, high roughness will reduce the wear resistance and fatigue strength of the coating, limiting its comprehensive application [23]. Generally, the current density has a significant impact on the roughness of the coating, as a high current density leads to a high gas evolution rate and gas resistance, causing an increase in coating roughness. Rapheal et al. conducted experiments using current densities of 30, 60, and 120 mA/cm^2, and the results showed a significant increase in coating roughness, while the corrosion resistance decreased due to the increased roughness [24]. Therefore, researchers often add different particles or organic substances to the electrolyte solution to reduce the gas evolution of the coating. Wu et al. [25] added glycerol as an additive to the electrolyte solution, reducing the mean square roughness of the coating from 424 nm to 174 nm. The corrosion current density of the coating decreased by two orders of magnitude to 10^{-7} J/(A·cm^2).

3.3. The Influence of Coating Thickness on Corrosion

Thicker coatings typically have higher corrosion resistance [26]. By extending the reaction time, it is possible to achieve a higher coating thickness. Liu et al. [27] were able to increase the coating thickness even further by using sodium citrate as an electrolyte, resulting in a tenfold decrease in corrosion current density. Generally, the outer layer of the coating contains more defects and holes, so the delay in corrosion is more dependent on the thickness of the outer layer than the inner layer. The thickness of the coating growth depends on various factors, such as the substrate, electrical parameters, and electrolyte solution used. The presence of different alloying elements in the substrate results in variations in the growth rate and uniformity of the coating during the filming process [14]. For example, the presence of the AlMn phase and β-Mg can impact the coating growth [17]. Additionally, the roughness of the substrate also plays a significant role in the thickness of the coating. As shown in Figure 10, Yoo et al. [28] demonstrated that a substrate with a roughness of 0.5 μm had a considerable improvement in thickness compared to a substrate with a roughness of 2.5 μm. This not only reduced the residual stress within the coating but also decreased the corrosion current density from 10^{-7} J/(A·cm^2) to 10^{-9} J/(A·cm^2). A lower substrate roughness leads to more uniform surface discharge, thereby increasing the coating growth rate and enhancing the mechanical properties of the coating.

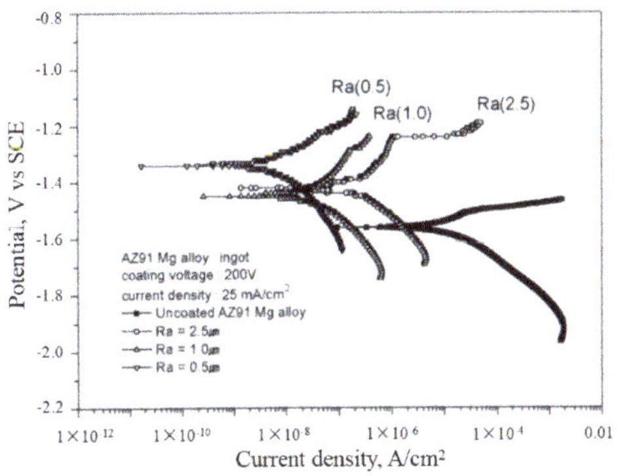

Figure 10. Tafel curves of oxide layers on AZ91 Mg alloys with different surface roughnesses [28] (Reprinted/adapted with permission from Elsevier, 2010).

The improvement in coating thickness in the PEO reaction process is more noticeable with changes in electrical parameters and electrolyte solution. Changes in electrical parameters will affect the rate of gas generation and particle movement during the coating growth process. The longer the reaction time, the thicker the coating becomes, and the corrosion resistance of the coating shows a trend of initial improvement followed by a decline with increasing coating thickness. Generally, the thickness of PEO coatings ranges between 20 and 55 μm. When the reaction time exceeds 60 min, the coating thickness increases significantly, but excessively long reaction times can cause coating ablation [29]. As illustrated in Figure 11, Chen et al. [30] conducted tests on the electrochemical properties of coatings at different times and discovered that with an extension of reaction time, the electrical resistance of the coating decreased due to the increase in coating defects. Moreover, the increase in coating thickness, to some extent, obstructed the entry of corrosion solution. Adding other electrolytes to enhance the average film-forming rate can effectively increase the thickness. For instance, adding sodium citrate to improve the conductivity of the solution can enhance the adhesion of the inner coating, thus facilitating more electrolytes to react with overflowed Mg^{2+} and increasing the thickness of the coating [27].

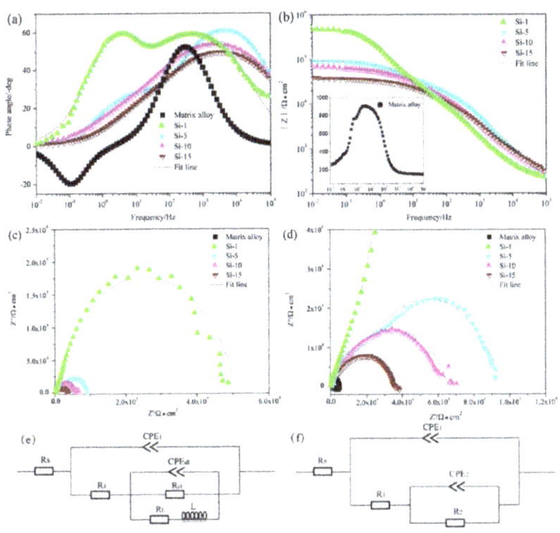

Figure 11. EIS curve of coating prepared by silicate electrolyte at different reaction times: (**a**,**b**) Bode plots, (**c**,**d**) Nyquist plots, and (**e**,**f**) the equivalent circuits employed to fit the spectra [30] (Reprinted/adapted with permission from Elsevier, 2018).

3.4. The Influence of Coating Density on Corrosion

High-density coatings can increase the difficulty for corrosive liquids to reach the inner layer, which is usually reflected in the resistance value of the coating per unit thickness. The resistance value of the coating is divided into an inner layer resistance value and an outer layer resistance value (Figure 5). Better density makes the outer layer resistance value increase by 1–2 orders of magnitude. This improvement has a significant impact on the corrosion resistance time in a 3.5% NaCl solution [16]. The inner layer structure of the coating with a resistance value higher than 8–10 times that of the outer layer is crucial. Due to the challenges of increasing the thickness of the inner layer, reducing defects in the inner layer is essential. Exterior defects are generally not connected to the inner layer, but the presence of electrolytes can lead to defects in the passive film generated in the initial reaction stages or due to excessive reaction time, causing partial dissolution of the passive film and a decrease in the resistance value of the inner layer. This, in turn, can lead to defects connecting the inner and outer layers, resulting in a decrease in the coating's corrosion resistance [31]. Gaps in the substrate structure may lead to incomplete formation of the passive film due to electrolytes and arc discharges. Han et al. [32] suggest that incomplete film formation can cause exterior defects to extend to the inner layer, decreasing the inner layer resistance value by one order of magnitude. Zhao et al. [33] found that adding substances like graphite oxide can create additional thin layers between the inner and outer layers, isolating the inner layer from the outer layer. The high surface area, strong van der Waals forces, and pi-pi stacking properties of this thin layer can reduce the corrosion current density to the level of 10^{-9} orders of magnitude. Additionally, Tran et al. [34] indicated that increasing current density or frequency can affect the flow of ions in a single-arc discharge reaction. Changes in frequency can lead to overflowing of Mg^{2+} ions and alterations in the return flow of anions, impacting the movement of molten material on the surface and affecting the coating's density. Dong et al. [35] developed a high-frequency discharge PEO technology that significantly shortened the single discharge time by increasing the number of power supply pulses. This technology can increase the discharge frequency of the coating, reduce the risk of breakdown, greatly enhance the coating's density, and significantly improve corrosion resistance (Figure 12).

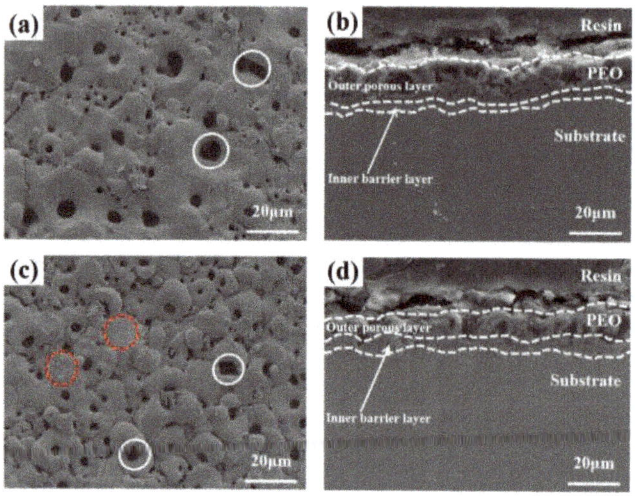

Figure 12. *Cont.*

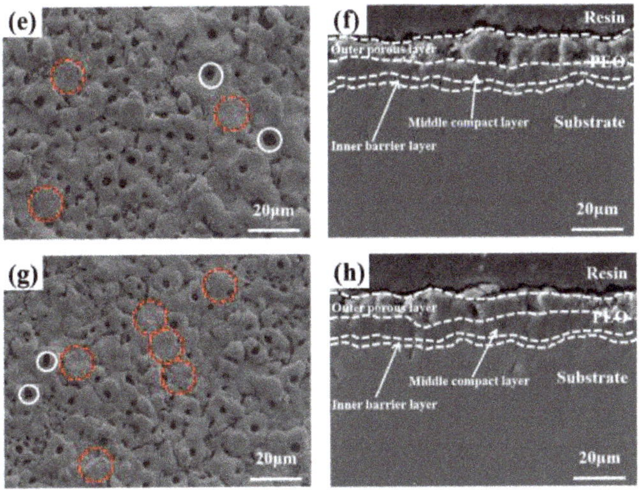

Figure 12. SEM images of (**a**,**b**) 0.5 kHz, (**c**,**d**) 5 kHz, (**e**,**f**) 10 kHz, and (**g**,**h**) 20 kHz [35] (Reprinted/adapted with permission from Elsevier, 2022).

3.5. The Influence of Coating Micropores and Defects on Corrosion

During the PEO reaction process, splashing of molten oxides is unavoidable, resulting in the formation of micro-pores on the PEO coatings. These pores accelerate the infiltration of corrosive liquids, triggering corrosion reactions that ultimately lead to the breakdown of the coating. Porosity and pore size are typically used to describe the micro-pores of the coating. As the current density increases or the reaction time lengthens, the resistance of the coating also increases. This results in a longer duration of high-density arc glow, causing the pore size and number to increase. However, as the thickness of the coating increases, current density decreases, leading to a decrease in arc glow. This transition from granular to blocky surface molten material results in larger pore sizes that no longer increase in size and gradually decrease in number. This reduces the entry of corrosive liquids and improves the corrosion resistance of the coating. It is important to note that higher current density can also lead to coating defects, which can in turn reduce the corrosion resistance of the coating. Chen et al. [36] have shown that high current density can result in outward growth of the coating at a much faster rate than inward growth. This leads to deeper pore sizes and more corrosion sites, ultimately reducing the corrosion resistance of the coating.

The pore ratio often changes due to variations in sparks. The increase in thickness in the later stage of the reaction results in larger individual sparks and a decrease in the number of sparks, causing the overall porosity of the coating to initially increase and then decrease, with a porosity rate between 0.2% and 10%. In addition to conventional methods for controlling the corrosion medium channel, sealing methods are becoming more popular. Common sealing methods are primarily divided into three types. The simplest approach is a two-step method to seal the pores, relying on different reactions in different electrolyte solutions to create a sealing effect by generating substances of varying sizes. For example, Dou et al. [37] first used a silicate solution to react for 300 s, followed by immersion in a phosphate solution to achieve sealing (Figure 13). This resulted in a decrease in corrosion current density from 10^{-6} J/(A·cm^2) to 10^{-8} J/(A·cm^2). The two-step method not only seals the underlying coating but also prevents film formation due to electrical conductivity. By utilizing different solidification times of the reaction products, substances with faster solidification times accumulate near or within the pores, effectively blocking them to achieve a sealing effect. Materials such as zirconium oxide [38] and zinc oxide [32] can be used for this purpose (Figure 14). As the reaction progresses, the leached Mg^{2+} reacts with the electrolyte solution, depositing into the interior, while high-melting-point substances from the surface reaction solidify into block-shaped solids or sol-gel particles, combining with a high porosity rate to effectively seal the pores [39].

The second method involves using different current modes to achieve the same effect as the current output mode (Figure 15) [40,41]. A high current density increases coating porosity significantly. By gradually reducing the current density, the surface arc weakens while the overall voltage remains stable. The molten material shifts from block to granule, and the low energy lowers the temperature of the generated substances, creating an effect similar to differences in melting points and achieving pore sealing. It is important to note that using a stepwise reduction in current or gradually decreasing pressure may not consistently alter the size of the arc, resulting in less noticeable effects. The third method involves a composite sealing approach. Despite the sealing methods mentioned above, PEO coatings may still have micropores and cracks that require additional sealing. Through composite sealing technology, PEO coatings can undergo secondary treatment, significantly enhancing corrosion resistance. Techniques such as electrophoresis, chemical immersion, and gel can be employed for sealing the coating, enabling it to withstand neutral salt spray tests for over 1000 h [42]. Additionally, the composite method can improve other properties, like wear resistance and impact resistance. However, this method is complex and costly, with limitations in its applicability.

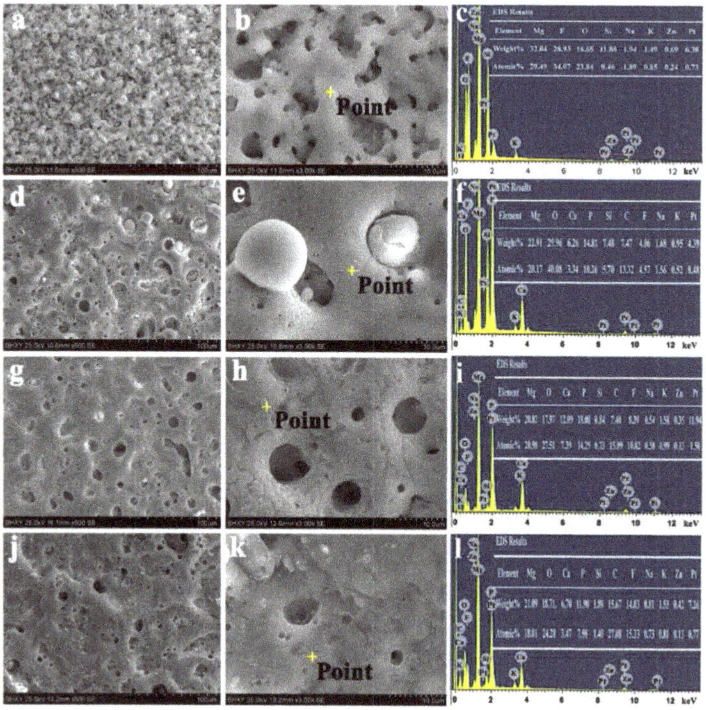

Figure 13. Surface morphology and the responding EDS analyses of PEO coatings produced using the first-step and two-step processes: CSA (a–c), CSB-350 (d–f), CSB-400 (g–i), and CSB-450 (j–l) [37] (Reprinted/adapted with permission from Elsevier, 2018).

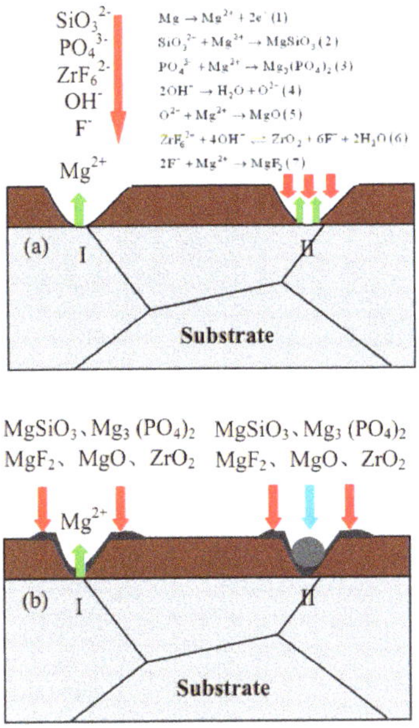

Figure 14. Sealing of zirconia in PEO coating on Mg alloy: (**a**) plasma discharge process and (**b**) formation process and structure of coating [38] (Reprinted/adapted with permission from Elsevier, 2015).

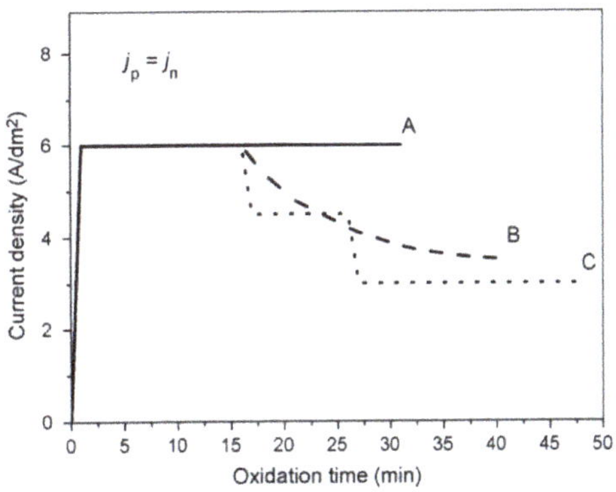

Figure 15. Schematic diagram of different current modes [40] (Reprinted/adapted with permission from Elsevier, 2007). (**A**) constant current density; (**B**) decaying freely current density in the later stages; (**C**) stepped decreasing of current density.

The generation of cracks is mainly due to the discharge holes formed by spark discharge passing through stress accumulation, the rapid cooling and solidification of molten material producing residual stress, and the release of residual stress generated by adjacent spark reactions. Snnivasan et al. [43] indicated that stress corrosion cracks make coatings more susceptible to exposure to corrosive liquids, thereby accelerating the corrosion process. Furthermore, there are ways such as uneven element distribution leading to large differences in the rate of molten material deposition and the generation of cracks. Adding appropriate phosphates to silicate solutions can effectively improve the corresponding conditions, increase the compactness of coatings, reduce arc size, and thereby reduce cracks. Luo et al. [44] showed that adding 5 g/L of sodium hexametaphosphate to PEO electrolytes can reduce the corrosion current density of the coating to 10^{-8} J/(A·cm^2). However, excessive phosphate can also affect the coating generation rate, leading to increased sparks and an increase in the corrosion current density to 10^{-6} J/(A·cm^2). In addition, using ultrasound can improve the concentration of sparks on the coating, effectively increasing the reaction area, dispersing the molten material more evenly, reducing stress generation, and making the reaction more uniform, thereby reducing coating stress and decreasing the corrosion current density [45]. An et al. [46] used EDTA-2Na as an additive, as its hydrophilic groups (-COOH and -NH_2) preferentially react with Mg^{2+}, reducing the solid–liquid interface tension, uniformly adsorbing anions, and decreasing spark intensity with subsequent breakdown discharge, ultimately reducing cracks and decreasing the corrosion current density to 10^{-6} J/(A·cm^2). Similarly, organic compounds such as glycerol have the same effect on improving cracks [25].

4. Improvement of the Coating Phase Composition

The structure of the coating primarily serves to restrain and slow down the immersion of corrosion solution and the spread of corrosion, while also reducing the stress that remains during the preparation process of the PEO coating. The main cause of corrosion is the dissolution of the coating and the damage inflicted by corrosion products on the coating. The phase composition and unique structure that form within the coating are crucial in inhibiting the dissolution and corrosion reactions of the coating. Transition metal compounds are commonly utilized in the electrolyte, relying on arc discharge to deposit insoluble substances onto the coating in order to alter the phase composition of the surface layer and inhibit corrosion. Simultaneously, transition metals create composite coatings on PEO ceramic films through processes of reaction, fusion, and solidification.

4.1. The Influence of PEO Electrolyte on Corrosion

Silicates are often utilized in conjunction with fluorides or hydroxides for the creation of coatings, typically containing hard-to-melt glassy materials like $MgSiO_3$ and Mg_2SiO_3. These materials, which are insoluble in water, act as the primary passivators during the initial stages of film development. Boosting the amount of sodium silicate can effectively lower the arc initiation voltage, thereby decreasing the initial oxidation time. On the other hand, prolonging the PEO reaction time enhances the formation of $MgSiO_3$. The dominant phase in the coating is MgO, and elevating the silicate content in the electrolyte can raise the percentage of $MgSiO_3$ in the coating. This results in a salt spray corrosion time of up to 250 h [47]. Nevertheless, due to the rapid rate of silicate film formation, the resulting coating tends to have poor density, often showing Mg-Al phases and α-Mg in surface analysis (in the matrix or electrolyte containing Al).

When preparing a phosphate coating, the arc voltage is higher, and the reaction time in the first stage is longer. Phosphate salts, such as sodium phosphate, can effectively promote the generation of electric sparks [44], but the growth rate of the coating is lower [29]. The sparks are usually finer than those formed during silicate film formation. The addition of Al elements can typically increase the growth rate of the coating. Coatings prepared with phosphate rely on the lower solubility product constant of magnesium phosphate to improve the corrosion resistance of the coating. Typically, Ca, Zn, and other elements are added to form compounds with even lower solubility product constants [48], while also reducing the surface MgO content to ultimately achieve corrosion resistance. Phosphates are often added as additives to silicate electrolyte solutions to improve the density of the coating. Phosphates easily enter the coating interior through the discharge channel, resulting in a more uniform distribution of elements in the coating (Figure 16) [11]. The discharge in the coating becomes more intense, promoting the conversion of magnesium oxide to magnesium silicate. Phosphates and other ions in the solution form compounds, ultimately reducing the MgO/Mg_2SiO_3 ratio in the coating and improving its corrosion resistance.

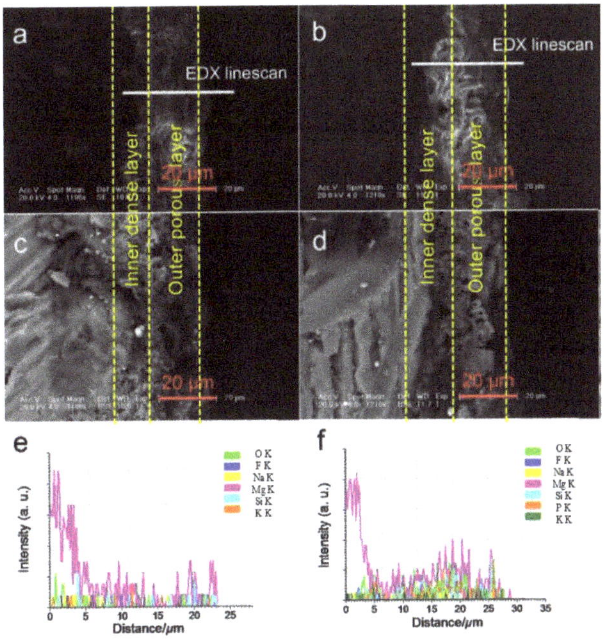

Figure 16. SEM and EDS of coating prepared from alkaline silicate solution: (**a**,**c**,**e**) no phosphate and (**b**,**d**,**f**) with phosphate [11] (Reprinted/adapted with permission from Elsevier, 2019).

4.2. The Influence of Added Compounds on Corrosion

In recent years, by adding compounds to the coating, such as powders and particles, or forming corresponding composite coatings through the process of melting and solidification after reaction, the coating has been conducive to hindering the substitution of Cl for O in MgO during the corrosion process, thereby improving the corrosion resistance of the coating [17]. Currently, silicate electrolyte solutions are mostly used, and the reaction produces a glass-like solid phase, which enhances the toughness of the coating, reduces cracks and internal stresses, and facilitates the particles to melt into the coating [20]. For example, adding zinc sulfate to the electrolyte can form $[Zn(OH)_4]^{2-}$ on the coating, increasing the anodic resistance and making it more difficult for the coating to breakdown. The larger reaction energy promotes the transformation of $[Zn(OH)_4]^{2-}$ on the coating into ZnO and reacts with the overflowed Mg^{2+} from the substrate to form the Mg-O-Zn structure [49]. Li et al. [50] found that Zr^{4+} ions have the same effect. Tu et al. [51] found that W forms a bright material band at the bottom of the coating, depositing W_xO_y oxide with the inner layer, and the addition of W results in a greater spark density on the surface, forming an insulating Mg-O-Al layer. Among many methods for preparing composite PEO coatings, the most direct way is to add particles or powder substances, so that they directly form alloy compounds with molten magnesium oxide during the reaction process, thereby improving the corrosion resistance of the coating. For example, by adding micrometer-sized sodium silicate particles, the particles form silicon dioxide in the discharge process, entering the PEO coating along with the cracks and accumulating in the corresponding positions, although it may sacrifice the uniformity of the coating to some extent [52]. The smaller the powder or particles, the more evenly distributed they are in the coating. Studies have shown that the ability of particles to enter the coating is related to the electric field force and the size of the pores, with larger pore sizes in the coating allowing for easier entry of particles, while the electric field force affects the solidification of molten substances and ion backflow, thereby affecting the entry and adsorption capacity of particles [33]. Yang et al. [53] doped high-content ZnO nanoparticles in the PEO coating of the AZ31B Mg alloy and found that the particles can plug the surface pores while ZnO is uniformly distributed in nano-crystalline form through re-melting behavior in the interior of the coating, resulting in a significant improvement in the corrosion resistance of the coating and increasing the neutral salt spray life to 2000 h (Figure 17).

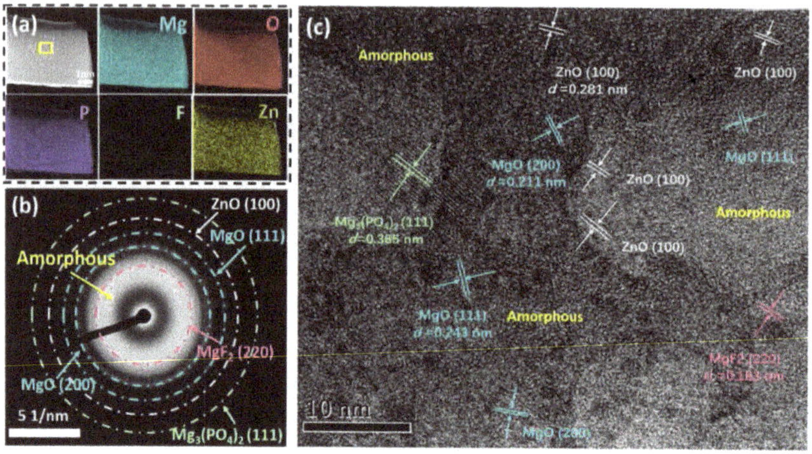

Figure 17. TEM observation on the cross-section of the PEO coating doped with ZnO nanoparticles: (**a**) EDS mapping; (**b**) SAED pattern, and (**c**) HRTEM image showing the compound as nanocrystalline and amorphous [53].

4.3. *The Influence of Self-Growing Insoluble Phases on the Corrosion*

The formation of self-growing insoluble phases more commonly uses phosphate as the electrolyte solution, primarily due to the fact that reactions in coatings prepared from silicates are more complex. This complexity leads to a lower content of corresponding silicates in the coating. Coatings containing only magnesium phosphate and magnesium oxide are not effective in providing corrosion resistance [13]. Therefore, it is essential to enhance the corrosion resistance of the coatings by adding other ions to form more insoluble phosphates or by reducing the surface content of MgO. By doing so, not only the structure of the coating is improved, but the onset of the first stage of corrosion is also delayed, ultimately enhancing the coating's corrosion resistance. A commonly employed method is to add elements such as Ca and Zn to form a phosphate of low solubility constant on the surface, thereby increasing the coating's density, as depicted in Figure 18 [48,54]. This technique effectively inhibits the corrosion of the coating. However, it is imperative to further improve the internal structure to address existing defects and stress-induced pitting. Electrochemical results indicate that PEO coatings doped with zinc phosphate exhibit exemplary corrosion resistance, as no noticeable pitting behavior was observed even after 4200 h of immersion in neutral salt spray, as shown in Figure 19 [48]. Given the specific requirement of an alkaline environment for the PEO electrolyte of Mg alloy, there are limited doping elements that can generate insoluble phosphate phases in situ. In the future, valuable insights and knowledge can be derived from these studies to further advance research in this area.

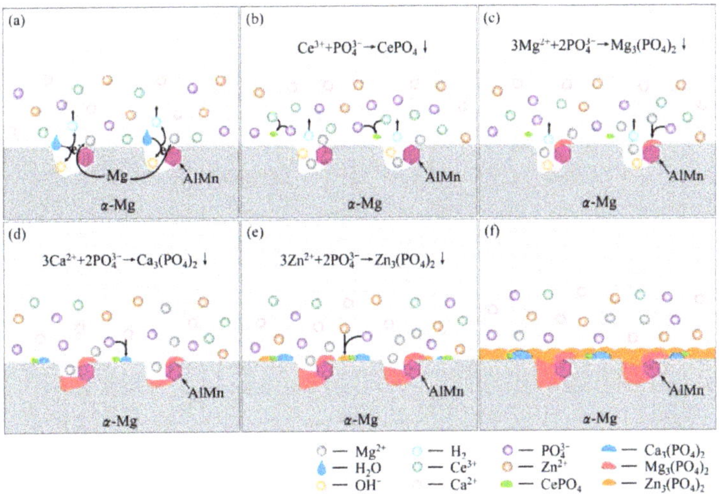

Figure 18. Schematic diagram of formation of calcium phosphate and zinc phosphate on Mg alloy: (**a**) dissolution of α-Mg matrix neighboring AlMn-phase and hydrogen release, (**b**) formation of the nucleus of cerium phosphate, (**c**) nuclei formation of magnesium phosphate, (**d**) formation of calcium phosphate, (**e**) formation of the zinc phosphate nucleus, and (**f**) coalescence of zinc phosphate nuclei and formation of crystalline zinc phosphate coating [54] (Reprinted/adapted with permission from Elsevier, 2022).

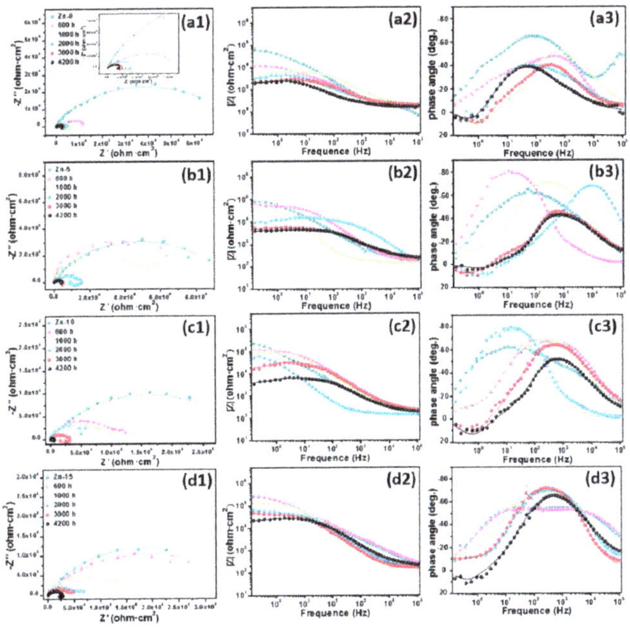

Figure 19. EIS curves for the PEO coatings after different salt spraying times with a 5 wt% NaCl solution: (**a**) Zn-0, (**b**) Zn-5, (**c**) Zn-10, and (**d**) Zn-15 [48] (Reprinted/adapted with permission from Elsevier, 2016).

Additionally, the inclusion of aluminate in the phosphate system results in improved corrosion resistance. This is evidenced by a reduction in the MgO content within the coating and the increased formation of the insoluble compound $MgAl_2O_4$. Figure 20 illustrates that as the aluminum content rises, the ratio of MgO to $Mg_3(PO_4)_2$ in the coating decreases. However, the variance in the $Mg_3(PO_4)_2$ content is not considered to be significant [55].

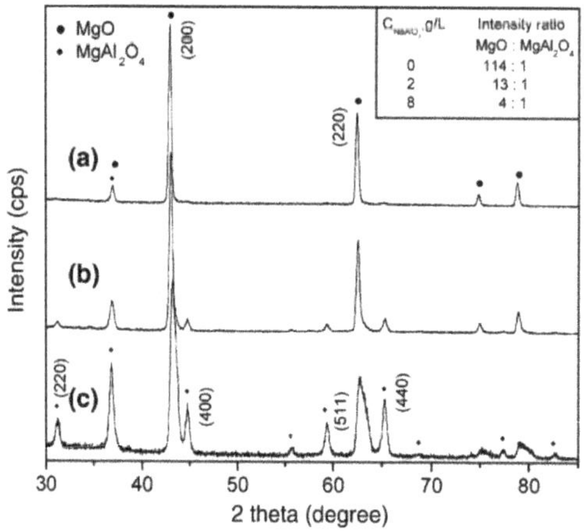

Figure 20. XRD of PEO coating containing aluminate: (**a**) 0 g/L, (**b**) 2 g/L, and (**c**) 8 g/L [55] (Reprinted/adapted with permission from Elsevier, 2005).

5. Prospect

Over the past 30 years, research on corrosion-resistant PEO coatings on Mg alloy has made significant progress; however, there are still some issues in the theoretical research and application of PEO technology that need further improvement.

(1) PEO mechanism. PEO technology is a complex and comprehensive technology that involves various disciplines such as chemistry, electrochemistry, plasma chemistry, and thermochemistry. While researchers have proposed some discharge processes and film-forming mechanisms of PEO technology, there are limitations, and the specific discharge and film-forming processes of PEO still cannot be clearly explained. Therefore, it is necessary to analyze and improve the theoretical model of PEO from an energy perspective, combining thermodynamic and kinetic theories, as well as advanced spectroscopic methods, in order to better design corrosion-resistant coatings.

(2) Electrolyte and electrical parameters. The selection of electrolyte directly affects the coating growth process on the Mg alloy. For example, phosphate system coatings grow inward, while silicate coatings maintain outward growth. Additives also affect the discharge process, resulting in coatings with different characteristics. The choice of electrical parameters affects the strength of discharge and the speed of film formation. Therefore, it is necessary to establish a large database of PEO coatings on Mg alloy and continuously share research theories and results of existing electrolytes and electrical parameters to improve the reference value of relevant data and refine more effective electrolyte systems for preparing corrosion-resistant coatings.

(3) Power supply and energy consumption. Coatings prepared using bipolar pulse power sources are believed to be uniform and dense, resulting in optimal coating performance. Additionally, they can reduce the energy consumption of PEO processes. However, achieving uniform preparation of large samples is still a challenge, with the current maximum treatment area being only 4 m^2. To meet the requirements of preparing uniform corrosion-resistant coatings on a large scale, it is necessary to design and develop an efficient, energy-saving, high-power power supply system suitable for large-area processing.

(4) Strengthening of corrosion-resistant coatings. Various enhancement methods, including improving the crystallinity, density, thickness, roughness, micropores, and cracks of coatings, as well as the study of corrosion-resistant phases, are widely researched. There is relatively little research on the preparation of insoluble metal salts on Mg alloys, such as zinc phosphate and calcium phosphate, which have extremely low solubility product constants. It is necessary to uniformly mix these insoluble metal salts in PEO coatings on Mg alloys to significantly improve the corrosion resistance.

Author Contributions: Conceptualization, C.Y. and P.C.; methodology, W.W.; software, C.Y.; validation, C.Y., P.C. and L.S.; formal analysis, Y.Z. and W.W.; investigation, P.C. and W.W.; resources, C.Y.; data curation, C.Y.; writing—original draft preparation, C.Y.; writing—review and editing, P.C. and P.K.C.; visualization, P.K.C.; supervision, L.S.; project administration, P.C.; funding acquisition, C.Y. All authors have read and agreed to the published version of the manuscript.

Funding: This study was financially supported by the National funded postdoctoral researcher program (grant No. GZC20231545) and the China Postdoctoral Science Foundation (grant No. 2023M742224).

Informed Consent Statement: Not applicable.

Data Availability Statement: Data are contained within the article.

Conflicts of Interest: The authors declare no conflicts of interest.

References

1. Parande, G.; Manakari, V.; Sharma Kopparthy, S.D.; Gupta, M. A study on the effect of low-cost eggshell reinforcement on the immersion, damping and mechanical properties of Mg–zinc alloy. *Compos. Part B Eng.* **2020**, *182*, 107650. [CrossRef]
2. Srikanth, N.; Zhong, X.L.; Gupta, M. Enhancing damping of pure Mg using nano-size alumina particulates Materials. *Letters* **2005**, *59*, 3851–3855.
3. Hamid, Z.A.; El-khair, M.T.A.; Hassan, H.B. Synthesis and protection of AM50 Mg alloy and its composites using environmentally pretreatment electrolyte. *Surf. Coat. Technol.* **2011**, *206*, 1041–1050. [CrossRef]
4. Shao, Z.; Cai, Z.; Hu, R.; Wei, S. The study of electroless nickel plating directly on Mg alloy. *Surf. Coat. Technol.* **2014**, *249*, 42–47. [CrossRef]
5. Ji, R.; Peng, G.; Zhang, S.; Li, Z.; Li, J.; Fang, T.; Zhang, Z.; Wang, Y.; He, Y.; Wu, J. The fabrication of a CeO_2 coating via cathode plasma electrolytic deposition for the corrosion resistance of AZ31 Mg alloy. *Ceram. Int.* **2018**, *44*, 19885–19891. [CrossRef]
6. Ono, S.; Moronuki, S.; Mori, Y.; Koshi, A.; Liao, J.; Asoh, H. Effect of Electrolyte Concentration on the Structure and Corrosion Resistance of Anodic Films Formed on Mg through Plasma Electrolytic Oxidation. *Electrochim. Acta* **2017**, *240*, 415–423. [CrossRef]
7. Kuo, Y.-L.; Chang, K.-H. Atmospheric pressure plasma enhanced chemical vapor deposition of SiO_x films for improved corrosion resistant properties of AZ31 Mg alloys. *Surf. Coat. Technol.* **2015**, *283*, 194–200. [CrossRef]
8. Saei, E.; Ramezanzadeh, B.; Amini, R.; Kalajahi, M.S. Effects of combined organic and inorganic corrosion inhibitors on the nanostructure cerium based conversion coating performance on AZ31 Mg alloy: Morphological and corrosion studies. *Corros. Sci.* **2017**, *127*, 186–200. [CrossRef]
9. Huang, Q.; Liu, L.; Wu, Z.; Ji, S.; Chu, P.K. Corrosion-resistant plasma electrolytic oxidation coating modified by Zinc phosphate and self-healing mechanism in the salt-spray environment. *Surf. Coat. Technol.* **2020**, *384*, 125321. [CrossRef]
10. Luo, D.; Liu, Y.; Yin, X.; Wang, H.; Han, Z.; Ren, L. Corrosion inhibition of hydrophobic coatings fabricated by micro-arc oxidation on an extruded Mge5Sne1Zn alloy substrate. *J. Alloys Compd.* **2018**, *731*, 731–738. [CrossRef]
11. Li, Z.; Ren, Q.; Wang, X.; Kuang, Q.; Ji, D.; Yuan, R.; Jing, X. Effect of phosphate additive on the morphology and anti-corrosion performance of plasma electrolytic oxidation coatings on Mg-lithium alloy. *Corros. Sci.* **2019**, *157*, 295–304. [CrossRef]
12. Verdier, S.; Boinet, M.; Maximovitch, S.; Dalard, F. Formation, structure and composition of anodic films on AM60 Mg alloy obtained by DC plasma anodizing. *Corros. Sci.* **2005**, *47*, 1429–1444. [CrossRef]
13. Durdu, S.; Aytaç, A.; Usta, M. Characterization and corrosion behavior of ceramic coating on Mg by micro-arc oxidation. *J. Alloys Compd.* **2011**, *509*, 8601–8606. [CrossRef]
14. Liu, X.Y.; Luan, B.L. Corrosion and wear properties of PEO coatings formed on AM60B alloy in $NaAlO_2$ electrolytes. *Appl. Surf. Sci.* **2011**, *257*, 9135–9141. [CrossRef]
15. Duan, H.; Yan, C.; Wang, F. Effect of electrolyte additives on performance of plasma electrolytic oxidation films formed on Mg alloy AZ91D. *Electrochim. Acta* **2007**, *52*, 3785–3793. [CrossRef]
16. Gu, Y.; Bandopadhyay, S.; Chen, C.-F.; Guo, Y.; Ning, C. Effect of oxidation time on the corrosion behavior of micro-arc oxidation produced AZ31 Mg alloys in simulated body fluid. *J. Alloys Compd.* **2012**, *543*, 109–117. [CrossRef]
17. Zhang, X.; Zhang, Y.; Lv, Y.; Dong, Z.; Hashimoto, T.; Zhou, X. Enhanced corrosion resistance of AZ31 Mg alloy by one-step formation of PEO/Mg-Al LDH composite coating. *Corros. Commun.* **2022**, *6*, 67–83. [CrossRef]

18. Guo, H.X.; Ying, M.A.; Wang, J.S.; Wang, Y.S.; Hao, Y. Corrosion behavior of micro-arc oxidation coating on AZ91D Mg alloy in NaCl solutions with different concentrations. *Trans. Nonferrous Met. Soc. China* **2012**, *22*, 1786–1793. [CrossRef]
19. Liang, J.; Srinivasan, P.B.; Blawert, C.; Störmer, M.; Dietzel, W. Electrochemical corrosion behaviour of plasma electrolytic oxidation coatings on AM50 Mg alloy formed in silicate and phosphate based electrolytes. *Electrochim. Acta* **2009**, *54*, 3842–3850. [CrossRef]
20. Hussein, R.; Northwood, D.; Nie, X. Processing-Microstructure Relationships in the Plasma Electrolytic Oxidation (PEO) Coating of a Magnesium Alloy. *Mater. Sci. Appl.* **2014**, *5*, 124–139. [CrossRef]
21. Dehnavi, V.; Binns, W.J.; Noël, J.J.; Shoesmith, D.W.; Luan, B.L. Growth behaviour of low-energy plasma electrolytic oxidation coatings on a Magnesium alloy. *J. Magnes. Alloys* **2018**, *6*, 229–237. [CrossRef]
22. Hussein, R.O.; Nie, X.; Northwood, D.O. A Yerokhin and A Matthews, Spectroscopic study of electrolytic plasma and discharging behaviour during the plasma electrolytic oxidation (PEO) process. *J. Phys. D Appl. Phys.* **2010**, *43*, 105203. [CrossRef]
23. Krishna, L.R.; Poshal, G.; Jyothirmayi, A.; Sundararajan, G. Relative hardness and corrosion behavior of micro arc oxidation coatings deposited on binary and ternary Mg alloys. *Mater. Des.* **2015**, *77*, 6–14. [CrossRef]
24. Rapheal, G.; Kumar, S.; Scharnagl, N.; Blawert, C. Effect of current density on the microstructure and corrosion properties of plasma electrolytic oxidation (PEO) coatings on AM50 Mg alloy produced in an electrolyte containing clay additives. *Surf. Coat. Technol.* **2016**, *289*, 150–164. [CrossRef]
25. Wu, D.; Liu, X.; Lu, K.; Zhang, Y.; Wang, H. Influence of $C_3H_8O_3$ in the electrolyte on characteristics and corrosion resistance of the microarc oxidation coatings formed on AZ91D Mg alloy surface. *Appl. Surf. Sci.* **2009**, *255*, 7115–7120. [CrossRef]
26. Kazanski, B.; Kossenko, A.; Zinigrad, M.; Lugovskoy, A. Fluoride ions as modifiers of the oxide layer produced by plasma electrolytic oxidation on AZ91D Mg alloy. *Appl. Surf. Sci.* **2013**, *287*, 461–466. [CrossRef]
27. Liu, F.; Shan, D.; Song, Y.; Han, E.-H. Effect of additives on the properties of plasma electrolytic oxidation coatings formed on AM50 Mg alloy in electrolytes containing K_2ZrF_6. *Surf. Coat. Technol.* **2011**, *206*, 455–463. [CrossRef]
28. Yoo, B.; Shin, K.R.; Hwang, D.Y.; Dong, H.L.; Dong, H.S. Effect of surface roughness on leakage current and corrosion resistance of oxide layer on AZ91 Mg alloy prepared by plasma electrolytic oxidation. *Appl. Surf. Sci.* **2010**, *256*, 6667–6672. [CrossRef]
29. Yang, C.; Sheng, L.Y.; Zhao, C.C.; Wu, D.; Zheng, Y.F. Regulating the ablation of nanoparticle-doped MAO coating on Mg alloy by MgF_2 passivation layer construction. *Mater. Lett.* **2024**, *355*, 135559. [CrossRef]
30. Chen, D.; Wang, R.; Huang, Z.; Wu, Y.; Zhang, Y.; Wu, G.; Li, D.; Guo, C.; Jiang, G.; Yu, S.; et al. Evolution processes of the corrosion behavior and structural characteristics of plasma electrolytic oxidation coatings on AZ31 Mg alloy. *Appl. Surf. Sci.* **2018**, *434*, 326–335. [CrossRef]
31. Lee, S.-J.; Do, L.H.T. Effects of copper additive on micro-arc oxidation coating of LZ91 Mg-lithium alloy. *Surf. Coat. Technol.* **2016**, *307*, 781–789. [CrossRef]
32. Han, H.; Wang, R.; Wu, Y.; Zhang, X.; Wang, D.; Yang, Z.; Su, Y.; Shen, D.; Nash, P. An investigation of plasma electrolytic oxidation coatings on crevice surface of AZ31 Mg alloy. *J. Alloys Compd.* **2019**, *811*, 152010. [CrossRef]
33. Zhao, J.; Xie, X.; Zhang, C. Effect of the Graphene Oxide Additive on the Corrosion Resistance of the Plasma Electrolytic Oxidation Coating of the AZ31 Mg Alloy. *Corros. Sci.* **2016**, *114*, 146–155. [CrossRef]
34. Tran, Q.-P.; Chin, T.-S.; Kuo, Y.-C.; Jin, C.-X.; Trung, T.; Tuan, C.V.; Dang, D.Q. Diamond powder incorporated oxide layers formed on 6061 Al alloy by plasma electrolytic oxidation. *J. Alloys Compd.* **2018**, *751*, 289–298. [CrossRef]
35. Liu, X.H.; Liu, L.; Dong, S.; Chen, X.B.; Dong, J.B. Towards dense corrosion-resistant plasma electrolytic oxidation coating on Mg-Gd-Y-Zr alloy by using ultra-high frequency pulse current. *Surf. Coat. Technol.* **2022**, *447*, 128881. [CrossRef]
36. Chen, W.-W.; Wang, Z.-X.; Sun, L.; Lu, S. Research of growth mechanism of ceramic coatings fabricated by micro-arc oxidation on Mg alloys at high current mode. *J. Magnes. Alloys* **2015**, *3*, 253–257. [CrossRef]
37. Dou, J.; Wang, C.; Gu, G.; Chen, C. Formation of silicon-calcium-phosphate-containing coating on Mg-Zn-Ca alloy by a two-step micro-arc oxidation technique. *Mater. Lett.* **2018**, *212*, 37–40. [CrossRef]
38. Cui, X.-J.; Liu, C.-H.; Yang, R.-S.; Li, M.-T.; Lin, X.-Z. Self-sealing micro-arc oxidation coating on AZ91D Mg alloy and its formation mechanism. *Surf. Coat. Technol.* **2015**, *269*, 228–237. [CrossRef]
39. Yang, W.; Wang, J.; Xu, D.; Li, J.; Chen, T. Characterization and formation mechanism of grey micro-arc oxidation coatings on Mg alloy. *Surf. Coat. Technol.* **2015**, *283*, 281–285. [CrossRef]
40. Liang, J.; Hu, L.; Hao, J. Improvement of corrosion properties of microarc oxidation coating on Mg alloy by optimizing current density parameters. *Appl. Surf. Ence* **2007**, *253*, 6939–6945. [CrossRef]
41. Zhang, Y.; Wu, Y.; Chen, D.; Wang, R.; Li, D.; Guo, C.; Jiang, G.; Shen, D.; Yu, S.; Nash, P. Micro-structures and growth mechanisms of plasma electrolytic oxidation coatings on aluminium at different current densities. *Surf. Coat. Technol.* **2017**, *321*, 236–246. [CrossRef]
42. Pezzato, L.; Rigon, M.; Martucci, A.; Brunelli, K.; Dabalà, M. Plasma Electrolytic Oxidation (PEO) as pre-treatment for sol-gel coating on aluminum and Mg alloys. *Surf. Coat. Technol.* **2019**, *366*, 114–123. [CrossRef]
43. Srinivasan, P.B.; Blawert, C.; Dietzel, W. Effect of plasma electrolytic oxidation treatment on the corrosion and stress corrosion cracking behaviour of AM50 Mg alloy. *Mater. Sci. Eng. A* **2008**, *494*, 401–406. [CrossRef]
44. Luo, H.; Cai, Q.; Wei, B.; Yu, B.; Li, D.; He, J.; Liu, Z. Effect of $(NaPO_3)_6$ concentrations on corrosion resistance of plasma electrolytic oxidation coatings formed on AZ91D Mg alloy. *J. Alloys Compd.* **2008**, *464*, 537–543. [CrossRef]

45. Li, S.; Wu, Y.; Yamamura, K.; Nomura, M.; Fujii, T. Improving the grindability of titanium alloy Ti–6Al–4V with the assistance of ultrasonic vibration and plasma electrolytic oxidation. *CIRP Ann.—Manuf. Technol.* **2017**, *66*, 345–348. [CrossRef]
46. An, L.; Ma, Y.; Liu, Y.; Sun, L.; Wang, S.; Wang, Z. Effects of additives, voltage and their interactions on PEO coatings formed on Mg alloys. *Surf. Coat. Technol.* **2018**, *354*, 226–235. [CrossRef]
47. Liang, J.; Hu, L.; Hao, J. Characterization of microarc oxidation coatings formed on AM60B Mg alloy in silicate and phosphate electrolytes. *Appl. Surf. Sci.* **2007**, *253*, 4490–4496. [CrossRef]
48. Zeng, R.-C.; Hu, Y.; Zhang, F.; Huang, Y.-D.; Wang, Z.-L.; Li, S.-Q.; Han, E.-H. Corrosion resistance of cerium-doped zinc calcium phosphate chemical conversion coatings on AZ31 Mg alloy. *Trans. Nonferrous Met. Soc. China* **2016**, *26*, 472–483. [CrossRef]
49. Li, H.; Lu, S.; Qin, W.; Wu, X. In-situ grown MgO-ZnO ceramic coating with high thermal emittance on Mg alloy by plasma electrolytic oxidation. *Acta Astronaut.* **2017**, *136*, 230–235. [CrossRef]
50. Qin, W.; Li, H.; Lu, S.; Wu, X. Influence of Zr^{4+} ions on solar absorbance and emissivity of coatings formed on AZ31 Mg alloy by plasma electrolytic oxidation. *Surf. Coat. Technol.* **2015**, *269*, 220–227.
51. Tu, W.; Cheng, Y.; Wang, X.; Zhan, T.; Han, J.; Cheng, Y. Plasma electrolytic oxidation of AZ31 Mg alloy in aluminate-tungstate electrolytes and the coating formation mechanism. *J. Alloys Compd.* **2017**, *725*, 199–216. [CrossRef]
52. Lu, X.; Blawert, C.; Mohedano, M.; Scharnagl, N.; Zheludkevich, M.L.; Kainer, K.U. Influence of electrical parameters on particle uptake during plasma electrolytic oxidation processing of AM50 Mg alloy. *Surf. Coat. Technol.* **2016**, *289*, 179–185. [CrossRef]
53. Yang, C.; Huang, J.; Cui, S.; Fu, R.; Sheng, L.; Xu, D.; Tian, X.; Zheng, Y.; Chu, P.K.; Wu, Z.Z. NaF assisted preparation and the improved corrosion resistance of high content ZnO doped plasma electrolytic oxidation coating on AZ31B alloy. *J. Magnes. Alloys*, 2023; *in press*. [CrossRef]
54. Yang, C.; Cai, H.; Cui, S.H.; Huang, J.; Zhu, J.Y.; Wu, Z.C.; Ma, Z.Y.; Fu, R.K.Y.; Sheng, L.Y.; Tian, X.B.; et al. A zinc-doped coating prepared on the Mg alloy by plasma electrolytic oxidation for corrosion protection. *Surf. Coat. Technol.* **2022**, *433*, 128148. [CrossRef]
55. Liang, J.; Guo, B.; Tian, J.; Liu, H.; Zhou, J.; Liu, W.; Xu, T. Effects of $NaAlO_2$ on structure and corrosion resistance of microarc oxidation coatings formed on AM60B Mg alloy in phosphate–KOH electrolyte. *Surf. Coat. Technol.* **2005**, *199*, 121–126. [CrossRef]

coatings

MDPI

Article

Anticorrosion Method Combining Impressed Current Cathodic Protection and Coatings in Marine Atmospheric Environment

Peichang Deng [1], Juyu Shangguan [2], Jiezhen Hu [2,*], Huan Huang [2] and Lingbo Zhou [2]

[1] College of Chemistry and Environment, Guangdong Ocean University, Zhanjiang 524088, China
[2] College of Mechanical Engineering, Guangdong Ocean University, Zhanjiang 524088, China
[*] Correspondence: jiezhen0520@163.com

Abstract: In this study, a new anticorrosion method combining impressed current cathodic protection (ICCP) with coatings that can be applied to marine atmospheric environments is proposed. As the corrosion medium fills the cracks and pores of the coating, an electrolyte film layer is inevitably formed on the metal surface. Therefore, a graphene conductive coating with excellent chemical inertness and shielding performance is selected as the intermediate coating to form an electrolytic cell system with a metal substrate serving as the cathode and a graphene coating serving as the auxiliary anode. By studying the surface corrosion morphologies and electrochemical signals of the coating samples at different protection potentials and coating thicknesses, the optimal potential is determined to be 0.6 V, and the optimal coating thickness is determined to be 20 μm. The samples protected by the joint method have lower corrosion rates and better anticorrosion performance than those protected by the coatings alone.

Keywords: anticorrosion method; impressed current cathodic protection; coatings; marine atmospheric environment; graphene conductive coating

1. Introduction

To resist marine corrosion to the environment, marine structures use a variety of corrosion control methods based on coating protection methods [1,2]. Coatings, as the primary option for steel structure corrosion protection, effectively delay metal corrosion by covering the metal surface with a protective film, which acts as a barrier against external media. However, the presence of nanoscale pores in the coating itself and of micrometer- and millimeter-sized pores formed during painting and use can increase the air and water permeability of the coating [3,4]. The thin liquid film on the surface gradually penetrates the coatings, leading to failure of the coating and the inability to provide long-term protection [5]. Moreover, ICCP is a widely used anticorrosion technology in marine engineering that is mainly used to protect ships, pipelines, reinforced concrete, and other structures from corrosion [6–8]. ICCP has the advantages of a low cost, remote monitoring ability, and greenness/environmental protection ability. However, due to the lack of macroscopic continuous conductive electrolytes in the atmosphere, ICCP is rarely used in marine atmospheric environments [9]. Therefore, overcoming the limitations of individual protective techniques and meeting the diverse requirements remain pressing issues in the field of marine engineering.

Combining ICCP with coatings in an atmospheric environment and in a seawater environment can inhibit the generation of corrosion products in coating defects. These defects can delay coating failure, thereby improving the corrosion protection effect of marine engineering equipment in the atmospheric environment. Against this background, researchers have investigated combined corrosion prevention techniques. Refait [10] studied coupons alternately immersed in seawater and exposed to the atmosphere placed in the tidal zone. He attempted to create a localized environment to maintain the liquid electrolyte around the metal. However, this method is difficult to apply on a large scale in the atmospheric environment and is similar to traditional cathodic protection techniques. The key to achieving cathodic protection in the atmospheric environment is to create an ionic conductive state. Ge [11] applied a specifically formulated coating containing the solid electrolyte sodium alumina as the ionically conducting component. The coating not only isolates metal substrates from the atmospheric environment but also provides an electrical path for impressed currents. However, the preparation of coatings is relatively complex, which has prevented this technology from becoming widespread.

This paper proposes an anticorrosion method to enhance the corrosion protection of marine engineering equipment in the atmospheric environment. This approach can overcome the limitations of traditional ICCP and compensate for the weaknesses of coatings, which are not able to provide effective long-term protection. This protection system is based on the principle of using a microporous channel-filled electrolyte to combine ICCP with coatings. Graphene is composed of sp^2-bonded carbon atoms densely packed in a honeycomb lattice. It presents a variety of properties, such as gas impermeability, chemical (acid/base/salt) resistance, a large aspect ratio, thermal stability, eco-friendliness, surface flexibility, and excellent mechanical properties [12]. Thus, graphene powder is chosen as the functional filler to prepare a graphene conductive coating. The graphene coating, as the intermediate coating of the sandwich composite coatings, has great anticorrosion performance, which is mainly due to the following two aspects. On the one hand, the graphene lattice can form a corrosion barrier on the metal surface, which can effectively prevent the penetration of corrosive media, thereby improving the corrosion resistance and the lifespan of the coating [13–15]. On the other hand, when coating cracks and pores appear, the corrosive medium permeates through the pores to the protected substrate surface, forming an electrolyte membrane layer [16,17]. This layer provides a pathway between the metal substrate and graphene conductive coating [18]. By applying ICCP to the substrate, an electrolytic cell system is formed, with the protected metal substrate serving as the cathode and graphene serving as the auxiliary anode. Since the auxiliary anodes are uniformly distributed on the metal surface, the potential distribution is relatively uniform. Cathodic protection can also be achieved with less electrolyte.

In this study, the anticorrosion performance of this joint method is evaluated by analyzing the surface corrosion morphologies and electrochemical signals of scratched and scratch-free samples exposed to a marine atmosphere for 20 days. Moreover, the coating degradation behaviors and metal corrosion states under different protection potential and coating thickness parameters are investigated to understand and improve the protection mechanisms of the novel method.

2. Materials and Methods

2.1. Materials and Specimens

Graphene powder with a diameter of 3~15 μm was purchased from Shenzhen Juzhan Materials Technology Co., Ltd. (Shenzhen, China). Waterborne epoxy resin and deionized water were supplied by Guangdong Kecheng Laboratory Equipment Co., Ltd. (Zhanjiang, China). Epoxy universal primer and polyurethane antifouling topcoat were supplied by Fuzhou Allgall Technology Co., Ltd. (Fuzhou, China).

The values of the corrosion development trend constants of Q235 carbon steel ranged from 0.7 to 1.5. It was greatly affected by environmental changes and could intuitively demonstrate the influences of changing the protection mechanism on the corrosion of the metal. Thus, the 150 mm × 100 mm × 5 mm long Q235 carbon steel plate (Table 1) was used as the metal material for research in this study.

Table 1. Chemical composition of material.

	Element wt (%)							
Material	C	Si	Mn	P	S	Cr	Ni	Cu
Q235	0.15	0.19	0.5	0.023	0.015	≤0.3	≤0.3	≤0.3

First, the Q235 carbon steel was sequentially wet abraded with 280#, 500#, 800#, and 1000# grit carbide metallographic sandpapers, cleaned with anhydrous ethanol and acetone, and then air-dried. Next, the wire was welded to the substrate as a cathode lead. Then, the surface was evenly coated with the epoxy universal primer, graphene composite coating, and polyurethane antifouling topcoat. The graphene composite coating was manufactured by mixing graphene powder, waterborne epoxy resin, and deionized water at a ratio of 1:1:2. After mixing and stirring, the graphene powder was completely dispersed through ultrasonic waves, as shown in Figure 1. When the graphene coating was not dry, a conductive carbon fiber was attached to one side of the layer, and a wire was connected to the other side as an anode lead. Finally, a power supply was connected in series between the substrate and the graphene coating.

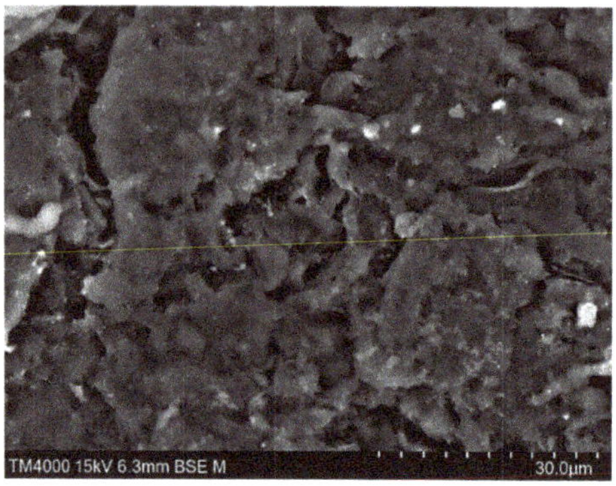

Figure 1. SEM image of graphene/epoxy coating.

Because the coating always has a certain air permeability and water permeability, it cannot be completely isolated from the outside world [19,20]. The permeation of the electrolytes allows for the formation of pathways between the metal substrate and the graphene coating. Figure 2 shows the structure and working principle of the joint method.

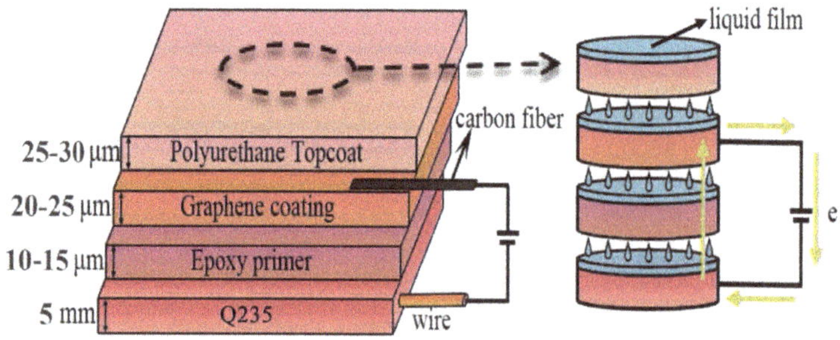

Figure 2. Structure and principle of the joint method.

2.2. Marine Atmospheric Exposure Site

Figure 3 shows that the samples were placed in the exposure field of Zhanjiang city, Donghai Island, Guangdong Province. A 20-day experiment in a marine atmospheric environment was conducted. Zhanjiang city is located in the southernmost part of mainland China, with a longitude of 109°31′~110°55′ E and a latitude of 20°~21°35′ N. This city includes the entire Leizhou Peninsula and part of the northern peninsula. The climate of this city is a maritime monsoon climate, with an average annual temperature of approximately 22.8 °C and an annual rainfall of approximately 2000 mm.

Figure 3. Experimental site.

2.3. Electrochemical Characterization

The electrochemical tests used a traditional three-electrode system (Figure 4). In this system, we used a silver chloride electrode serving as the reference electrode, a 2 cm × 2 cm platinum plate serving as the counter electrode, a sample serving as the working electrode, and a 3.5 wt% NaCl solution serving as the electrolyte. The test area was 19.6 mm^2.

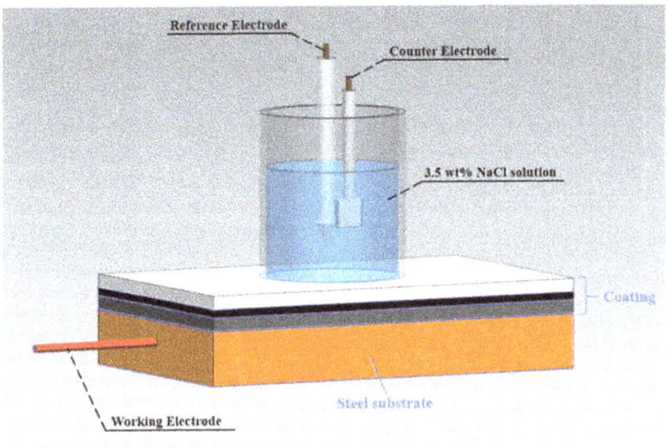

Figure 4. Schematic diagram of the coating sample test system.

EIS was carried out at the open-circuit potential (OCP), and the amplitude of the AC excitation signal was 10 mV in the frequency range of 10^5~10^{-2} Hz. Then, the acquired EIS data were fitted with an equivalent circuit using Zview 2.7 software. To minimize the influences of the surrounding environment on the measurement signals, the samples were placed in a Faraday cage during the measurement process.

The open-circuit potential test time was 15 min. The scanning interval for the potentiodynamic polarization test was 0.05 V with a scanning rate of 0.1 mV/s.

3. Results and Discussion

3.1. Determination of the Optimum Protection Potential

In this experiment, an impressed voltage system of two coating samples connected in series was adopted, and one coating sample surface was carved with an X-shaped scratch. A potentiometer was connected in series in this circuit to adjust the voltage. The voltages of each group of circuits were fixed at 0 V, 0.6 V, 1.2 V, and 1.8 V, indicating that the impressed protection voltages of a single coating sample were 0 V, 0.3 V, 0.6 V, and 0.9 V. After the samples were retrieved, the surface morphologies of the scratch samples were recorded with the infinite-focus three-dimensional microscope and camera. Then, the impedances of the scratch-free samples were recorded by an electrochemical workstation and Tait cell kit.

Figure 5 shows the surface corrosion morphologies of the scratched samples before and after peeling off the coating under different protection voltages. The impressed voltages of the samples from left to right were 0 V, 0.3 V, 0.6 V, and 0.9 V. According to the image before peeling off the coating, the corrosion products that accumulated at the scratches of the sample with an impressed voltage of 0.6 V were the least abundant and did not spread to the coating. The corrosion degree of the sample with an impressed voltage of 0.3 V was the second highest. For the samples with impressed voltages of 0 V and 0.9 V, the corrosion products were highly enriched on the scratch. The corrosion products of the samples in the marine atmospheric environment diffuse with the aid of water, so there were many obvious corrosion products on the coating surfaces around the scratches. After the coating was stripped by a muffle furnace at a high temperature, the corrosion of the substrate under the coating could be detected by the infinite-focus three-dimensional microscope. The samples without an impressed voltage not only exhibited severe corrosion at the scratch but the substrate also had uniform corrosion. The corrosion rust spots were uniformly distributed over a large area of the substrate. Obviously, when the impressed voltage was 0 V or 0.9 V, the surface of the sample substrate had the most corrosion marks, followed by the surface of the sample with an impressed voltage of 0.3 V. The surface of the sample with an impressed voltage of 0.6 V had the fewest corrosion marks. This finding demonstrated that the best protective effect of the substrate was approximately 0.6 V for the scratched coating sample.

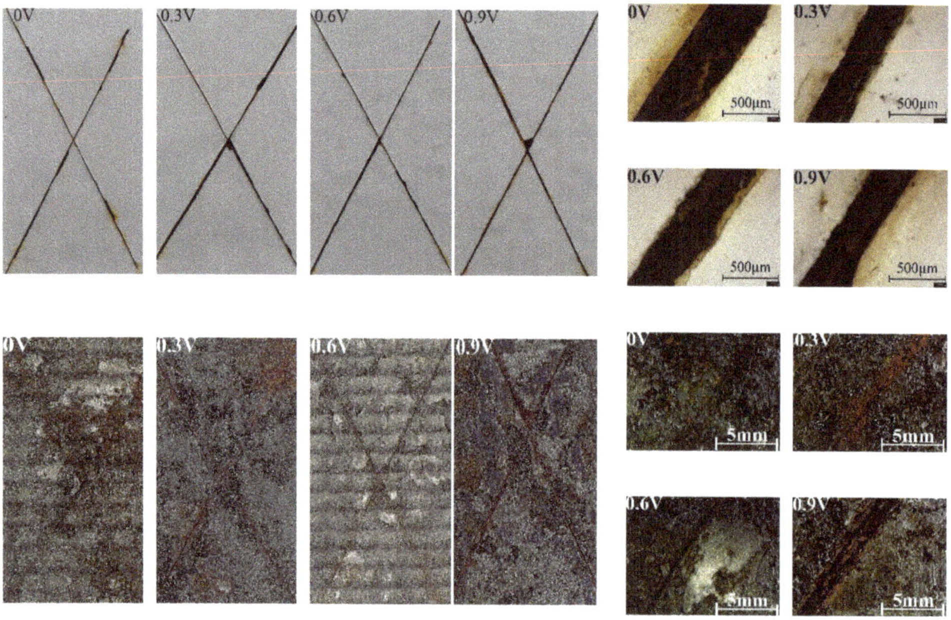

Figure 5. Surface morphologies of the scratched samples before and after peeling off the coating under different protection voltages.

Figure 6 shows the EIS images of the scratch-free samples under different protection voltages. In the proposed system, the metal substrate and the graphene conductive coating underwent galvanic corrosion under impressed voltage. Although the metal substrate received electrons, the graphene coating used as an auxiliary anode suffered some deterioration during use, resulting in an overall reduction in impedance [21]. The Nyquist diagram (Figure 6a) shows that the impedances of the samples with impressed voltages were lower than the sample without impressed voltages. Moreover, the impedance of the impressed voltage samples did not linearly change with the impressed voltage. When the impressed voltage was 0.6 V, the impedance was closest to the sample without impressed voltages and much greater than the samples with impressed voltages of 0.6 V and 0.9 V. Thus, a high impressed voltage effectively protected the metal substrate better and increased the degradation of the graphene coating.

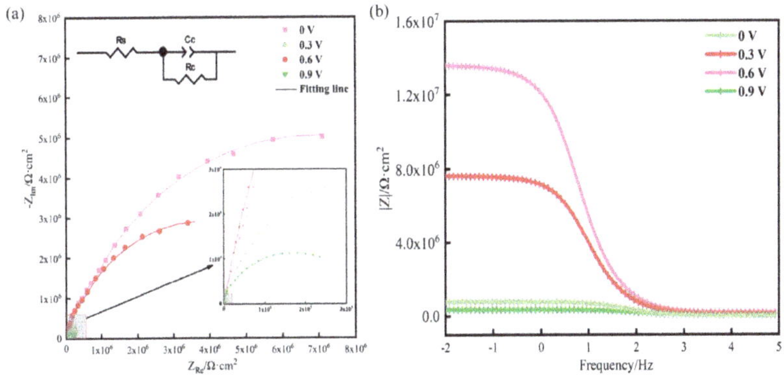

Figure 6. Nyquist (**a**) and Bode (**b**) impedance plots of scratch-free samples under different protection voltages.

Typically, the impedance at a low frequency (|Z|0.01 Hz) is one of the main parameters used to evaluate the corrosion resistance of a coating sample. This phenomenon is directly related to the corrosion resistance of the coating sample. A high impedance would indicate an enhanced corrosion resistance of the coating. According to the Bode diagram (Figure 6b), when the frequency was 0.01 Hz, the barrier ability decreased in the following order: 0.6 V > 0.3 V > 0 V > 0.9 V. Thus, the most suitable voltage was approximately 0.6 V when the coating was intact. This voltage would not damage the coating protection technology and would cause the least interference.

3.2. Determination of the Optimal Graphene Coating Thickness

In this experiment, the thicknesses of the graphene coatings in the five coating samples were controlled to 10 μm, 15 μm, 20 μm, 25 μm, and 30 μm. The dry film thickness of each coating was measured and averaged over five measurements to ensure that the margin of error was maintained at ±2 μm. Each coating sample was carved with an X-shaped scratch. Then, the samples were split into two groups, one with an impressed voltage of 0.6 V and the other without impressed voltage. The samples were tested for their open-circuit potential, and potentiodynamic polarization testing was performed by an electrochemical workstation and the Tait cell kit after retrieval.

Figure 7 shows the open-circuit potentials of the scratched samples with different coating thicknesses. In general, the greater the risk of water and corrosive media diffusing into the interface between the coating and the metal, the more susceptible the coating is to corrosion. The open-circuit potentials of the samples with impressed voltages were greater than the samples without impressed voltages. Thus, it can be observed that the presence of ICCP could enhance the corrosion resistance of the coating samples, and the novel anticorrosion method could function in a marine atmospheric environment. In addition, the figure shows that the overall open-circuit potential tended to increase as the thickness of the graphene coating increased. At thicknesses of 10~20 μm, the open-circuit potential moved significantly in a positive direction. At thicknesses of 20~30 μm, the change in the open-circuit potential became flat. Therefore, it was not possible to improve the protective performance of the combined anticorrosion system by further increasing the thickness.

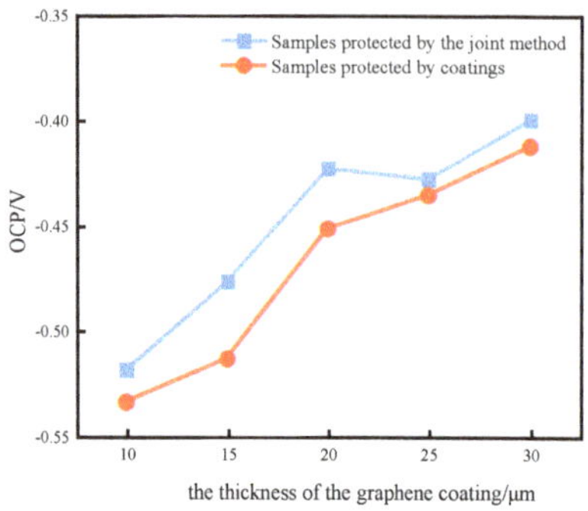

Figure 7. Open-circuit potentials of the scratched coating samples with different thicknesses.

Figure 8 shows the potentiodynamic polarization curves and the corrosion current density (I_{corr}) variations. As the thickness of the graphene coating increased, the I_{corr} values of the samples without impressed voltages increased, ranging from 1.36×10^{-4} mA/cm^2 to 9.87×10^{-4} mA/cm^2, while the I_{corr} values of the samples with impressed voltages significantly decreased, ranging from 2.15×10^{-5} mA/cm^2 to 5.88×10^{-5} mA/cm^2. Because the graphene coating effectively inhibited corrosion, its atomic lattice could prevent the intercalation of oxidizing species and the penetration of corrosive media by virtue of its impermeability [22]. The thicker the coating was, the more difficult it was for corrosion products to reach the metal substrate. As a result, the I_{corr} tended to decrease overall. However, if defects were present, the graphene layer lost the benefit of shielding protection [23]. Cathodic protection delivered a current to the scratched area in this case, reducing the number of active sites or forming a protective film. Moreover, the electrolyte membrane layer, which was present in the coating cracks and pores, protected the scratch-free site. As a result, the metal at the scratch had a low corrosion rate and was in a protected state [24]. Overall, the samples with impressed voltages had better corrosion resistance than the samples without impressed voltages.

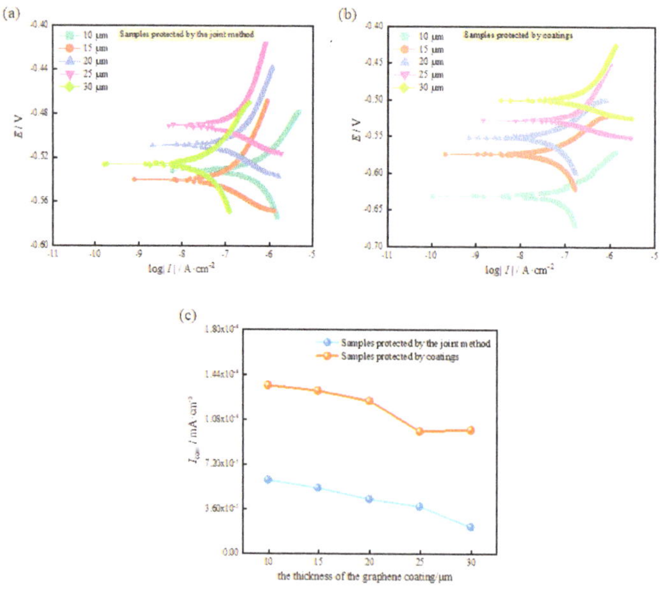

Figure 8. Potentiodynamic polarization curves (**a**,**b**) and corrosion current density variations (**c**) in scratched coating samples at different thicknesses.

4. Conclusions

(1) In this paper, a novel corrosion prevention method that effectively combines the advantages of coatings and ICCP (impressed current cathodic protection) technology is proposed. The method constructs a sandwich composite coating system. The microscopic electrolyte membrane layer provides a pathway between the metal substrate and the conductive graphene coating, allowing ICCP to improve the corrosion protection of equipment in the marine atmospheric environment.

(2) Wherever the coatings are stripped before and after, the scratched sample with an impressed voltage of 0.6 V has the least corrosion, while the samples with impressed voltages of 0 V and 0.9 V have the most severe corrosion. The graphene conductive coating serving as the auxiliary anode suffers to an extent with increasing impressed voltage. However, the resistance of the scratch-free sample with an impressed voltage of 0.6 V is similar to the sample without an impressed voltage, and the protection effect is the greatest. The results show that 0.6 V can effectively protect the carbon steel substrate for a long time.

(3) As the thickness of the graphene coating increases, it becomes increasingly difficult for the corrosion products to reach the metal substrate; thus, the open-circuit potentials shift in the positive direction. Moreover, the scratched samples with impressed voltages have higher open-circuit potentials and lower corrosion current densities than those without impressed voltages. This further confirms that the method provides excellent corrosion protection. Considering the efficiency of corrosion prevention methods and the cost of application, the protective characteristics of the combined anticorrosion system are ensured by the optimal thickness of the graphene coating of 20 μm.

Author Contributions: Conceptualization, P.D. and J.H.; methodology, P.D.; software, H.H. and L.Z.; validation, J.S., H.H. and L.Z.; formal analysis, J.S.; investigation, L.Z.; resources, J.H.; data curation, J.S. and L.Z.; writing—original draft preparation, J.S.; writing—review and editing, J.S.; visualization, P.D.; supervision, H.H.; project administration, J.H.; funding acquisition, J.H. All authors have read and agreed to the published version of the manuscript.

Funding: The authors were sincerely grateful for the financial support from the Natural Science Foundation of Guangdong Province (No. 2021A1515012129) and the Science & Technology Development Foundation of Zhanjiang (No. 2022A01029).

Institutional Review Board Statement: Not applicable.

Informed Consent Statement: Not applicable.

Data Availability Statement: Data are contained within the article.

Conflicts of Interest: The authors declare no conflicts of interest.

References

1. Li, X.; Zhang, D.; Liu, Z.; Li, Z.; Du, C.; Dong, C. Materials science: Share corrosion data. *Nature* **2015**, *527*, 441–442. [CrossRef] [PubMed]
2. Deshpande, P.; Kolekar, A.; Bhopale, A.; Kalendova, A.; Kohl, M. ICCP of mild steel in association with zinc based paint coating. *Mater. Today Proc.* **2022**, *50*, 1660–1665. [CrossRef]
3. Wu, W.; Zeng, Z.; Cheng, X.; Li, X.; Liu, B. Atmospheric corrosion behavior and mechanism of a Ni-Advanced weathering steel in simulated tropical marine environment. *J. Mater. Eng. Perform.* **2017**, *26*, 6075–6086. [CrossRef]
4. Price, J.S.; Figueira, B.R.; Soucek, D.M. Corrosion protection systems and fatigue corrosion in offshore wind structures: Current status and future perspectives. *Coatings* **2017**, *7*, 25. [CrossRef]
5. Zhang, H.; Sun, W.; Wang, L.; Wang, J.; Wang, S.; Liu, G. A mechanically and chemically stable superhydrophobic coating for preventing marine atmospheric corrosion. *Surf. Interfaces* **2021**, *27*, 101537. [CrossRef]
6. Mitolo, M.; Pettinger, A. Interactions between cathodically protected pipelines and grounding systems. *IEEE Trans. Ind. Appl.* **2016**, *52*, 3694–3698. [CrossRef]
7. Guo, B.; Qiao, G.; Li, D.; Ou, J. Multi-species reactive transport modeling of electrochemical corrosion control in saturated concrete structures including electrode reactions and thermodynamic equilibrium. *Constr. Build. Mater.* **2021**, *278*, 122228. [CrossRef]

8. Feng, R.; Wang, J.; Zhu, J.; Dong, Z. Fatigue behavior of corroded reinforced concrete continuous beams with multi-intervention system. *Eng. Struct.* **2021**, *231*, 111748. [CrossRef]
9. Anwar, S.M.; Sujitha, B.; Vedalakshmi, R. Light-weight cementitious conductive anode for ICCP of steel reinforced concrete application. *Constr. Build. Mater.* **2014**, *71*, 167–180. [CrossRef]
10. Refait, P.; Jeannin, M.; Sabot, R.; Antony, H.; Pineau, S. Corrosion and cathodic protection of carbon steel in the tidal zone: Products, mechanisms and kinetics. *Corros. Sci.* **2015**, *90*, 375–382. [CrossRef]
11. Ge, D.; Zhang, Z.; Zhang, C.; Xu, N. Novel technique of impressed current cathodic protection in atmospheric environments. *Br. Corros. J.* **2013**, *24*, 265–268.
12. Ashok, S.S.K.; Shahid, B.; Ramesh, K.; Ramesh, S. New perspectives on graphene/graphene oxide based polymer nanocomposites for corrosion applications: The relevance of the graphene/polymer barrier coatings. *Prog. Org. Coat.* **2021**, *154*, 106215.
13. Bunch, J.S.; Verbridge, S.S.; Alden, J.S.; Van Der Zande, A.M.; Parpia, J.M.; Craighead, H.G.; McEuen, P.L. Impermeable atomic membranes from graphene sheets. *Nano Lett.* **2008**, *8*, 2458–2462. [CrossRef] [PubMed]
14. Berry, V. Impermeability of graphene and its applications. *Carbon* **2013**, *62*, 1–10. [CrossRef]
15. Papageorgiou, G.D.; Kinloch, A.I.; Young, J.R. Mechanical properties of graphene and graphene-based nanocomposites. *Prog. Mater. Sci.* **2017**, *90*, 75–127. [CrossRef]
16. Chen, Y.; Wu, L.; Yao, W.; Wu, J.; Xiang, J.; Dai, X.; Wu, T.; Yuan, Y.; Wang, J.; Jiang, B.; et al. Development of metal-organic framework (MOF) decorated graphene oxide/MgAl-layered double hydroxide coating via microstructural optimization for anticorrosion micro-arc oxidation coatings of magnesium alloy. *J. Mater. Sci. Technol.* **2022**, *130*, 12–26. [CrossRef]
17. Aidin, B.K.; Benyamin, Y.; Masoud, M. Effect of ZnO pore-sealing layer on anticorrosion and in-vitro bioactivity behavior of plasma electrolytic oxidized AZ91 magnesium alloy. *Mater. Lett.* **2020**, *258*, 126779.
18. Yang, Z.; Che, J.; Zhang, Z.; Yu, L.; Hu, M.; Sun, W.; Gao, W.; Fan, J.; Wang, L.; Liu, G. High-efficiency graphene/epoxy composite coatings with outstanding thermal conductive and anticorrosion performance. *Compos. Part A* **2024**, *181*, 108152. [CrossRef]
19. Cui, G.; Bi, Z.; Zhang, R.; Liu, J.; Yu, X.; Li, Z. A comprehensive review on graphene-based anti-corrosive coatings. *Chem. Eng. J.* **2019**, *373*, 104–121. [CrossRef]
20. Zhang, F.; Ju, P.; Pan, M.; Zhang, D.; Huang, Y.; Li, G.; Li, X. Self-healing mechanisms in smart protective coatings: A review. *Corros. Sci.* **2018**, *144*, 74–88. [CrossRef]
21. Quezada-Renteria, A.J.; Chazaro-Ruiz, F.L.; Rangel-Mendez, R.J. Poorly conductive electrochemically reduced graphene oxide films modified with alkyne chains to avoid the corrosion-promoting effect of graphene-based materials on carbon steel. *Carbon* **2020**, *167*, 512–522. [CrossRef]
22. Zhang, R.; Yu, X.; Yang, Q.; Cui, G.; Li, Z. The role of graphene in anticorrosion coatings: A review. *Constr. Build. Mater.* **2021**, *294*, 123613. [CrossRef]
23. Zhou, S.; Yao, W.; Wang, Z.; Ma, L.; Lu, Z.; Hou, C. The first-principles calculations to explore the mechanism of oxygen diffusion on vacancy defective graphene in marine environment. *Appl. Surf. Sci.* **2020**, *525*, 136585. [CrossRef]
24. Wang, X.; Wang, Z.; Chen, Y.; Song, X.; Yang, Y. Effect of a DC Stray Current on the corrosion of X80 pipeline steel and the cathodic disbondment behavior of the protective 3PE coating in 3.5% NaCl solution. *Coatings* **2019**, *9*, 29. [CrossRef]

coatings

MDPI

Article

Effect of Co Contents on Microstructure and Cavitation Erosion Resistance of NiTiAlCrCo$_x$N Films

Hongjuan Yan *, Fangying Cheng, Lina Si, Ye Yang, Zhaoliang Dou and Fengbin Liu

School of Mechanical and Materials Engineering, North China University of Technology, Beijing 100144, China; fangy0111@163.com (F.C.); silina_thu@163.com (L.S.); yangye@ncut.edu.cn (Y.Y.); dzl@ncut.edu.cn (Z.D.); fbliu@ncut.edu.cn (F.L.)

* Correspondence: yanhongj@sina.com

Abstract: In order to investigate the effect of Co contents on the structure and cavitation erosion property, NiTiAlCrCo$_x$N films were prepared by the magnetron sputtering system. The X-ray diffractometer (XRD), the scanning electron microscope (SEM) and the energy dispersive spectrometer (EDS) were used to characterize the structure and morphology of the films. The nanoindenter and the scratch tester were used to analyze the mechanical properties of the films. Cavitation erosion experiments were carried out by the ultrasonic vibration cavitation machine. The results show that NiTiAlCrCo$_x$N films with different Co contents have a simple face-centered cubic (FCC) structure and show a preferred orientation on the (200) crystal plane. The diffraction angle on the (200) crystal plane decreases and the interplanar spacing increases with the increase in Co content in NiTiAlCrCo$_x$N films. NiTiAlCrCo$_x$N films exhibit a typical columnar crystalline structure. With the increase in Co content, the nanohardness of the films increases and the elastic modulus of the films decreases, while the mass loss of cavitation erosion monotonously increases except for the film with a 1.2 Co molar ratio. The NiTiAlCrCo$_{1.4}$N film has a minimum hardness of 13.264 GPa, a maximum elastic modulus of 253.22 GPa and a minimum mass loss of 0.72 mg in the cavitation erosion experiment. The NiTiAlCrCo$_{1.4}$N film exhibits the best cavitation corrosion resistance because the addition of the Co element enhances the solid solution strengthening effect and the NiTiAlCrCo$_{x1.4}$N film with the biggest elastic modulus has better elasticity to reduce the micro jet impact.

Keywords: NiTiAlCrCo$_x$N film; Co contents; cavitation erosion; hardness

1. Introduction

Cavitation erosion is one of the main factors causing the failure of key parts such as ship propellers and turbine blades [1]. Coating is widely used to improve the cavitation erosion resistance of parts [1,2]. The high-entropy alloy film is a new material with high hardness, corrosion resistance [1,2], oxidation resistance [3], wear resistance [4,5] and corrosion resistance [6–9]. Nonmetallic elements such as C, N and O were doped into high-entropy alloys to form high-entropy nitride or carbide films, which have better properties than high-entropy alloy films and can be used to improve the cavitation erosion resistance of parts [10–13]. In high-entropy ceramic films, the metal elements share cation positions and the nonmetallic elements occupy anionic positions, which forms a new material system and has a unique microstructure [11,14]. Researchers studied the effect of deposition parameters on film properties [15–20]. With the increase in bias potential on the substrate, the hardness of (TiZrHfVNb)N coatings increases. With the increase in working gas pressure, the hardness of (TiZrHfVNb)N decreases [12]. With the increase in the N_2:Ar flow ratio, (AlCrTiZrV)N high-entropy alloy nitride film exhibits the preferred orientation on the (200) crystal plane. The hardness and modulus firstly increase and then decrease [21,22]. Our group studied the effect of a N_2:Ar flow ratio on the cavitation erosion resistance of NiTiAlCrN films. The results show that, when the N_2:Ar flow ratio is 1:1, the NiTiAlCrN films have the best cavitation erosion resistance [23]. The researchers also studied the effect of the elements on the structure and properties of HEAs [24–29]. The CoCrFeNiAl coating shows a BCC structure accompanied by a small amount of FCC and $AlCrO_3$ phase. The CoCrFeNiMn coating shows an FCC structure and a large amount of $MnCr_2O_4$ phase. The wear resistance of CoCrFeNiMn coating is better than the one of CoCrFeNiAl coating [26]. With the increase in x, the structures of $CoCrFeNiAl_xMn_{(1-x)}$ high-entropy alloy (HEA) coatings changes from an FCC structure to dual-phase FCC + BCC structure to BCC structure. The $CoCrFeNiAl_{0.8}Mn_{0.2}$ HEA coating with an FCC + BCC structure has the best corrosion resistance [27]. The $FeCoCr_xNiAl$ HEA coatings have a dual phase of FCC and BCC. The $FeCoCr_{1.5}NiAl$ coating has the highest hardness and the best wear resistance and corrosion resistance because the Cr element promotes the formation of a hard phase and a dense oxide film is formed in 3.5 wt.% NaCl solution [28]. With the increase in Si contents, the (AlCrTiZrMo)-Si_x-N high-entropy films with Si contents change from crystal to amorphous phases and the hardness and modulus first increase and then decrease [29]. Therefore, the content of the element can change the structure and improve the properties of the films. The cavitation erosion can cause phase transformations in Co alloys and Co shows superior cavitation erosion resistance in 304 and 316 stainless steel [30]. Therefore, Co plays a crucial role in the cavitation erosion of films.

The N_2:Ar flow ratio has been determined in our previous research. On this basis, the $NiTiAlCrCo_xN$ films with different Co contents were deposited by a magnetron sputtering system. The effect of Co contents on the microstructure, nanohardness, elastic modulus and cavitation erosion resistance of the $NiTiAlCrCo_xN$ films are studied.

2. Materials and Methods

2.1. Materials

The 304 stainless steel (Juncheng Co., Ltd., Tianjin, China) is a widely used chromium–nickel stainless steel. Therefore, it is selected as the substrate, which is mirror-polished. The element contents of 304 stainless steel are shown in Table 1. The dimensions of the substrate are Φ20 mm × 3 mm.

Table 1. Element contents of 304 stainless steel.

Elements	Fe	Cr	Ni	Mn	Si	C	S	P
Contents/wt.%	67~71	17~19	8~11	≤2.0	≤1.0	≤0.08	≤0.03	≤0.035

According to the definition of the high-entropy alloys, the content of every element ranges from 5 at.% to 35 at.%. Therefore, $NiTiAlCrCo_x$ alloys with a Co molar ratio of 0.6, 0.8, 1, 1.2 and 1.4 are selected as targets. The contents of Ni, Ti, Al and Cr in $NiTiAlCrCo_x$ targets are in equimolar ratios. The $NiTiAlCrCo_x$ targets are fabricated by powder metallurgy technology with a temperature of 900 °C and a pressure of 40 MPa. The purity of the targets is 99.99%. The dimensions of the targets are Φ50.4 mm × 4 mm.

2.2. Film Deposition

The $NiTiAlCrCo_xN$ films with different Co contents were deposited on 304 stainless steel by the magnetron sputtering system. The vacuum degree of the chamber is pumped to 3×10^{-3} Pa. The target was pre-sputtered for 15 min to clear impurities and oxide in the target surface. The surface impurities and oxide of the substrate were etched by Ar^+. The nitrogen argon flow ratio was 3:4. In order to strengthen the adhesive strength between the substrate and $NiTiAlCrCo_xN$ film, the TiN layer was deposited for 60 min on the substrate. The $NiTiAlCrCo_x$ target was controlled by a DC power of 110 W. The $NiTiAlCrCo_xN$ films with different Co contents were deposited by changing the $NiTiAlCrCo_x$ targets with a different Co molar ratio. The deposition time was 180 min and the thickness of $NiTiAlCrCo_xN$ film is about 2 μm.

2.3. Film Characterization

The microstructures of the NiTiAlCrCo$_x$N films were analyzed by Rigaku Ultima IV X-ray diffraction (Tokyo, Japan) with Cu-Kα, a wavelength of 0.154 06 nm, a current of 40 mA, a voltage of 40 kV, a test step of 0.02°, a scanning speed of 8°·min^{-1} and an angle range from 10° to 80°. The surface and cross-section morphologies of NiTiAlCrCo$_x$N films and the wear track were analyzed with a Cart Zeiss Sigma-300 scanning electron microscope (SEM). The chemical compositions of NiTiAlCrCo$_x$N films and the wear track were analyzed with the Ultim Max energy spectrometer (Oberkochen, Germany) (EDS). The nanohardness and elastic modulus of NiTiAlCrCo$_x$N films were measured with the Anton Parr UNHT nanoindenter (Graz, Austria) with a Berkovich indenter (Graz, Austria), which has a curvature radius of the tip of 100 nm, a maximum load of 40 mN and loading and unloading rates of 20 mN·min^{-1}, with an indentation depth of 300 nm. A total of 5 points were selected to test the nanohardness and elastic modulus. The adhesive force of the NiTiAlCrCo$_x$N film was measured with a WS-2005 automatic scratch meter (Zhongke Kaihua Technology Co., Ltd., Lanzhou, China) with a load of 30 N, a loading rate of 30 N·min^{-1} and a scratch length of 3 mm. The scratch test was repeated 3 times in every sample. The cavitation erosion experiment was carried out with the ultrasonic vibration cavitation machine with a power of 1200 kW and an amplitude of 25 µm, which is shown in Figure 1. The diameter of the vibrating head was Φ20 mm. The 3.5 wt.% NaCl solution was selected as the cavitation erosion medium. The distance between the sample surface and the vibrating head was 0.5 mm. The ice was added into the circulating water in the bath to keep the samples at 0 °C. The sample was taken out and the mass loss of the sample was measured with a high-precision electronic balance for every 2 h of the cavitation erosion experiment. The total duration of the cavitation erosion experiment was 12 h. The cavitation rate is defined as the mass loss per hour. The cavitation erosion experiment was repeated 3 times for every NiTiAlCrCo$_x$N film.

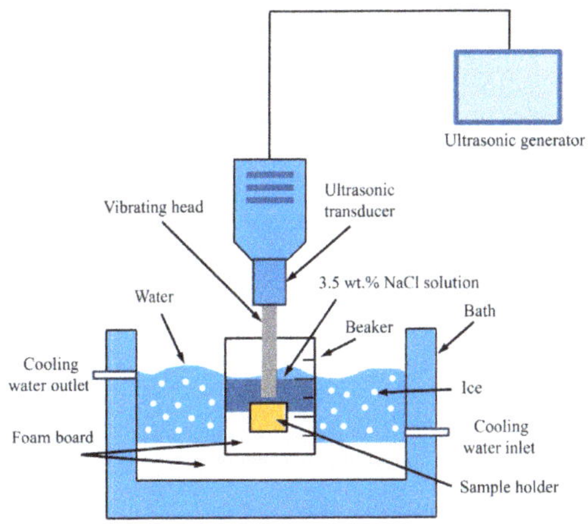

Figure 1. Diagram of the ultrasonic cavitation erosion machine.

3. Results and Discussion

3.1. Film Structure

Figure 2 shows the XRD patterns of NiTiAlCrCo$_x$N films with different Co contents. The films exhibit a face-centered cubic (FCC) structure and have a preferred orientation on the (200) crystal plane. The phases consist of TiN, AlN, CrN, Co_3Ti and $Co_{5.47}N$. When the molar ratio of Co in the targets is bigger than 1.0, the diffraction peak of (111) plane disappears. The diffraction angle on the (200) crystal plane shifts to the bigger angle and the interplanar spacing decreases with the increase in Co contents.

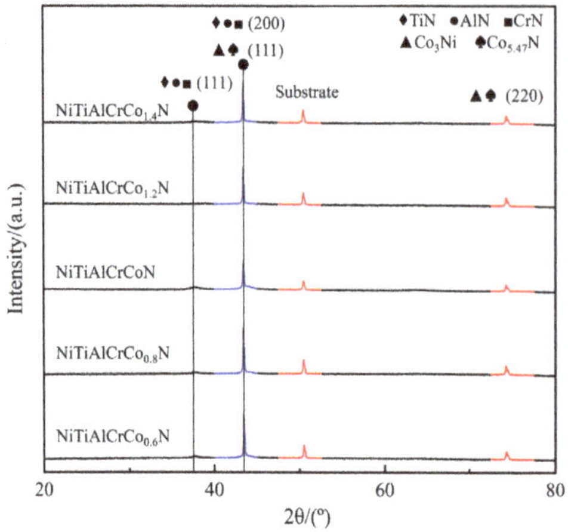

Figure 2. XRD patterns of NiTiAlCrCo$_x$N films.

Table 2 shows the diffraction angle, the interplanar spacing and the full-width half of the maximum (FWHM) on the (200) crystal plane. Except for the NiTiAlCrCoN film with equimolar ratios, the diffraction angle decreases and the interplanar spacing increases with the increasing of Co content. But the NiTiAlCrCoN film with equimolar ratios has the maximum diffraction angle and FWHM and the minimum interplanar spacing, which means that the NiTiAlCrCoN film has higher crystallinity and a finer grain size. The reason for this is that the film with equimolar ratios has the lattice distortion effect and the slow diffusion effect.

Table 2. Diffraction angle, interplanar spacing and FWHM.

Co Content	Diffraction Angle 2θ/(°)	Interplanar Spacing *d*/nm	FWHM *B*/rad
NiTiAlCrCo$_{0.6}$N	43.399	2.0833	0.119
NiTiAlCrCo$_{0.8}$N	43.342	2.0859	0.152
NiTiAlCrCoN	43.381	2.0842	0.193
NiTiAlCrCo$_{1.2}$N	43.340	2.0860	0.122
NiTiAlCrCo$_{1.4}$N	43.338	2.0861	0.115

With of the increase in Co content, the FWHM of NiTiAlCrCo$_x$N films first increases and then decreases, which means that the Co element can improve the peak quality. When the Co element is in an equimolar ratio, the peak quality of the XRD pattern is the best. The NiTiAlCrCoN film with equimolar ratios is more likely to generate multicomponent crystals in an alloy structure, which greatly increases entropy value and easily generates a crystal structure. Due to the different sizes of each atom in the site, the lattice position changes and the lattice distortion is intensified, which reduces the diffraction peak intensity and increases the FWHM of the NiTiAlCrCoN film on the (200) crystal plane. The NiTiAlCrCo$_x$N films with unequal molar ratios have a lower entropy which weakens the unique microstructure caused by the "cocktail" effect. Therefore, the NiTiAlCrCoN film with equimolar ratios exhibits specificity in XRD patterns.

3.2. Morphology

Figure 3 shows the elements map, the surface and the cross-section morphologies of the NiTiAlCrCo$_x$N films with different Co contents. The NiTiAlCrCo$_x$N films exhibit the typical columnar crystalline structure which is perpendicular to the substrate. The morphologies are sequentially the NiTiAlCrCo$_x$N layer, the TiN transition layer and the substrate from top to bottom. The interfaces between layers are clear. The surface is smooth, flat, with no pores and no peel.

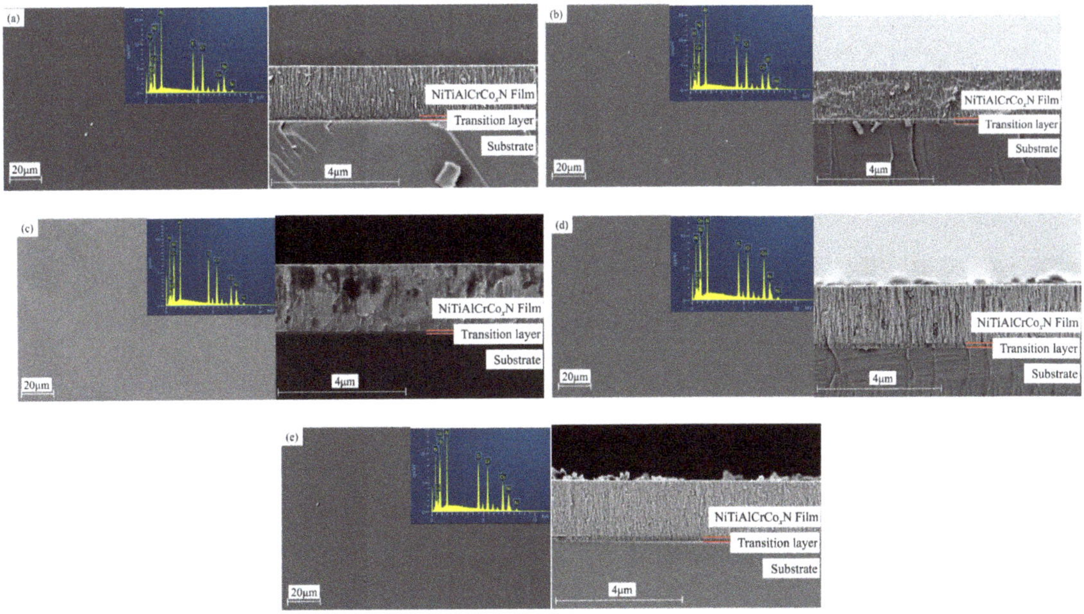

Figure 3. Element map, surface and cross-section morphologies of NiTiAlCrCo$_x$N films. (**a**) NiTiAlCrCo$_{0.6}$N; (**b**) NiTiAlCrCo$_{0.8}$N; (**c**) NiTiAlCrCoN; (**d**) NiTiAlCrCo$_{1.2}$N; (**e**) NiTiAlCrCo$_{1.4}$N.

Table 3 and Figure 4 show the element contents of NiTiAlCrCo$_x$N films with different Co contents, which was observed by EDS. The N contents in NiTiAlCrCo$_x$N films are about (39 $\pm$ 0.6) in percentage. The Co contents in NiTiAlCrCo$_x$N films increase from 9.4% to 22.53% with an increase in the Co molar ratio in the NiTiAlCrCo$_x$ targets. The other elements such as Ni, Ti, Al and Cr in NiTiAlCrCo$_x$N films are approximately equal.

Table 3. Element contents in NiTiAlCrCo$_x$N films.

	Ni	Ti	Al	Cr	Co	N
NiTiAlCrCo$_{0.6}$N	14.85	10.92	11.56	14.59	9.4	38.68
NiTiAlCrCo$_{0.8}$N	13.86	10.7	11.36	13.82	10.72	39.54
NiTiAlCrCoN	12.15	10.64	12.16	13.29	11.88	39.88
NiTiAlCrCo$_{1.2}$N	11.52	9.6	10.86	11.37	17.89	38.76
NiTiAlCrCo$_{1.4}$N	10.35	8.21	9.05	10.74	22.53	39.12

Figure 5 shows the element distribution of point scanning, which was taken five points from top to bottom in the cross-section of the NiTiAlCrCo$_x$N film. In the NiTiAlCrCo$_{0.6}$N film, the metal elements are more easily concentrated near the substrate and the N element increases from bottom to top, which increases the nitride content near the surface. With the increase in Co content in the NiTiAlCrCo$_{1.4}$N film, the metal elements increase and the N element decreases near the surface. The solid solution strengthening effect increases significantly between metal elements and nitrides near the substrate, resulting in an increase in nitrides near the substrate.

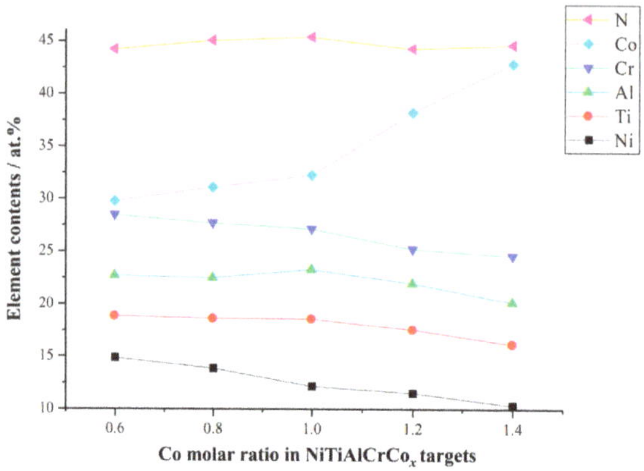

Figure 4. Element contents in NiTiAlCrCo$_x$N films.

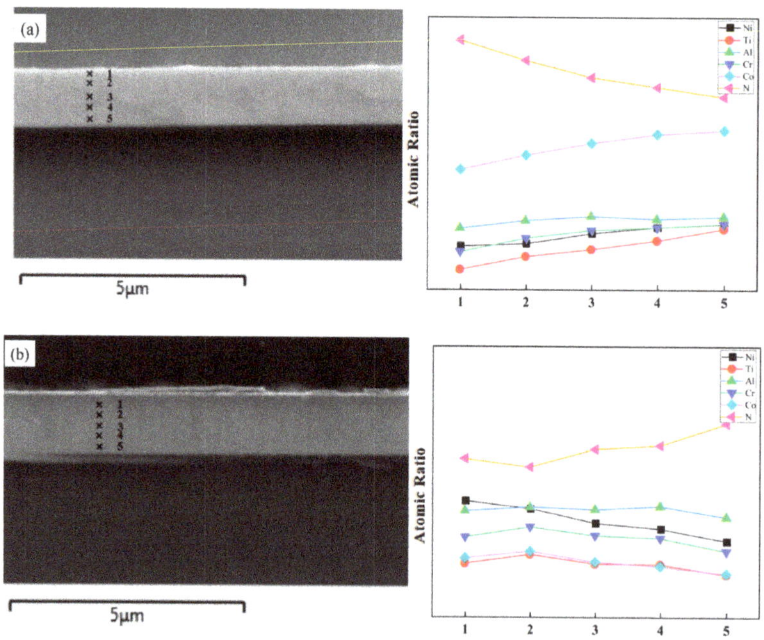

Figure 5. Element distribution in cross-section of NiTiAlCrCo$_x$N films (the numbers of 1, 2, 3, 4, 5 in cross-section are the position of the point scanning). (**a**) NiTiAlCrCo$_{0.6}$N; (**b**) NiTiAlCrCo$_{1.4}$N.

3.3. Mechanical Properties

Figure 6 shows the average nanohardness and the elastic modulus of NiTiAlCrCo$_x$N films. With the increase in Co content, the nanohardness of the NiTiAlCrCo$_x$N films decreases and the elastic modulus of the NiTiAlCrCo$_x$N films increases, expect for the ones of the NiTiAlCrCo$_{1.2}$N film. When the Co molar ratio is 1.4, the film has a minimum hardness of 13.264 GPa and a maximum elastic modulus of 253.22 GPa. When the Co molar ratio is 0.6, the film has a maximum hardness of 14.178 GPa and a minimum elastic modulus of 229.40 GPa.

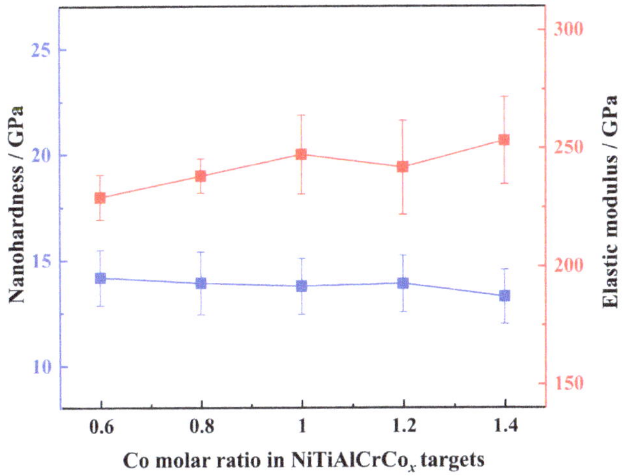

Figure 6. Nanohardness and elastic modulus of NiTiAlCrCo$_x$N films.

When the Co molar ratio is lower, the elements with larger atomic radii concentrate near the surface of the film but AlN, TiN and CrN concentrate near the surface, which causes the higher nanohardness of the film. With the increase in the Co element, the solid solution phase increases near the transition layer and AlN, TiN and CrN concentrate near the substrate of the film, which decreases the nanohardness of the film.

Figure 7 shows the adhesive force of NiTiAlCrCo$_x$N films. The adhesive force firstly increases and then decreases with the increase in Co content. When the Co molar ratio is 1.0, the film has a maximum adhesive force of 24.2 N. The reason is that AlN, TiN and CrN are concentrated near the substrate and have a better adhesive force with a TiN transition layer.

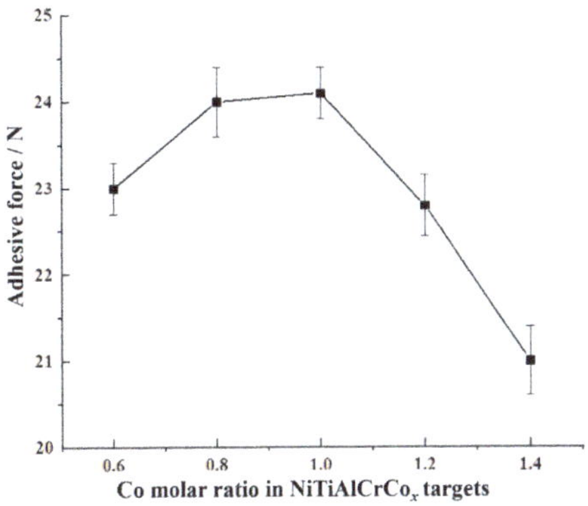

Figure 7. Adhesive force of NiTiAlCrCo$_x$N films.

3.4. Cavitation Erosion Resistance

Figure 8 shows the relationship between the mass loss and the cavitation erosion time of NiTiAlCrCo$_x$N films with different Co contents. The mass loss increases monotonically with the cavitation erosion time. The accumulative mass loss firstly decreases and then increases with the increase in Co content. When the Co content has a molar ratio of 1.4, the accumulative mass loss is the minimum of 0.72 mg and the cavitation erosion rate is of 0.12 mg/h.

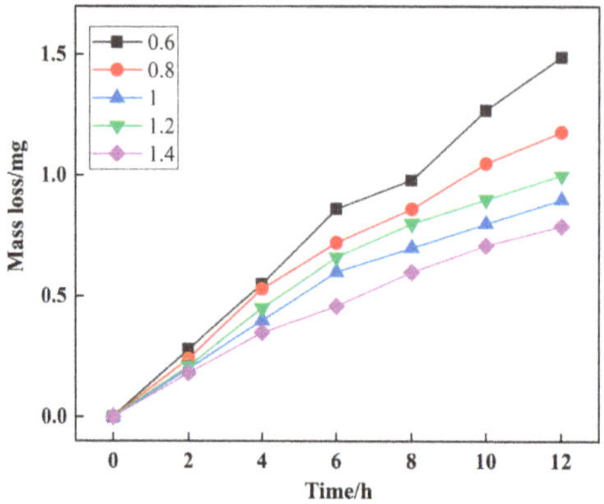

Figure 8. Mass loss curve of NiTiAlCrCo$_x$N films.

There are two reasons for the improvement in the cavitation erosion resistance of the films. First, when the Co content has a 1.4 molar ratio, the addition of the Co element enhances the solid solution strengthening effect [30]. Secondly, the film with the biggest elastic modulus has better elasticity to reduce the micro jet impact, which improves the cavitation erosion resistance of the film.

Figure 9 shows the surface and cavitation pits of NiTiAlCrCo$_x$N films after 12 h of cavitation erosion experiment. The film with a 0.6 Co molar ratio peels off. When the Co molar ratio is greater than 0.6, the films have no peeling, cracking or plastic deformation, and there are a few cavitation pits in the surface of NiTiAlCrCo$_x$N films. There is no rupture and spallation in NiTiAlCrCo$_x$N films, which implies a different cavitation mechanism from TiAlN and AlTiN films [31]. When the Co molar ratio is 1.4, the size of the cavitation pit is the minimum of 2.202 μm, which is consistence with the result seen in Figure 7.

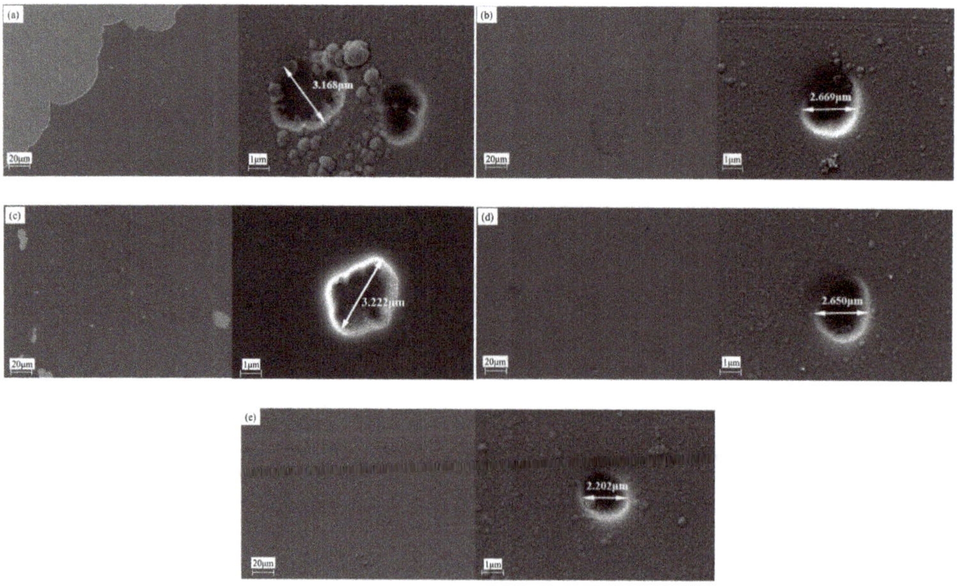

Figure 9. Surface and cavitation pit after 12 h cavitation erosion experiment. (**a**) NiTiAlCrCo$_{0.6}$N; (**b**) NiTiAlCrCo$_{0.8}$N; (**c**) NiTiAlCrCoN; (**d**) NiTiAlCrCo$_{1.2}$N; (**e**) NiTiAlCrCo$_{1.4}$N.

Figure 10 shows the element distribution of the cavitation pits of films with equimolar ratios. In the cavitation pit, Ni, Ti, Al, Co and N elements disappear and Cr and Fe elements appear, which means that the film undergoes breakdown and that the substrate of 304 stainless steel is exposed. The O element appears in the pit edge, which means that oxidations occur during the cavitation erosion. The Al, Co, N and O elements increase near the cavitation pits, which means that AlN, $Co_{5.47}N$ enriches near the surface of the films and reacts with O_2 to form Al_2O_3 and Co_2O_3 to resist the impact of micro jets. The addition of the Co element enhances the solid solution strengthening effect of the cavitation pits, which improves the cavitation erosion resistance of the films.

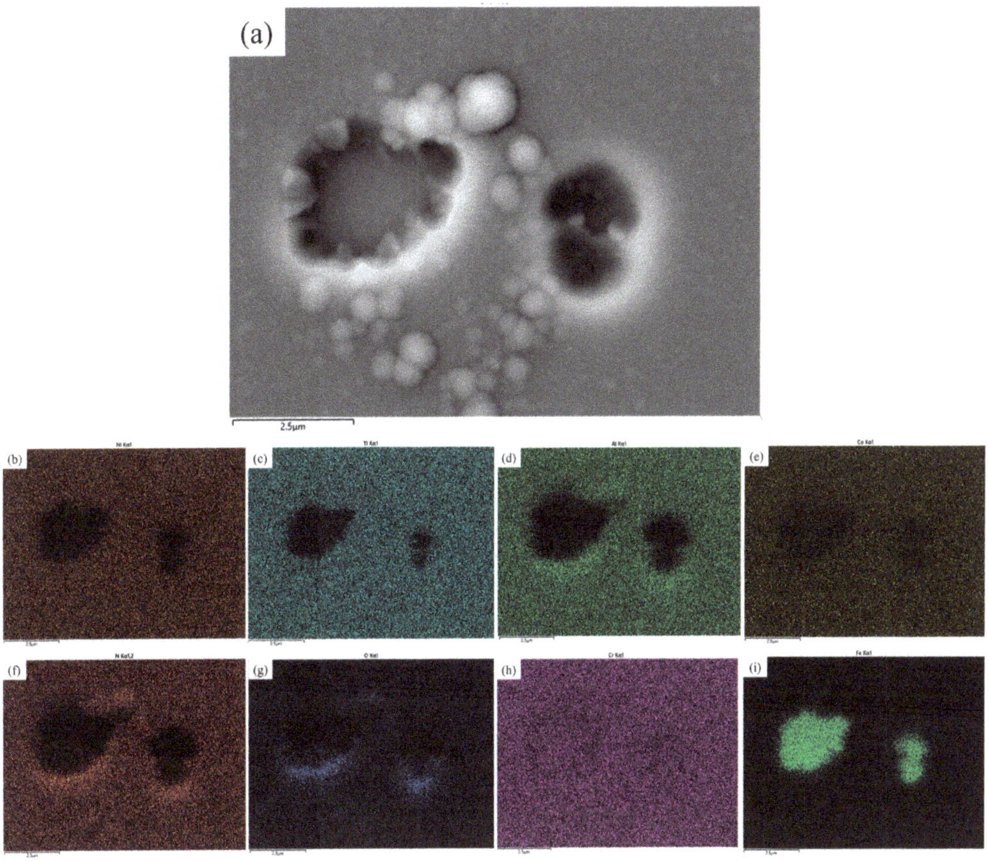

Figure 10. Element distribution of $NiTiAlCrCo_{0.6}N$ film cavitation pits. (**a**) Surface of cavitation pit; (**b**) Ni; (**c**) Ti; (**d**) Al; (**e**) Co; (**f**) N; (**g**) O; (**h**) Cr; (**i**) Fe.

4. Conclusions

The $NiTiAlCrCo_xN$ films with different Co contents were deposited on 304 stainless steel substrates by the magnetron sputtering system. The effect of Co content on microstructure and cavitation erosion resistance of the $NiTiAlCrCo_xN$ films was studied.

(1) The $NiTiAlCrCo_xN$ films with different Co contents have a simple face-centered cubic structure, and the preferred orientation appears on the (200) crystal plane. With the increase in Co contents, the interplanar spacing first increases and then decreases. The NiTiAlCrCoN film with equimolar ratios has the minimum interplanar spacing due to the lattice distortion effect and the slow diffusion effect.

(2) With the increase in Co content, the nanohardness of the $NiTiAlCrCo_xN$ films decreases and the elastic modulus of the $NiTiAlCrCo_xN$ films increases, expect for the ones of the $NiTiAlCrCo_{1.2}N$ film. The $NiTiAlCrCo_{1.4}N$ film has the lowest nanohardness of 13.264 GPa, and the highest elastic modulus of 253.22 GPa.

(3) The $NiTiAlCrCo_xN$ films have no peeling, cracks and plastic deformation, and there are few cavitation pits on the surface of the films, except for the $NiTiAlCrCo_{0.6}N$ film. The $NiTiAlCrCo_{x1.4}N$ film exhibits the minimum mass loss of cavitation erosion. There are two reasons for the improvement in the cavitation erosion resistance. Firstly, the addition of the Co element enhances the solid solution strengthening effect. Secondly, the $NiTiAlCrCo_{x1.4}N$ film with the biggest elastic modulus has better elasticity to reduce the micro jet impact, which improves the cavitation erosion resistance of the film.

Author Contributions: H.Y., F.C. and F.L. proposed the idea. F.C., L.S. and Y.Y. carried out the experiments. H.Y., F.C. and Z.D. analyzed the experimental results. H.Y. and F.C. wrote the main manuscript text. All authors have read and agreed to the published version of the manuscript.

Funding: This research was funded by National Natural Science Foundation of China (Grant No. U23A2025, 52305328), Natural Science Foundation of Beijing (Grant No. 3212003) and Yuyou Team of North China University of Technology (Grant No. 22XN746).

Institutional Review Board Statement: Not applicable.

Informed Consent Statement: Not applicable.

Data Availability Statement: All data are included in this published article.

Conflicts of Interest: The authors declare no conflicts of interest.

References

1. Wang, Y.; Hao, E.; An, Y.; Chen, J.; Zhou, H. Effects of microstructure and mechanical properties on cavitation erosion resistance of NiCrWMoCuCBFe coatings. *Appl. Surf. Sci.* **2021**, *547*, 149125. [CrossRef]
2. Wei, Z.; Wu, Y.; Hong, J.; Cheng, J.; Qiao, L.; Cheng, J.; Zhu, S. Ultrasonic cavitation erosion behaviors of high-velocity oxygen-fuel (HVOF) sprayed AlCoCrFeNi high-entropy alloy coating in different solutions. *Surf. Coat. Technol.* **2021**, *409*, 126899. [CrossRef]
3. Guan, Y.; Cui, X.; Chen, D.; Su, W.; Zhao, Y.; Li, J.; Li, X.; Feng, L.; Jin, G. Microstructure and properties analysis of FeCoNiAlCu dual-phase high-entropy alloy coating by laser cladding. *Surf. Coat. Technol.* **2023**, *467*, 129695. [CrossRef]
4. Luo, D.; Zhou, Q.; Huang, Z.; Li, Y.; Liu, Y.; Li, Q.; He, Y.; Wang, H. Tribological Behavior of High Entropy Alloy Coatings: A Review. *Coatings* **2022**, *12*, 1428. [CrossRef]
5. Huang, Y.; Hu, Y.; Zhang, M.; Mao, C.; Tong, Y.; Zhang, J.; Li, K.; Wang, K. On the enhanced wear resistance of laser-clad $CoCrCuFeNiTi_x$high-entropy alloy coatings at elevated temperature. *Tribol. Int.* **2022**, *174*, 107767. [CrossRef]
6. Xu, J.; Peng, S.; Li, Z.; Jiang, S.; Xie, Z.H.; Munroe, P.; Lu, H. Remarkable cavitation erosion–corrosion resistance of CoCrFeNiTiMo high-entropy alloy coatings. *Corros. Sci.* **2021**, *190*, 109663. [CrossRef]
7. Rong, Z.; Wang, C.; Wang, Y.; Dong, M.; You, Y.; Wang, J.; Liu, H.; Liu, J.; Wang, Y.; Zhu, Z. Microstructure and properties of $FeCoNiCr_x$ (x = Mn, Al) high-entropy alloy coatings. *J. Alloys Compd.* **2022**, *921*, 166061. [CrossRef]
8. Liu, Y.; Xiang, D.; Wang, K.; Yu, T. Corrosion of Laser Cladding High-Entropy Alloy Coatings: A Review. *Coatings* **2022**, *12*, 1669. [CrossRef]
9. Wu, X.; Lv, Y. Study on the Corrosion Resistance of Laser Clad Al0.7FeCoCrNiCux High-Entropy Alloy Coating in Marine Environment. *Coatings* **2022**, *12*, 1855. [CrossRef]
10. Lan, L.W.; Wang, X.J.; Guo, R.P.; Yang, H.J.; Qiao, J.W. Effect of environments and normal loads on tribological properties of nitrided $Ni_{45}(FeCoCr)_{40}(AlTi)_{15}$ high-entropy alloys. *J. Mate. Sci. Technol.* **2020**, *42*, 85–96. [CrossRef]
11. Straumal, B.; Rabkin, E.; Lopez, G.A.; Korneva, A.; Kuzmin, A.; Gornakova, A.; Straumal, A.; Baretzky, B. Grain boundary wetting phenomena in high entropy alloys containing nitrides, carbides, borides, silicides, and hydrogen: A review. *Crystals* **2021**, *11*, 1540. [CrossRef]
12. Pogrebnjaka, A.D.; Yakushchenkoa, I.V.; Abadiasb, G.; Chartier, P.; Bondar, O.V.; Beresnev, V.M.; Takeda, Y.; Sobol', O.V.; Oyoshi, K.; Andreyev, A.A.; et al. The effect of the deposition parameters of nitrides of high entropy alloys (TiZrHfVNb)N on their structure, composition, mechanical and tribological properties. *J. Super. Mater.* **2013**, *35*, 356–368. [CrossRef]
13. Lu, K.; Zhu, J.; Ge, W.; Hui, X. Progress on New Preparation Methods, Microstructures, and Protective Properties of High-Entropy Alloy Coatings. *Coatings* **2022**, *12*, 1472. [CrossRef]
14. Wang, Y.; Liu, S.; Dong, C. Research process and challenge of high entropy ceramic metarials. *J. Mater. Eng.* **2024**, *52*, 83–100.

15. Thorhallsson, A.I.; Fanicchia, F.; Davison, E.; Paul, S.; Davidsdottir, S.; Olafsson, D.I. Erosion and Corrosion Resistance Performance of Laser Metal Deposited High-Entropy Alloy Coatings at Hellisheidi Geothermal Site. *Materials* **2021**, *14*, 3071. [CrossRef]
16. Bao, Y.; Guo, L.; Zhong, C.; Song, Q.; Yang, K.; Jiang, Y.; Wang, Z. Effects of WC on the cavitation erosion resistance of $FeCoCrNiB_{0.2}$ high entropy alloy coating prepared by laser cladding. *Mater. Today Commun.* **2021**, *26*, 102154. [CrossRef]
17. Wu, C.L.; Zhang, S.; Zhang, C.H.; Zhang, H.; Dong, S.Y. Phase evolution and cavitation erosion-corrosion behavior of FeCoCrAlNiTix high entropy alloy coatings on 304 stainless steel by laser surface alloying. *J. Alloys Compd.* **2017**, *698*, 761–770. [CrossRef]
18. Baufeld, B. Effect of deposition parameters on mechanical properties of shaped metal deposition parts. *Proc. Inst. Mech. Eng. B-J. Eng.* **2012**, *226*, 126–136. [CrossRef]
19. Dai, C.D.; Fu, Y.; Guo, J.X.; Du, C.W. Effects of substrate temperature and deposition time on the morphology and corrosion resistance of $FeCoCrNiMo_{0.3}$ high-entropy alloy coating fabricated by magnetron sputtering. *Int. J. Miner. Metall. Mater.* **2020**, *27*, 1388–1397. [CrossRef]
20. Beresnev, V.M.; Nyemchenko, U.S.; Srebniuk, P.O.; Lytovchenko, S.V.; Sobol, O.V. Study of influence physical and technological parameters of deposition on the structure, physical and mechanical properties of vacuum arc coatings (Mo + Ti6%Si)N. In Proceedings of the 2016 International Conference on Nanomaterials: Application & Properties (NAP), Lviv, Ukraine, 14–19 September 2016.
21. Chen, L.; Li, W.; Liu, P.; Zhang, K.; Ma, F.; Chen, X.; Zhou, H.; Liu, X. Microstructure and mechanical properties of $(AlCrTiZrV)N_x$high-entropy alloy nitride films by reactive magnetron sputtering. *Vacuum* **2020**, *181*, 109706. [CrossRef]
22. Xu, Y.; Li, G.; Xia, Y. Synthesis and characterization of super-hard AlCrTiVZr high-entropy alloy nitride films deposited by HiPIMS. *Appl. Surf. Sci.* **2020**, *523*, 146529. [CrossRef]
23. Yan, H.; Liu, Y.; Mi, Z.; Si, L.; Dou, A.; Liu, F. Effect of nitrogen-argon flow ratio on cavitation resistance of NiTiAlCrN coating. *J. Func. Mater.* **2023**, *54*, 1007–1011.
24. Novikov, V.; Stepanov, N.; Zherebtsov, S.; Salishchev, G. Structure and properties of high-Entropy nitride coatings. *Metals* **2022**, *12*, 847. [CrossRef]
25. Zhou, C.; Wang, A.; Wang, G.; Zhang, J.; Liu, W. Investigation on the microstructure, wear and corrosion resistance of $FeCoNiCrMo_x$high-entropy alloy coatings deposited on 40Cr by laser cladding. *J. Mater. Sci.* **2022**, *57*, 18615–18639. [CrossRef]
26. Xiao, J.; Li, T.; Wu, Y.; Chen, J.; Zhang, C. Microstructure and tribological properties of plasma-sprayed CoCrFeNi-based high-entropy alloy coatings under dry and oil-lubricated sliding conditions. *J. Therm. Spray. Technol.* **2021**, *30*, 926–936. [CrossRef]
27. Sun, S.; Liu, H.; Hao, J.; Yang, H. Microstructural evolution and corrosion behavior of $CoCrFeNiAl_xMn_{(1-x)}$ dual-phase high-entropy alloy coatings prepared by laser cladding. *J. Alloys Compd.* **2021**, *886*, 161251. [CrossRef]
28. Ben, Q.; Zhang, Y.; Sun, L.; Wang, L.; Wang, Y.; Zhan, X. Wear and corrosion resistance of FeCoCrxNiAl high-entropy alloy coatings fabricated by laser cladding on Q345 welded joint. *Metals* **2022**, *12*, 1428. [CrossRef]
29. Yu, W.; Li, W.; Liu, P.; Zhang, K.; Ma, F.; Chen, X.; Feng, R.; Liaw, P.K. Silicon-content-dependent microstructures and mechanical behavior of $(AlCrTiZrMo)-Si_x-N$ high-entropy alloy nitride films. *Mater. Design.* **2021**, *203*, 109553. [CrossRef]
30. Szala, M.; Chocyk, D.; Skic, A.; Kamiński, M.; Macek, W.; Turek, M. Effect of nitrogen ion implantation on the cavitation erosion resistance and Cobalt-based solid solution phase transformations of HIPed Stellite 6. *Materials* **2021**, *14*, 2324. [CrossRef] [PubMed]
31. Szala, M.; Walczak, M.; Pasierbiewicz, K.; Kamiński, M. Cavitation erosion and sliding wear mechanisms of AlTiN and TiAlN flims deposited on stainless steel substrate. *Coatings* **2019**, *9*, 340. [CrossRef]

coatings

MDPI

Article

Performance of a Composite Inhibitor on Mild Steel in NaCl Solution: Imidazoline, Sodium Molybdate, and Sodium Dodecylbenzenesulfonate

Xue Wang [1,2], Changfeng Fan [2,*], Lanyi Sun [1], Hongyan Shang [1], Duxin Zhang [1], Nan Xu [3,4], Bin Wang [3,4] and Jianbin Xu [3,4]

[1] College of Chemical Engineering, China University of Petroleum, Qingdao 266580, China
[2] Shandong Institute of Petroleum and Chemical Technology, Dongying 257061, China
[3] Key Laboratory of Oilfield Produced Fluid New Demulsifier, Dongying 257061, China
[4] Dongying Jinmei Chemical Co., Ltd., Dongying 257061, China
* Correspondence: fancf2015@hotmail.com

Abstract: Mild steel corrosion is a significant challenge in oil and gas exploitation. Inhibitors are frequently employed to minimize the corrosive impact on mild steel. Mixing corrosion inhibitors is an effective method in reducing the dosage of toxic compounds and expanding the potential applications of inhibitors in NaCl solutions. Herein, a mixed corrosion inhibitor composed of imidazoline (IM), sodium molybdate, and sodium dodecylbenzenesulfonate (SDBS) for mild steel in a 3.5 wt% NaCl solution are investigated by orthogonal experimental design and electrochemical measurement. The imidazoline compound was synthesized and identified using Fourier transform infrared (FTIR) spectroscopy. The inhibitory effect is improved by higher concentrations of sodium molybdate and is further enhanced with the addition of 10 mg/L of SDBS. The electrochemical impedance spectroscopy indicates that the combination of IM (100 mg/L), sodium molybdate (50 mg/L), and SDBS (100 mg/L) results in excellent performance with electrochemical impedance (1.8 kohm·cm^2). The mild steel surfaces after electrochemical measurement were analyzed using scanning electron microscopy (SEM). The information can contribute to the development of corrosion inhibitors with high performance or to understand the influence of mixing inhibitors on corrosion processes of mild steels.

Keywords: mild steel; mixed corrosion inhibitor; orthogonal test; electrochemical measurement

1. Introduction

Mild steel is a crucial structural component for manufacturing equipment and pipelines due to its excellent mechanical properties, cost-effectiveness, wide availability, and high recyclability. However, mild steel is continuously exposed to corrosion during its service life in chloride-containing environments, resulting in performance degradation and significant economic losses [1–3]. Various corrosion inhibitors have been developed to reduce the corrosion rate of mild steel [4,5] and thereby extend the life of the equipment [6,7] in response to this problem. However, there are serious health and environmental hazards associated with conventional high-performance corrosion inhibitors that contain metal salts like tungstates and chromates along with a variety of chemical compounds [8,9]. This contradicts the increasing demand for eco friendly production. Therefore, it is crucial to develop effective corrosion inhibitors that are both high-performing and environmentally friendly [10–13].

Among the available organic compounds, IM and its derivatives are widely recognized as corrosion inhibitors, especially for controlling mild steel pipeline corrosion. The IM-based inhibitors are regarded as green corrosion inhibitors, stemming from their excellent performance, low toxicity, biodegradability, and easy availability [14]. Nevertheless, these inhibitors encounter certain challenges, such as issues related to stability during storage, emulsification in production water, and cost-effectiveness [15]. At the same time, sodium dodecylbenzenesulfonate (SDBS), characterized as a highly active surfactant, exhibits an excellent ability to adsorb onto steel surfaces for corrosion protection in aqueous solutions [16]. It can be utilized independently or in conjunction with other chemicals. Sodium molybdate, recognized as a green and non-toxic oxidative corrosion inhibitor, exhibits limited corrosion inhibition efficiency when applied in isolation [17]. Its efficacy is typically enhanced through combination with other corrosion inhibitors. It is thus expected that the combination of IM, SDBS, and sodium molybdate would produce improved corrosion inhibition, as compared to their uses independently, while overcoming the potential problems associated with the use of IM alone [18]. Many studies in the literature consistently show that the incorporation of molybdate into corrosive media increases the adsorption capacity of organic cations [19]. This enhancement is enabled by the establishment of a linking bridge between the negatively charged metal surface and the inhibitor cation, as evidenced in various preceding investigations. Xie et al. investigated the corrosion inhibition performance of imidazoline inhibitors combined with uracil, potassium thiocyanate, sodium dodecyl sulfate (SDS), cetyltrimethylammonium bromide (CTAB), and other substances on A3 carbon steel in a 6% hydrochloric acid solution [20]. Meanwhile, Meng et al. studied the corrosion inhibition efficiency of sodium molybdate at a concentration of 100 mg/L combined with 10 mg/L of SDS, showing an improvement of approximately 17% compared to using sodium molybdate alone at the same concentration [21]. It is reasonable to believe that the combination of inhibitors involving IM, SDBS, and sodium molybdate will improve corrosion protection beyond the performance that can be achieved when applied individually. However, there is limited literature on the significant impact of each corrosion inhibitor when multiple inhibitors are combined.

This study was conducted to investigate the corrosion behavior of mild steel immersed in a 3.5 wt% NaCl solution at ambient temperature, focusing specifically on the impact of three distinct inhibitors IM, SDBS, and sodium molybdate. Corrosion properties were evaluated by dynamic polarization (DP) and electrochemical impedance spectroscopy (EIS) measurements. Scanning electron microscopy (SEM) and energy dispersive spectrometer analysis (EDS) were carried out to gain insight into the surface morphology and element composition. Additionally, Fourier transform infrared spectroscopy (FTIR) was applied to characterize the synthesized IM. This study shed light on the combination and the significant impact of these inhibitors in corrosion inhibition on mild steel and provides an insight on the development of innovative corrosion protection strategies.

2. Experimental

2.1. Sample Preparation and Synthesis of Imidazoline

The chemical composition of the mild steel is listed in Table 1. All samples were cut into 30 mm × 30 mm × 1 mm. Prior to each experiment, samples were wet-ground successively with 240, 500, 800, 1000, and 2000 grit silicon carbide (SiC) paper, cleaned with ethanol and deionized water, and dried with a blower.

Table 1. Chemical composition of mild steel used (in wt%).

Element	C	Si	Mn	S	P	Fe
Content	0.2	0.3	0.7	0.042	0.042	Bal.

Figure 1 shows the synthesis process of water-soluble imidazoline quaternary ammonium corrosion inhibitors. Figure 2 illustrates the chemical structure of imidazoline. The imidazoline intermediate is initially prepared via a bifurcated synthesis method with fatty acids and diethylene triamine as precursors. This intermediate is then quaternized by adding benzyl chloride, which results in the formation of water-soluble imidazoline-based corrosion inhibitors.

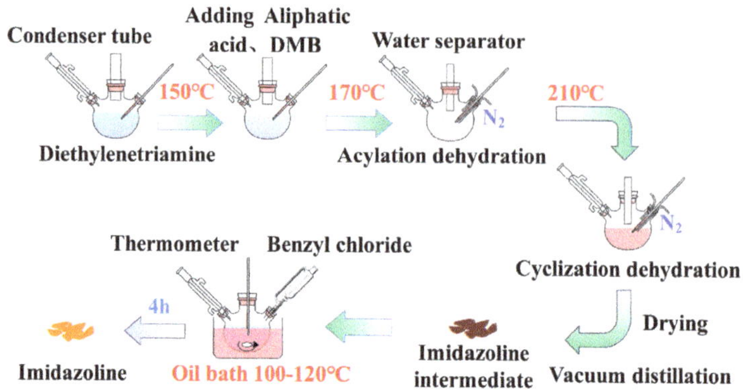

Figure 1. Synthesis process of water-soluble imidazoline quaternary ammonium corrosion inhibitors.

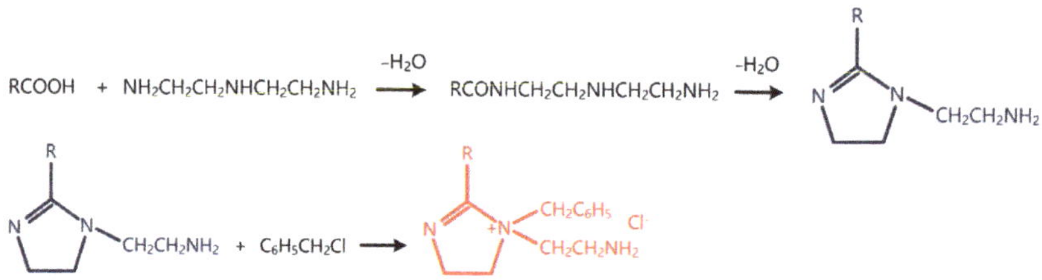

Figure 2. Chemical structure of imidazoline.

A predetermined amount of diethylene triamine is added to a triple-necked flask equipped with a phase separator, reflux condenser, and thermometer. The flask is then heated to about 150 °C to synthesize the fatty-acid-type imidazoline intermediate. Specified quantities of fatty acids and 5 mL of xylene are then added under the protective shield of a N_2 atmosphere, followed by rapid heating to 170 °C to initiate the acylation dehydration process. This process is sustained until no water can be observed exiting the phase separator.

The temperature is raised to 210 °C via a step-by-step, program-controlled heating technique, which aids in cyclization dehydration. Heating is stopped and the system is cooled in the safe N_2 atmosphere after no more water is visible, and the water content in the phase separator matches the theoretical volume.

The remaining diethylene triamine and residual xylene are removed under low pressure, and the mixture is then dried in a vacuum oven to produce a reddish-brown, viscous imidazoline intermediate.

The acylation and cyclization dehydration procedures involved in this reaction occur when the amino groups on the diethylene triamine attack the fatty acid carbonyl carbon atoms. This reaction also results in dehydration and the formation of an alkyl amide intermediate. Excess amine is employed to accelerate the reaction process and prevent the formation of diamides. In the intermediate alkyl amide structure, the amine group (-NH) and the amide carbonyl subsequently undergo dehydration and cyclization.

The best conditions for synthesizing the intermediate can be determined by assessing how the temperature, reaction time, and raw material ratio influence the synthesis reaction. A temperature of 230 °C, a molar ratio of 1:1.2, and a reaction duration of 7 h are optimal for synthesizing the imidazoline intermediate (AI) from fatty acid and diethylene triamine.

The imidazoline intermediate is placed into a flask with three necks and heated to between 100 and 120 °C in an oil bath, followed by treatment with a comparable quantity of benzyl chloride to promote the quaternization reaction and improve the imidazoline's water solubility. After four hours of reaction, a group of water-soluble imidazoline quaternary salts is generated.

2.2. Orthogonal Test

The additive concentrations in the inhibitors were evaluated using an orthogonal test, shown in Table 2. An $L_9(3^3)$ orthogonal array obtained based on the Taguchi method was employed to estimate the optimum operating conditions by reducing the trials of the experiment. The Taguchi method is a structured approach to experimental design and analysis, involving several key steps identifying quality characteristics and design parameters, choosing the number of factor levels, selecting an orthogonal array, conducting experiments according to the array, performing variance analysis, determining optimal factor levels, and confirming parameters through a test.

Table 2. Factors and levels by orthogonal test.

Level	Factor		
	IM (mg/L)	SDBS (mg/L)	Na_2MoO_4 (mg/L)
1	0	0	0
2	20	10	50
3	100	100	100

The orthogonal test was used to evaluate the additive concentrations of the inhibitors included in the compound formula in order to evaluate the anti-corrosion properties of each inhibitor in addition to the potential effects of the inhibitor combination.

In the study, the corrosion impedance measured through EIS measurement was utilized as the evaluation criteria. The effects of each factor at their respective levels were determined by averaging the inhibition efficiencies across all experiments. These values, denoted as I, II, and III, were used to calculate the mean and range values (R) for each factor. The corrosion inhibition performances showed varying degrees of improvement within the specified range as the inhibitor quantities increased. The inhibition mechanism studies were subsequently carried out through SEM observation.

2.3. Surface Characterization

FTIR spectra were obtained using a Nicolet iS50 (Thermo Fisher Scientific, Waltham, MA, USA) with a 32-scan data accumulation, covering a range of 400–4000 cm^{-1} at a spectral resolution of 4.0 cm^{-1}. The steel specimens were analyzed for morphology using a scanning electron microscope (SEM, SUS3800, Hitachi, Tokyo, Japan). Energy dispersive spectroscopy (EDS, Xplore30, Oxford Instruments, Oxfordshire, UK) was utilized to analyze the element distribution.

2.4. Electrochemical Measurements

Electrochemical measurement was executed with an electrochemical workstation (Reference 3000, Gamry, Warminster, PA, USA) controlled by Echem Software (Version 5.61). Prior to testing, the test solution was deaerated by passing nitrogen at a rate of 100 mL/min for 24 h. The three-electrode cell configuration consisted of the mild steel sample (working electrode); a silver/silver chloride (Ag/AgCl), which acted as the reference electrode; and platinum mesh as a counter electrode. The exposed area was 1 cm^2. The open circuit potential (OCP) was first completed to achieve stability. Electrochemical impedance spectroscopy (EIS) measurements were performed at the Open Circuit Potential (OCP) across a frequency spectrum from 100 kHz to 10 mHz using a 20 mV signal amplitude. The EIS data were analyzed using the Zview software with a specific equivalent circuit model. Dynamic polarization (DP) measurements were performed using a scanning rate of 1 mV/s across a voltage range of ± 0.1 V relative to the OCP. For each sample, three parallel tests were performed under the same conditions.

3. Results and Discussion

Figure 3 shows the FTIR spectra of the synthesized imidazoline. The sharp band around 3380 cm^{-1} corresponds to NH stretching [22]. The absorption peaks observed at 2918 cm^{-1} and 2854 cm^{-1} are attributed to the asymmetric methyl CH_3 and symmetric methylene CH_2 vibrations in IM [23]. A prominent absorption peak is observed at 1605 cm^{-1}, which is associated with the C=N bond within the IM ring [24]. The peaks observed at 1460 cm^{-1} correspond to the stretching vibrations of the C-N bond [24]. The 1343 cm^{-1} peak is attributed to the N-H bending vibration. The peak at 1242 cm^{-1} is related to the NH groups [25]. The C-Cl stretching is responsible for the adsorption peaks observed at 701 cm^{-1} [26]. Typical molecular absorption bands can be observed, such as C-C-O stretching (1281 cm^{-1}), C-O-C deformation (964 cm^{-1}), and CH_2-bending (1151 cm^{-1}) [27]. The peak observed at 841 cm^{-1} is associated with the bending motion of HNC [28].

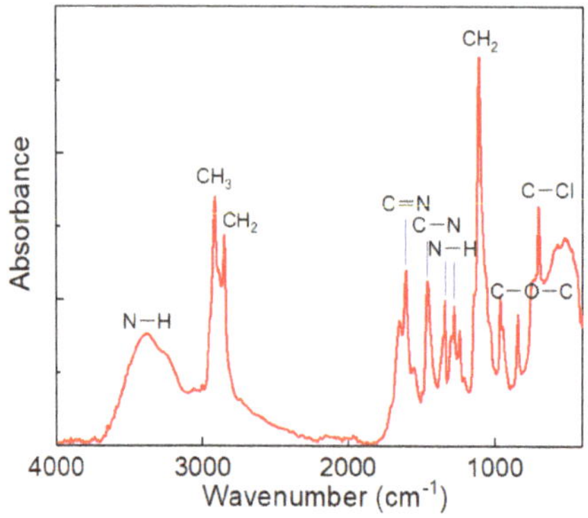

Figure 3. FTIR spectra of the corrosion inhibitor IM.

Figure 4 depicts polarization curves of mild steel immersed in a 3.5 wt% NaCl solution at ambient temperature, both with and without various concentrations of inhibitors, including IM, SDBS, and sodium molybdate. As shown in Figure 4a, the introduction of sodium molybdate slightly increases the current density from 73.29 $\mu A/cm^2$ to 114.82 $\mu A/cm^2$.

The inhibitor molecules initially adsorb onto the surface of the mild steel, where they block reaction sites, thereby hindering the process. Contrary to its intended inhibitive function, sodium molybdate exacerbates the corrosion of carbon steel, potentially attributed to insufficient quantities of molybdate ions inducing partial passivation of the steel surface, while other areas remain in an active state. Consequently, this leads to the formation of localized "active-passive" cells, thereby accelerating the corrosion of mild steel [17]. Furthermore, as the concentration of sodium molybdate increases, there is a slight reduction in the corrosion rate of carbon steel by NaCl, indicating that the exclusive use of sodium molybdate fails to provide effective protection against carbon steel in NaCl solutions.

The cooperative influence of IM, SDBS, and sodium molybdate on the polarization characteristics of mild steel is illustrated in Figure 4b,c. Current density and corrosion rates for different combinations of inhibitors are summarized in Table 3. The combination with IM (20 mg/L) and sodium molybdate (50 mg/L) shown in Figure 4b presents a higher current density 93.74 $\mu A/cm^2$. As shown in Figure 4c, the combined use of IM, SDBS, and sodium molybdate led to a decrease in corrosion current densities, resulting in reductions to 12.66, 8.95, and 6.97 $\mu A/cm^2$, respectively.

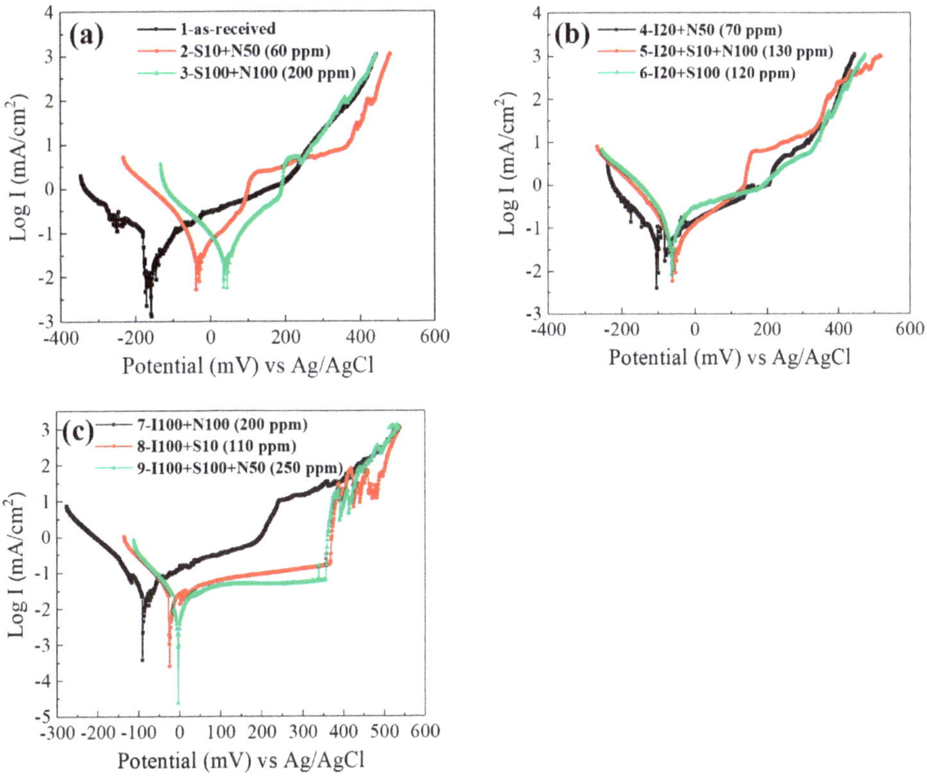

Figure 4. Potentiodynamic polarization curves for corrosion of carbon steel specimens exposed to a 3.5 wt% NaCl solution with various concentrations of IM, SDBS, and sodium molybdate. (**a**) As received, S10+N50, S100+N100; (**b**) I20+N50, I20+S10+N100, I20+S100; (**c**) I100+N100, I100+S10, I100+S100+N50.

Table 3. Current density and corrosion rates for different combinations of inhibitors.

No.	Combination	Current Density ($\mu A/cm^2$)	Corrosion Rate ($mg/cm^2 \cdot Y$)
1	As received	73.29	670.77
2	S10+N50	114.82	1050.86
3	S100+N100	54.10	495.14
4	I20+N50	93.74	757.93
5	I20+S10+N100	36.91	337.81
6	I20+S100	41.01	375.33
7	I100+N100	12.66	115.87
8	I100+S10	8.95	81.91
9	I100+S100+N50	6.97	63.79

Figure 5 shows Nyquist plots that depict the impedance characteristics of mild steel when exposed to a 3.5 wt% NaCl solution with different inhibitor concentrations of IM, SDBS, and Na_2MoO_4. To evaluate the inhibitory efficacy of these inhibitors on mild steel, specimens were subjected to corrosion solutions following an orthogonal test design.

Across the frequency spectrum, all electrochemical impedance spectroscopy (EIS) plots exhibit a depressed capacitive semicircle. The magnitude of this semicircle and the low-frequency impedance modulus show an increase, indicating enhanced corrosion inhibition with differing concentrations of IM, SDBS, and sodium molybdate.

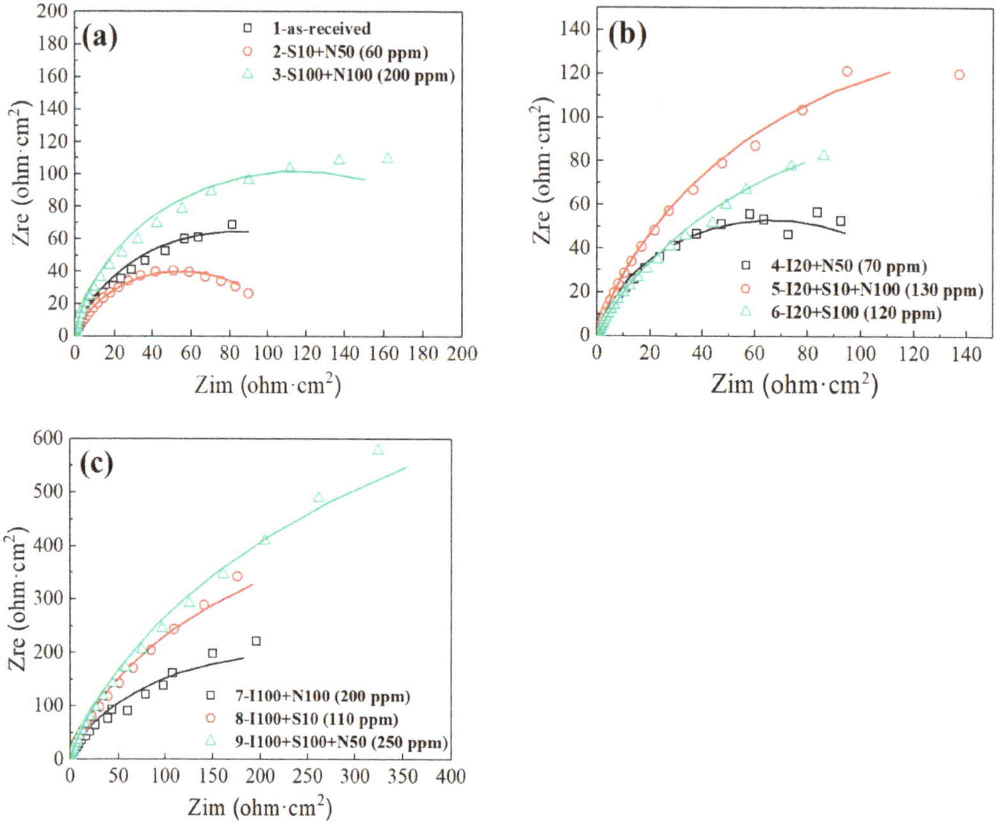

Figure 5. Nyquist plots for corrosion of mild steel specimens exposed to a 3.5 wt% NaCl solution with various concentrations of IM, SDBS, and sodium molybdate. (**a**) As received, S10+N50, S100+N100; (**b**) I20+N50, I20+S10+N100, I20+S100; (**c**) I100+N100, I100+S10, I100+S100+N50.

To evaluate the impedance data, an electrochemical equivalent circuit is employed, which consists of a parallel combination of a constant phase element (CPE) and charge-transfer resistance (R_{ct}), along with a solution resistance (R_s), as depicted in Figure 6. Typically, R_{ct} inversely correlates with the rate of steel corrosion and serves as an indicator for evaluating corrosion behavior. The impedance of the base material in the 3.5 wt% NaCl solution is around 172 ohm·cm^2. However, when 10 mg/L SDBS and 50 mg/L Na_2MoO_4 are present, it decreases to 110 ohm·cm^2, in line with the findings from the polarization curves shown in Figure 4a. With increasing concentrations of SDBS and Na_2MoO_4 to 100 mg/L each, the impedance rises to approximately 234 ohm·cm^2. The combination of IM (100 mg/L), SDBS (100 mg/L), and Na_2MoO_4 (50 mg/L) demonstrates the highest impedance with the value of 1.8 kohm·cm^2.

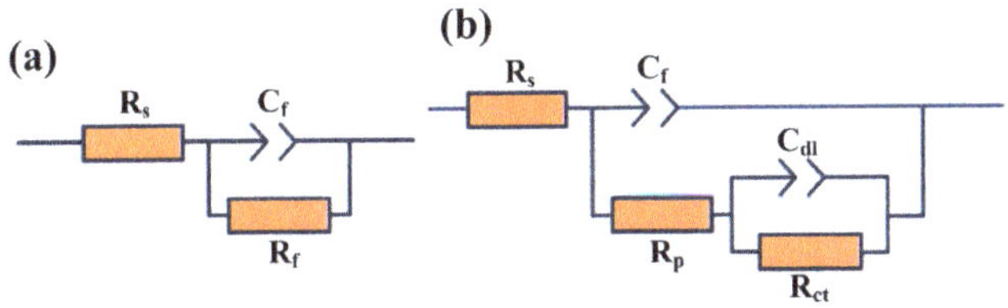

Figure 6. Equivalent circuit for the electrochemical impedance spectra. (**a**) R(RC) for substrate; (**b**) R(RC(RC)) for specimens exposed to a 3.5 wt% NaCl solution with various concentrations of IM, SDBS, and sodium molybdate.

EIS measurement was performed with the purpose of analyzing changes as a consequence of various concentrations of IM, SDBS, and sodium molybdate on mild steel. Two time constants for protective films are depicted according to the depressed arc in the Nyquist plot, whereas only one relaxation process can be determined on the substrate. The behavior of the impedance is generally explained using an equivalent circuit including a solution resistance (R_s), the pore resistance (R_p), and a constant phase element (CPE). Impedance refers to the overall resistance including the pore resistance and charge transfer resistance ($R_{total} = R_p + R_{ct}$). Figure 6 illustrates the equivalent circuits employed to analyze the EIS spectrum of both the substrate and protective films. Table 4 illustrates the equivalent circuit parameters of mild steel specimens exposed to a 3.5 wt% NaCl solution with various concentrations of IM, SDBS, and sodium molybdate. The CPE is utilized to simulate the dispersion behavior arising from various physical phenomena, such as local dielectric material inhomogeneities and surface roughness, offering a more accurate calibration. The CPE represents a shift from the capacitor; it is defined as:

$$Q = \frac{1}{C \cdot (j\omega)^{\alpha}} \quad (1)$$

where α serves as the CPE exponent that controls the phase shift and can indicate the surface's roughness or heterogeneity level ($0 < \alpha < 1$). Additionally, j^2 is defined as an imaginary number equal to -1, and ω represents the angular frequency ($\omega = 2\pi f$, with f denoting the frequency). The different resistances for mild steel in a 3.5 wt% NaCl solution in the presence of inhibitors are listed in Table 5. The response of each factor to its individual level was calculated by averaging the impedance in all experiments at each level for each factor, as indicated by the I, II, and III values. Based on the mean values, range values R were calculated to estimate the influence of each factor. The corrosion inhibition performances improved in different degrees during the selected range with the increased concentrations of all inhibitors. The concentration of the inhibitors with 100 mg/L, 100 mg/L, and 50 mg/L for IM, SDBS, and Na_2MoO_4 exhibits the most excellent corrosion inhibition. The results from extreme R analysis in Table 5 show that among the factors, the concentration of IM has the greatest impact on the impedance, followed by that of SDBS, and the concentration of Na_2MoO_4 has the least influence.

Analysis of variance (ANOVA) calculations were performed to investigate the effect of the selected factors on the inhibition efficiencies. The results are summarized in Table 6. Herein, D.F. stands for degrees of freedom and sum of sq. represents the sum of squares. All the factors have two degrees of freedom. The F-ratio is a tool to demonstrate whether the parameter has a significant effect or not. The F-ratio is a statistical measure used to determine the significance of the factors on the inhibition efficiencies. The higher the F-ratio indicated, the greater the significance of all the three factors. The p-ratio, which is defined as a ratio of the parameters' sum of square to the total sum of square, indicates the contributions of these parameters. It can be noticed that IM is a significant factor ($F = 20.283$, $p = 0.047 < 0.05$), indicating the presence of a main effect. IM exhibits a differential impact on the impedance. Variance analysis can be used to examine particular differences in more

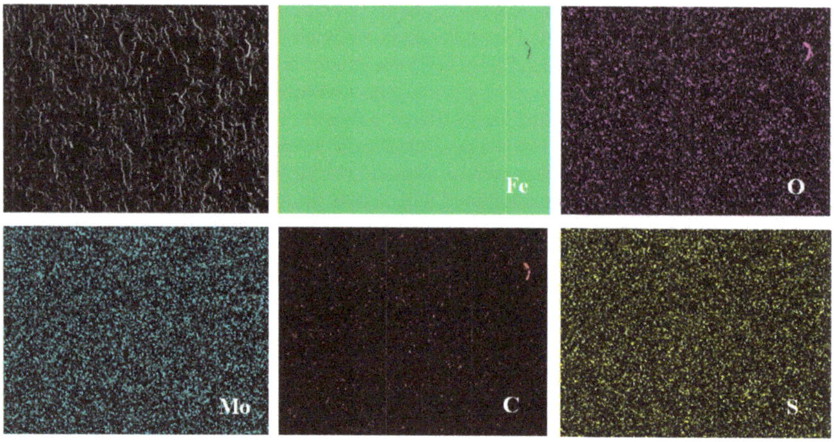

Figure 9. EDS elemental mapping of mild steel after 24 h immersion in a 3.5 wt% NaCl solution with inhibitors (IM, SDBS, and Na_2MoO_4).

Characterizations demonstrate the formation of the inhibitor adsorption film on the steel surface, which is also indicated by the capacitive semicircle measured in the whole frequency range in EIS plots. It is widely acknowledged that the effectiveness of the inhibitor greatly depends on the structure and integrity of the inhibitor film that is formed. A more compact, uniform, and dense film is generated on the steel surface when the solution contains IM, SDBS, and Na_2MoO_4, compared to the films formed with only IM, SDBS, or Na_2MoO_4 individually.

This work also confirms that according to the R value of the orthogonal test, for the inhibitors IM, SDBS, and Na_2MoO_4, the inhibition performance of IM is much better than that of SDBS and Na_2MoO_4 when they are used individually. According to the orthogonal experiment analysis, IM exhibits a differential impact on the impedance by EIS measurement, while SDBS and Na_2MoO_4 do not show significant effects. It is thus expected that, when the three inhibitors are co-present in the solution, the inhibitor IM would dominate the enhanced corrosion inhibition performance.

Figure 10 shows schematically the mechanisms of inhibitors IM and SDBS on corrosion inhibition to the steel. When IM and SDBS are included in the solution, they will be ionized into the corrosion-inhibition cations and anions, respectively. When employed separately, the inhibitor molecules adhere to the steel surface to create a protective layer of inhibitor films. The molecular interaction parameters for inhibitors IM and SDBS are negative, which means that repulsive forces exist among the adsorbed inhibitor molecules (Figure 10a,b) [18].

SDBS is a common anionic surfactant that possesses a relatively high molecular weight. It can attach to the surface of carbon steel via van der Waals forces and electrostatic attraction, leading to the creation of a physically adsorbed film. However, at low concentrations of SDBS, the formed adsorption film can only achieve partial coverage. Moreover, the adsorption film is characterized by unevenness and lack of compactness, which can potentially give rise to the formation of multiple localized micro-cells, thereby accelerating the corrosion process on mild steel.

Sodium molybdate acts as a mildly oxidizing corrosion inhibitor that produces a gradual and non-dense protective layer. It often results in the formation of micropores, which can cause localized corrosion on carbon steel. The formation of an $FeMoO_4$ solid effectively hinders the dissolution of activated Fe, thereby mitigating the diffusion and propagation of corrosion. As is shown in Figure 10c, there could be a competitive adsorption between MoO_4^{2-} and Cl^- at the defects of the passivation film on the metal surface, according to the literature [17]. The presence of MoO_4^{2-} weakens the adsorption of

Cl^- ions, thereby enhancing the passivation film's capability to withstand corrosion and partially inhibiting the occurrence of corrosion.

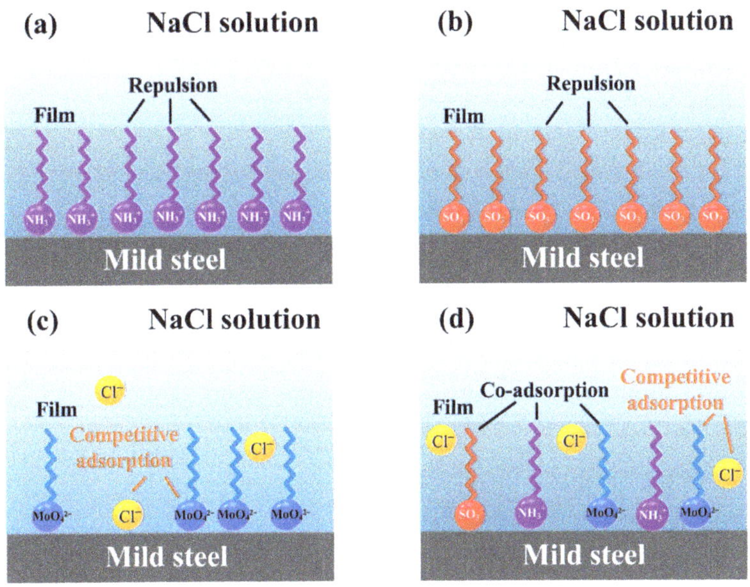

Figure 10. The schematic diagram of the mechanisms for corrosion inhibition by inhibitors: (**a**) repulsive forces among the NH_3^+; (**b**) repulsive forces among the SO_3^-; (**c**) competitive adsorption between Cl^- and among the MoO_4^{2-}; (**d**) Co-adsorption and competitive adsorption with the application of different inhibitors in NaCl solution.

Figure 10d illustrates that when the inhibitors IM, SDBS, and MoO_4^{2-} are employed simultaneously, the anions and cations are co-adsorbed on the steel surface, forming a dense and homogeneous film. SDBS, as an anionic surfactant, possesses a positive effect with IM for the film formation due to the co-adsorption of elements sulfur (from SDBS) and N (from IM) on the exposed Fe atoms. The formed inhibitor film is thus more compact and intact, achieving a better corrosion inhibition performance to the steel. With the addition of a small amount of SDBS, the surface tension of the solution undergoes a significant decrease. As the concentration of SDBS increases, the surface tension further decreases, thereby enhancing the wetting properties of the carbon steel surface by the solution. Consequently, the number of effective contact particles at the interface increases, promoting the film-forming reaction of sodium molybdate on the carbon steel surface. This results in accelerated film formation and the development of denser film layers. Furthermore, the RSO_3^- active groups present in SDBS possess certain adsorption capabilities. They compete with chloride ions and selectively adhere to the surface of molybdate oxide films, creating a double-layer protective film that fills the gaps in the film layer. The enhanced compactness of the passivation film improves the corrosion inhibition protection on the metal.

4. Conclusions

The study explores the inhibitory impact of imidazoline and its combined effect with SDBS and sodium molybdate on mild steel corrosion in a 3.5 wt% NaCl solution through orthogonal experimental design and electrochemical measurements. The results indicate that the composite inhibitors demonstrate significant corrosion inhibition properties. The combination of imidazoline (100 mg/L), sodium molybdate (50 mg/L), and SDBS (100 mg/L) results in an excellent corrosion inhibition with the value of 1.8 kohm·cm^2. The concentration of IM has the greatest impact on the impedance, followed by that of

SDBS, and the concentration of Na_2MoO_4 has the least influence. Insufficient quantities of molybdate ions can induce partial passivation of the steel surface and result in the accelerate of corrosion. The introduction of sodium molybdate with 10 mg/L marginally elevates the current density from 73.29 $\mu A/cm^2$ to 114.82 $\mu A/cm^2$. When the inhibitors IM, SDBS, and Na_2MoO_4 are used at the same time, the anions and cations are co-adsorbed on the steel surface, where the ions are arranged. A dense and homogeneous film can be formed, as indicated by the elemental analysis.

Author Contributions: Conceptualization, X.W.; methodology, X.W. and C.F.; investigation, X.W., D.Z. and N.X.; formal analysis, B.W.; data curation, X.W.; writing—original draft preparation, X.W.; writing—review and editing, X.W., C.F., H.S. and L.S.; funding acquisition, C.F., B.W. and J.X. All authors have read and agreed to the published version of the manuscript.

Funding: This work is supported by the Dongying Major Science and Technology Innovation Project (Science and Technology Development Guidance Plan) (2023ZDJH114) and the Shandong Provincial Natural Science Foundation (ZR2022QE186, ZR2021ME180, ZR2023QE139).

Institutional Review Board Statement: Not applicable.

Informed Consent Statement: Not applicable.

Data Availability Statement: The data supporting this study's findings are available from the corresponding author upon reasonable request.

Conflicts of Interest: Authors Nan Xu, Bin Wang and Jianbin Xu were employed by Dongying Jinmei Chemical Co., Ltd. The remaining authors declare that the research was conducted in the absence of any commercial or financial relationships that could be construed as a potential conflict of interest.

References

1. Behzadnasab, M.; Mirabedini, S.M.; Kabiri, K.; Jamali, S. Corrosion performance of epoxy coatings containing silane treated ZrO_2 nanoparticles on mild steel in 3.5% NaCl solution. *Corros. Sci.* **2011**, *53*, 89–98. [CrossRef]
2. Afshari, F.; Ghomi, E.R.; Dinari, M.; Ramakrishna, S. Recent advances on the corrosion inhibition behavior of Schiff base compounds on mild steel in acidic media. *ChemistrySelect* **2023**, *8*, e202203231. [CrossRef]
3. Emmanuel, J.K. Corrosion protection of mild steel in corrosive media, a shift from synthetic to natural corrosion inhibitors: A Review. *Bull. Natl. Res. Cent.* **2024**, *48*, 26. [CrossRef]
4. Shanmugapriya, R.; Ravi, M.; Ravi, S.; Ramasamy, M.; Maruthapillai, A.; J, A.S. Electrochemical and morphological investigations of Elettaria Cardamomum pod extract as a green corrosion inhibitor for mild steel corrosion in 1 N HCl. *Inorg. Chem. Commun.* **2023**, *154*, 110958. [CrossRef]
5. Kadhim, A.; Betti, N.; Al-Bahrani, H.; Al-Ghezi, M.; Gaaz, T.; Kadhum, A.; Alamiery, A. A mini review on corrosion, inhibitors and mechanism types of mild steel inhibition in an acidic environment. *Int. J. Corros. Scale Inhib.* **2021**, *10*, 861–884.
6. Bidi, H.; Ebn Touhami, M.; Baymou, Y.; Chung, I.-M.; Lgaz, H.; Zehra, S. Toward the development of an innovative descaling and corrosion inhibiting solutions to protect mild steel equipment: An experimental and theoretical approach. *Chem. Eng. Commun.* **2019**, *207*, 632–651. [CrossRef]
7. Singh, A.; Ansari, K.R.; Chauhan, D.S.; Quraishi, M.A.; Lgaz, H.; Chung, I.-M. Comprehensive investigation of steel corrosion inhibition at Macro/micro level by ecofriendly green corrosion inhibitor in 15% HCL medium. *J. Colloid. Interf. Sci.* **2020**, *560*, 225–236. [CrossRef] [PubMed]
8. Wang, Y.; Hu, J.; Zhang, L.; Cao, J.; Lu, M. Electrochemical and thermodynamic properties of 1-phenyl-3-(phenylamino)propan-1-one with Na_2WO_4 on N80 Steel. *Roy. Soc. Open. Sci.* **2020**, *7*, 191692. [CrossRef]
9. Dehghani, A.; Bahlakeh, G.; Ramezanzadeh, B.; hossein Mostafatabar, A.; Ramezanzadeh, M. Estimating the synergistic corrosion inhibition potency of (2-(3,4-)-3,5,7-trihydroxy-4H-chromen-4-one) and trivalent-cerium ions on mild steel in NaCl solution. *Constr. Build. Mater.* **2020**, *261*, 119923. [CrossRef]
10. Vakili Azghandi, M.; Davoodi, A.; Farzi, G.A.; Kosari, A. Water-base acrylic terpolymer as a corrosion inhibitor for SAE1018 in simulated sour petroleum solution in stagnant and hydrodynamic conditions. *Corros. Sci.* **2012**, *64*, 44–51. [CrossRef]
11. Cheng, F.; Tian, H. Novel inhibitors containing multi-functional groups for pipeline corrosion inhibition in oilfield formation water. *Corrosion* **2015**, *72*, 472–485. [CrossRef] [PubMed]
12. Hegazy, M.A.; Abdallah, M.; Ahmed, H. Novel cationic Gemini surfactants as corrosion inhibitors for carbon steel pipelines. *Corros. Sci.* **2010**, *52*, 2897–2904. [CrossRef]
13. Wei, H.; Heidarshenas, B.; Zhou, L.; Hussain, G.; Li, Q.; Ostrikov, K. (Ken) Green inhibitors for steel corrosion in acidic environment: State of art. *Mater. Today Sustain.* **2020**, *10*, 100044. [CrossRef]

14. Kousar, K.; Walczak, M.S.; Ljungdahl, T.; Wetzel, A.; Oskarsson, H.; Restuccia, P.; Ahmad, E.A.; Harrison, N.M.; Lindsay, R. Corrosion inhibition of carbon steel in hydrochloric acid: Elucidating the performance of an imidazoline-based surfactant. *Corros. Sci.* **2021**, *180*, 109195. [CrossRef]
15. Han, P.; Chen, C.; Yu, H.; Xu, Y.; Zheng, Y. Study of pitting corrosion of L245 steel in H_2S environments induced by imidazoline quaternary ammonium salts. *Corros. Sci.* **2016**, *112*, 128–137. [CrossRef]
16. Liu, J.; Wang, D.; Gao, L.; Zhang, D. Synergism between cerium nitrate and sodium dodecylbenzenesulfonate on corrosion of AA5052 aluminium alloy in 3 wt.% NaCl solution. *Appl. Surf. Sci.* **2016**, *389*, 369–377. [CrossRef]
17. Fang, J.; Xu, B. Development and developing direction of molybdate inhibitor. *Shangdong Chem. Ind.* **2008**, *37*, 17–19.
18. Qian, S.; Cheng, Y.F. Synergism of imidazoline and sodium dodecylbenzenesulphonate inhibitors on corrosion inhibition of X52 carbon steel in CO_2-Saturated Chloride Solutions. *J. Mol. Liq.* **2019**, *294*, 111674. [CrossRef]
19. Rui, Y.; Lu, M.; Liang, Y.; Wei, Y. Inhibition behavior of molybdate inhibitors for carbon steel in sea water. *Corros. Prot.* **2007**, *28*, 61–64.
20. Xie, F.; Xuan, H. Multi-site adsorption imidazoline quaternary ammonium salt corrosion inhibitor Synthesis and corrosion inhibition properties. *Chin. J. Chem.* **2011**, *28*, 94–100.
21. Meng, X.; Ji, D.; Ge, H. Effect of sodium dodecylbenzene sulfonate on corrosion inhibition and surface tension of carbon steel in 3.5% nacl-sodium molybdate solution. *Chem. World.* **2014**, *8*, 472–474.
22. He, Z.; Xiong, L.; Zhang, R.; Hu, J.; Lei, J.; Liu, J.; Shen, M. The synergistic inhibition effects of oleic imidazoline quaternary ammonium salt and sulfonate on Q235 Steel Corrosion in HCl media. *Surf. Topogr-Metrol.* **2019**, *7*, 045005. [CrossRef]
23. Scarano, D.; Bertarione, S.; Spoto, G.; Zecchina, A.; Otero Areán, C. FTIR spectroscopy of hydrogen, carbon monoxide, and methane adsorbed and co-adsorbed on zinc oxide. *Thin Solid Film.* **2001**, *400*, 50–55. [CrossRef]
24. Feng, L.; Yang, H.; Wang, F. Experimental and theoretical studies for corrosion inhibition of carbon steel by imidazoline derivative in 5% NaCl saturated $Ca(OH)_2$ solution. *Electrochim. Acta.* **2011**, *58*, 427–436. [CrossRef]
25. Rusu, E.; Drobota, M.; Barboiu, V. Structural investigations of amines treated polyester thin films by FTIR-ATR spectroscopy. *J. Optoelectron. Adv. Mater.* **2008**, *10*, 377–381.
26. Aiad, I.; Hafiz, A.; El-Awady, M.; Habib, A. Some imidazoline derivatives as corrosion inhibitors. *J. Surfactants Deterg.* **2010**, *13*, 247–254. [CrossRef]
27. Hermann, P.; Kästner, B.; Hoehl, A.; Kashcheyevs, V.; Patoka, P.; Ulrich, G.; Feikes, J.; Ries, M.; Tydecks, T.; Beckhoff, B.; et al. Enhancing the sensitivity of nano-FTIR spectroscopy. *Opt. Express.* **2017**, *25*, 16574. [CrossRef]
28. Raunier, S.; Chiavassa, T.; Marinelli, F.; Allouche, A.; Aycard, J.P. Reactivity of HNCO with NH_3 at low temperature monitored by FTIR spectroscopy: Formation of $NH_4^+OCN^-$. *Chem. Phys. Lett.* **2003**, *368*, 594–600. [CrossRef]

coatings

MDPI

Article

Lady's Mantle Flower as a Biodegradable Plant-Based Corrosion Inhibitor for CO_2 Carbon Steel Corrosion

Katarina Žbulj *, Gordana Bilić, Katarina Simon and Lidia Hrnčević

Faculty of Mining, Geology and Petroleum Engineering, University of Zagreb, 10000 Zagreb, Croatia; gordana.bilic@rgn.unizg.hr (G.B.); katarina.simon@rgn.unizg.hr (K.S.); lidia.hrncevic@rgn.unizg.hr (L.H.)
* Correspondence: katarina.zbulj@rgn.unizg.hr

Abstract: Due to issues with the corrosion problem in the petroleum industry and the use of less ecologically acceptable corrosion inhibitors, great emphasis, within research on corrosion inhibitors, is now being put on green corrosion inhibitors (GCIs). In this study, Lady's mantle flower extract (LMFE) has been observed as a plant-based GCI for carbon steel in a simulated CO_2-saturated brine solution. The effectiveness of the inhibitor in static and flow conditions has been determined using potentiodynamic polarization with Tafel extrapolation and electrochemical impedance spectroscopy (EIS). In static conditions, the inhibitor has been tested at concentrations from 1 g/L to 5 g/L with an increment of 1 g/L per measurement, while, in dynamic (flow) conditions, the inhibitor has been tested at concentrations from 3 g/L to 6 g/L with an increment of 1 g/L per measurement. All measurements were performed at room temperature. EIS and potentiodynamic polarization methods showed that LMFE achieves maximum effectiveness in protecting carbon steel from corrosion when added at a concentration of 4 g/L in static conditions and at a concentration of 5 g/L in flow conditions. The test methods proved that the inhibitory effectiveness of LMFE is greater than 90% in both test conditions (static and flow). The inhibitor efficiency was attributed to the adsorption of LMFE on the carbon steel surface, which was demonstrated by Fourier transform infrared spectroscopy (FTIR) and scanning electron microscopy (SEM). A biodegradability of 0.96 and a toxicity of 19.34% for LMFE were determined. The conducted laboratory tests indicate that LMFE could be used as an effective corrosion inhibitor for CO_2 carbon steel corrosion.

Keywords: Lady's mantle flower extract; plant-based corrosion inhibitors; CO_2 corrosion; carbon steel; biodegradability; toxicity

1. Introduction

Since its beginning, the petroleum industry has been faced with the problem of corrosion, as most of the equipment and pipelines within the petroleum industry are mainly composed of carbon steel. During the production of hydrocarbons, depending on the type of reservoir, a certain amount of brine and other impurities (dissolved gases, sand, additives used in the production process, etc.) are also produced. As production progresses, i.e., as the production field matures, the amount of brine produced also increases. In addition to the aforementioned brine, the produced fluid may also contain carbon dioxide (CO_2) and hydrogen sulfide (H_2S) (among other impurities), which cause corrosion and damage equipment when dissolved in water (brine).

The flow of the produced fluid has a significant effect on the rate of corrosion and the processes that occur on the steel surface. If a protective layer of iron carbonate ($FeCO_3$;) is formed on a steel surface, or the steel surface is covered with a corrosion inhibitor's film, the effect of the flow on the corrosion rate becomes insignificant, since the $FeCO_3$; layer or the inhibitor's film represent the main resistance to corrosion. However, the presence of a turbulent flow, typical for hydrocarbon production and transportation, as well as the transportation of brine during its reinjection into the reservoir, could disrupt the formation of or mechanically damage the mentioned layer/film, which leads to a higher corrosion risk on a local basis [1].

Regarding its omnipresence, corrosion in the petroleum industry is a major problem. Its aftermaths can damage the equipment and, additionally, can influence the environment (i.e., potentially cause a spill). The mentioned environmental impact has a major economic impact. The International Measures of Prevention, Application, and Economics of Corrosion Technologies Study (NACE IMPACT report) gave estimates of the total costs caused by corrosion. It was reported that the overall cost of corrosion is approximately USD 2.5×10^{12}, or approximately 3.4% of the world's gross domestic product (GDP) [2].

Electrochemical corrosion will occur in systems where water is present. As mentioned earlier, production lines and flowlines are the most vulnerable parts of an oil and gas production and transportation system. The produced fluid that is transported through the flowlines has not yet been treated. This means that the fluid still contains brine and impurities such as CO_2. According to CONCAWE (CONservation of Clean Air and Water in Europe) data [3], there were fifteen leakage incidents caused by corrosion in a ten-year period (2012–2021). Corrosion has the largest share in spilled volume (1194.4 m^3), and thus has the largest negative environmental impact compared to spills caused by mechanical failure (spill volume 109.2 m^3) and operational causes (spill volume 63.1 m^3). The corrosion problem can be alleviated by the use of corrosion inhibitors, among other things. Since conventional corrosion inhibitors are considered to be harmful to the environment, they are to be replaced by just as effective, less toxic, and biodegradable so-called green corrosion inhibitors (GCI). Plant extracts, derived either from flowers, roots, leaves, or fruit, are currently being tested as GCI. Most of them have been tested in strong acidic [4–24] and neutral environments [25–28], with only a few tests being performed in CO_2-saturated media [29–33].

Nowadays, in addition to the fact that a corrosion inhibitor must meet the criteria of high effectiveness in corrosion protection, it must also be environmentally friendly. Such corrosion inhibitors certainly include plant extracts that are rich in organic compounds that protect the metal from corrosion through a specific binding mechanism on the metal surface. Therefore, numerous tests are focused on the discovery of new plant extracts that show high inhibitory effectiveness in the protection of metals from corrosion [4–33].

Although tests show that plant extracts achieve comparable and higher effectiveness in corrosion protection than organic inhibitors, not all plant extracts are effective in all media, nor is the use of certain extracts as corrosion inhibitors justified (in the case of plants that are simultaneously used as food), and not all extracts are equally available in all parts of the world. Basically, finding an effective extract to act as a corrosion inhibitor in specific conditions is not easy.

Tests of plant extracts from different parts of the world (cashew nut, Tridax procumbens, Chromolaena odorata, kuding, olive leaf, Jatropha curcas, Sida acuta, Coptis chinensis, berberine, licorice, tobacco leaf, pomelo, Calotropis procera, corn oil, Dimocarpus longan, Ginkgo biloba, and Momordica charantia) were carried out in systems that simulate real ones in the production and transport systems of the oil industry, with a brine solution saturated with CO_2 [29,34–36].

The aim of this work is to examine the action of Lady's mantle flower (LMF) extract as a corrosion inhibitor in a brine solution saturated with CO_2. Although it can be cultivated, Lady's mantle is a plant that grows wild in the temperate regions of Europe, all the way to Siberia, but also in Asia, North America, and Greenland. It blooms from May to October [37,38]. The plant is characteristic of the Croatian climate, which is the reason for its selection as a potential corrosion inhibitor. The tests were carried out in a brine solution saturated with CO_2, which, in its composition, is similar to natural aquifers in Croatia.

This article presents the inhibitory effect of Lady's mantle flower extract (LMFE). The inhibitory effect was determined using electrochemical methods (potentiodynamic polarization with Tafel polarization and electrochemical impedance spectroscopy (EIS)). In addition, the results of scanning electron microscopy (SEM) and Fourier transform infrared spectroscopy (FTIR) scans are presented to show surface characterization and determine how, and by which functional groups, the inhibitor was absorbed on the steel surface. Finally, biodegradability and toxicity tests were conducted to determine if LMFE is biodegradable and/or toxic.

2. Materials and Methods

Carbon steel was used as a metal sample for the measurements conducted in this study. The composition (wt.%) of the carbon steel was 0.32 C, 0.25 Si, 1.38 Mn, 0.016 P, 0.009 S, 0.24 Cr, <0.01 Mo, 0.02 Ni, 0.01 Cu, and balance Fe. It was determined using optical emission spectrometry on the Laboratory Equipment Companies' (LECO) Glow Discharge Atomic Emission spectrophotometer GDS 850 A. Prior to conducting each corrosion rate progression measurement, the carbon steel sample was ground with 300, 600, and 1200 grit paper. After grinding, the carbon steel sample was washed using distilled water and degreased with ethanol (96%). The area of the carbon steel sample exposed to the corrosive solution was 1 cm^2. Considering that the brine is corrosive media for the carbon steel in the petroleum industry, it was necessary to simulate the brine for the laboratory measurements. Simulated brine solution contained 30 g/L of NaCl, 0.1 g/L of $NaHCO_3$; and 0.1 g/L of $CaCO_3$. For the measurements, the commercially available powder of Lady's mantle flower extract, DARvitalis, Croatia, was used.

Prior to the start of the measurements, the prepared solution was saturated with carbon dioxide (CO_2) for 45 min and continued to be saturated during the measurements. LMFE was added at a concentration from 1 g/L to 5 g/L, which increased by 1 g/L per measurement in static conditions. For the measurements in flow conditions, the added concentration of LMFE was 3 g/L to 6 g/L, also with an increase of 1 g/L per measurement. All measurements were performed at room temperature.

After saturation, the carbon steel sample was immersed in the prepared solution and the electrode was stabilized for 1 h before the electrochemical measurement was performed. The measurements that were performed in static conditions were performed by using a three-electrode corrosion cell (see Figure 1a). For the measurements in flow conditions, a three-electrode flow-through corrosion cell was used (see Figure 1b). Two graphite rods were used as counter electrodes for the measurements in static conditions and a platinum electrode was used for the measurements in flow conditions. In both cases, a saturated calomel electrode (SCE) was used as the reference electrode. In maximum flow conditions of 50 cm^3/min, which is equal to a maximum flow rate of 400 cm/min for the flow cell used in this case, the measurements were conducted for a simulation of flow. The electrochemical measurements were performed, using the SP1 potentiostat and SmartManager software (ver6743), with the DC data analysis software IVMAN^{TM} 1.5 and Impedance Data Analysis Software ZMAN^{TM} 2.5 to analyze the data. All the measurements were taken three times, resulting in a reproducibility of less than 5% of relative standard deviation.

For potentiodynamic polarization with Tafel extrapolation, the corrosion potential was adjusted to ±250 mV opposite to the open-circuit potential after a 1 h stabilization of the electrode. Electrochemical impedance spectroscopy (EIS) was performed in the frequency range of 100 kHz to ~10 mHz. By repeating the measurements, the obtained results indicated a negligible change in the values after 10 mHz and had no effect on the results or the chosen equivalent electrical circuits. The characterization of the steel sample surface, with or without the corrosion inhibitor, was performed using the non-destructive EIS method. In addition to electrochemical measurements, surface methods (SEM and FTIR) were used to determine the presence of inhibitory film and the mode of the adsorption on the steel surface. To perform these methods, the carbon steel sample was immersed in the solution containing the most effective concentration of the LMFE for four hours, after which it was compared to the carbon steel samples, which were exposed to the solution without the addition of the extract. To determine the impact of the LMFE as a GCI on the environment, the biodegradability and toxicity of the LMFE were studied. The value of chemical oxygen demand (*COD*) was determined spectrophotometrically, and biochemical oxygen demand (*BOD*) was determined using the Winkler method. The bioluminescent bacterium *Vibrio fischeri* was used to determine the toxicity.

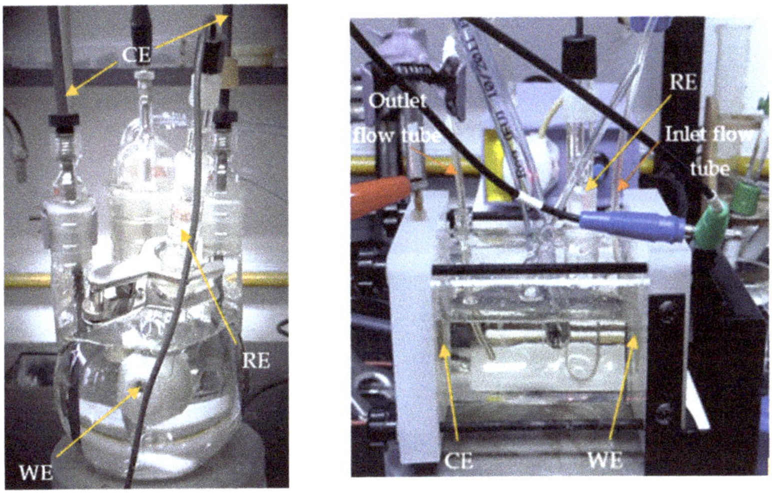

(**a**) (**b**)

Figure 1. Three-electrode corrosion cell used for measurements in static conditions (**a**); three-electrode flow-through corrosion cell used for measurements in flow conditions (**b**); CE—counter electrodes, RE—reference electrode, WE—working electrode.

3. Results and Discussion

3.1. Electrochemical Measurements

3.1.1. Potentiodynamic Polarization

To detect the efficiency of LMFE as a GCI, a potentiodynamic polarization measurement was conducted. Figures 2 and 3 show polarization curves that were scanned during the measurements in static and flow conditions. By applying Tafel extrapolation on the recorded polarization curves, some electrochemical parameters could be determined. These parameters are corrosion potential (E_{corr}), corrosion current (j_{corr}), anodic and cathodic Tafel slopes (β_a and β_c, respectively), and corrosion rate (v_{corr}) (see Tables 1 and 2). Inhibitor efficiency was calculated using Equation (1) [39], as follows:

$$IE(\%) = \frac{v^0_{corr} - v^{inh}_{corr}}{v^0_{corr}} \times 100 \quad (1)$$

where IE is the inhibitor efficiency (%), v^0_{corr} is the corrosion rate of the sample in the uninhibited solution (mm/year), and v^{inh}_{corr} is the corrosion rate of the sample in the inhibited solution (mm/year).

For the specified test conditions, the anodic corrosion reaction is the dissolution of iron from the steel, according to Equation (2), as follows:

$$Fe \rightarrow Fe^{2+} + 2e^- \quad (2)$$

Although there were disagreements about the cathodic corrosion reaction, it was determined that, under the tested conditions, the dominant cathodic reaction is the reduction of hydrogen ions produced by the dissociation of carbonic acid (the so-called "buffering effect", shown in Equation (3)) [40–42], as follows:

$$2H^+ + 2e^- \rightarrow H_2 \quad (3)$$

Figures 2 and 3 show the polarization curves scanned on carbon steel in static and flow conditions, while Tables 1 and 2 contain electrochemical parameters determined from these polarization curves.

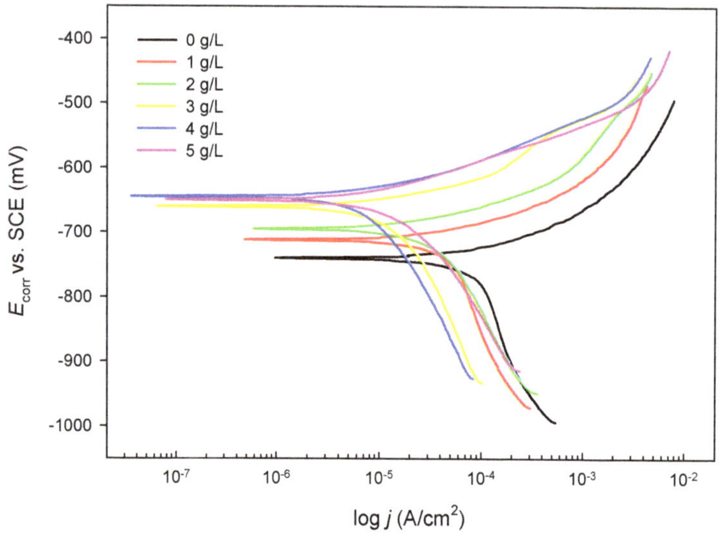

Figure 2. Polarization curves of carbon steel samples in the simulated brine solution saturated with CO_2 with different concentrations of LMFE as a corrosion inhibitor in static conditions (adapted from [43]).

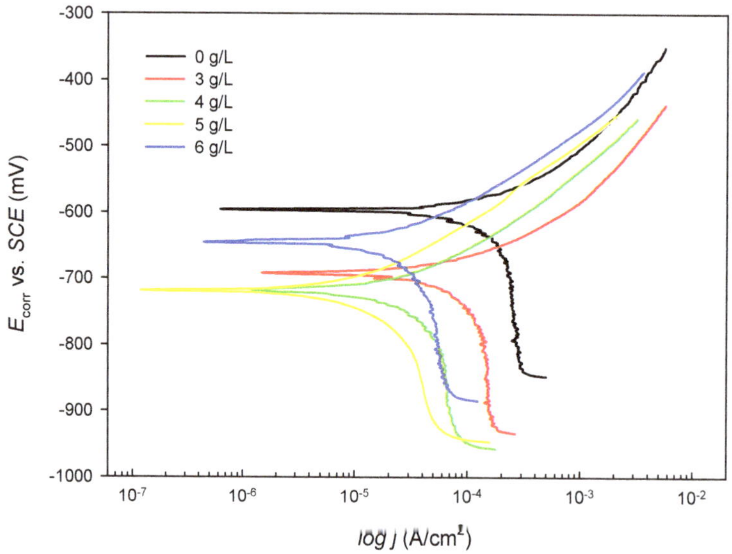

Figure 3. Polarization curves of carbon steel samples in the simulated brine solution saturated with CO_2 with different concentrations of LMFE as a corrosion inhibitor in flow conditions (adapted from [43]).

Table 1. Electrochemical polarization parameters for carbon steel samples in the simulated brine solution saturated with CO_2 with different concentrations of LMFE as a corrosion inhibitor in static conditions (adapted from [43]).

γ (g/L)	E_{corr} (mV)	j_{corr} (μA/cm^2)	$-\beta_k$ (V/dec)	β_a (V/dec)	v_{corr} (mm/year)	IE (%)	σ
0	−741	97.627	0.517	0.065	1.136	-	-
1	−699	46.199	0.375	0.054	0.538	52.64	1.49
2	−685	34.675	0.287	0.053	0.404	64.44	1.92
3	−662	13.370	0.275	0.053	0.156	86.27	2.51
4	−641	5.516	0.182	0.038	0.064	94.37	1.44
5	−640	12.352	0.188	0.057	0.144	87.32	1.59

Table 2. Electrochemical polarization parameters for carbon steel samples in the simulated brine solution saturated with CO_2 with different concentrations of LMFE as a corrosion inhibitor in flow condition (adapted from [43]).

γ (g/L)	E_{corr} (mV)	j_{corr} (μA/cm^2)	$-\beta_k$ (V/dec)	β_a (V/dec)	v_{corr} (mm/year)	IE (%)	σ
0	−613	175.441	0.686	0.145	2.042	-	-
3	−683	86.783	0.626	0.085	1.010	50.54	1.49
4	−708	32.193	0.347	0.099	0.351	82.81	1.18
5	−713	8.845	0.158	0.084	0.103	94.96	2.77
6	−642	18.330	0.206	0.078	0.213	89.57	2.28

Potentiodynamic polarization in static conditions

Corrosion current densities, j_{corr}, and corrosion potentials, E_{corr}, were determined using the Tafel method, i.e., as the intersection of the extrapolated linear parts of the cathodic and anodic branches of the polarization curves in the $E_{corr} \pm 100$ mV range.

In the presence of LMFE, a shift of the cathodic and anodic part of the polarization curve towards lower values of the corrosion currents is observed, which implies lower values of the corrosion rate. Corrosion rates decrease (from 1.136 mm/year to 0.064 mm/year) with an increase in extract concentration (from 0 g/L to 5 g/L). In the presence of LMFE in the highest concentrations (5 g/L), a decrease in the effectiveness (increase in corrosion rate to 0.144 mm/year) of LMFE was observed, which is often mentioned in the literature for organic inhibitors at higher concentrations as the phenomenon of critical inhibitor concentration (CMC) and is associated with adsorption [44–46].

As can be seen in Figure 2, the anodic Tafel slope (β_a) in simulated brine solution saturated with CO_2 without the LMFE was 65 mV/dec, which indicates that, in the tested condition, anodic dissolution of the carbon steel sample occurs [47]. The cathodic Tafel slope in the simulated brine saturated with CO_2 without the LMFE was very high ($\beta_c = -517$ mV/dec). This high value is characteristic for saturated CO_2 solutions, and it is a result of the diffusion of dissolved CO_2 on the reduction process [48].

Regarding the polarization curves and the values of the slope of the anodic Tafel lines in the presence of LMF extract, it is evident that they change only slightly when compared to the uninhibited system. A significant increase in the anodic slope of the Tafel lines would indicate the effect of the extract on the reaction mechanism. Since only small decreases in the anodic Tafel slopes are evident, this indicates the adsorption of the extract on the surface of the steel and its inhibitory effect by blocking the surface, without affecting the reaction mechanism [49,50]. A more significant change in the anodic Tafel slope is observed at an extract concentration of 4 g/L, which also corresponds to the greatest reduction in the corrosion rate (0.064 mm/year), i.e., the greatest inhibitory effectiveness (94.37%).

The dominant cathodic reaction in the tested conditions is the reduction of hydrogen ions produced by the dissociation of carbonic acid [40]. The cathodic Tafel slopes in the presence of LMFE decrease compared to the uninhibited system, which indicates a reduction in the influence of the diffusion process on the cathodic reaction [45]. This change is especially noticeable at an extract concentration of 4 g/L.

The corrosion potential of a carbon steel sample in a simulated brine solution saturated with CO_2 without the LMFE was −741 mV, which is in accordance with the values of corrosion potential under similar test conditions [51].

In the presence of the LMFE, a shift of the corrosion potential (E_{corr}) in the direction of more positive values was observed. The shift in the positive direction increases with an increase in extract concentration, and, at the extract concentration of 4 g/L, it was −641 mV vs. SCE, which is 100 mV more positive compared to the uninhibited system (E_{corr} = −741 mV). The shift of the corrosion potential in the direction of more positive values indicates the preferential inhibition of the anodic reaction of iron dissolution and is attributed to the formation of a protective film on the steel surface [48].

Since the value of the corrosion potential at a concentration of LMFE of 4 g/L is greater than 85 mV compared to the uninhibited system, LMFE can be considered as an anodic corrosion inhibitor of carbon steel in brine solutions saturated with CO_2 in static conditions [39,52–56]. The polarization curves scanned at higher concentrations of the extract, 4 g/L and 5 g/L, show a more significant change in the shape of the anodic part of the curve, which also supports the action of the extract on the anodic process.

The inhibitory efficiencies, IEs, given in Table 1 show that the LMFE inhibits the corrosion of carbon steel in a brine solution saturated with CO_2 under static conditions. The highest inhibitory effectiveness was achieved by the LMFE at a concentration of 4 g/L and is 94.37%. The inhibitory effect is attributed to the adsorption of the LMFE on the steel surface.

Although the tested extract also showed an action on the cathodic reaction, the shift of the corrosion potential in the anodic direction greater than 85 mV, a significant decrease in the anodic current densities, in comparison with the decrease in the cathodic current densities in the presence of the extract, and an obvious change in the shape of the anodic branch of the polarization curves at concentrations of 4 g/L and 5 g/L extract, categorize the extract as an anodic corrosion inhibitor for carbon steel in a brine solution saturated with CO_2 under static conditions.

Potentiodynamic polarization in flow conditions

Scanned polarization curves for measurements conducted in flow conditions are shown in Figure 3. In Table 2, the parameters determined from the Tafel extrapolation of scanned curves are given.

The corrosion potential in the non-inhibited system (E_{corr} = −613 mV) takes on a more positive value compared to the non-inhibited system in static conditions (E_{corr} = −741 mV), which is consistent with the tests of the influence of flow on the corrosion potential [57].

In the presence of the extract, a shift of the corrosion potentials towards more negative values is observed (Figure 3), and the shift is more significant with an increase in the concentration of the extract, which indicates the inhibition of the cathodic reaction [58].

The change in corrosion potential in the cathodic direction for 3 g/L, 4 g/L, and 5 g/L of LMFE is greater than 85 mV (Table 2), which means that LMFE exhibits the behavior of a cathodic corrosion inhibitor in flow conditions.

In the presence of LMFE, a shift of the cathodic parts of the polarization curves towards lower current densities was observed when compared to the system without extract. This proves the effect of the extract on the cathodic reaction, such that it suppresses the reduction of hydrogen ions [59]. On the other hand, the anodic parts of the polarization curves show active corrosion for the uninhibited and inhibited systems, which means that the LMFE does not significantly affect the anodic reaction under flow conditions [58].

With an increase in the concentration of the extract (from 0 g/L to 5 g/L), a decrease in the rate of corrosion (from 2.042 mm/year to 0.103 mm/year) and an increase in the effectiveness of the extract (up to 94.96%) as a corrosion inhibitor can be observed. Even in flow conditions, at the highest tested concentration, 6 g/L, an increase in the corrosion current density and a decrease in effectiveness compared to a lower extract concentration were observed, which is also attributed to the CMC. LMF extract showed a maximum effectiveness of 94.96% as a corrosion inhibitor at a concentration of 5 g/L.

The values of the parameters obtained by Tafel extrapolation, shown in Table 2, indicate a slight change in the anodic Tafel slopes (maximum 0.067 V/dec) and a significant decrease in the cathodic Tafel slope (maximum 0.528 V/dec) in the presence of the extract compared to the uninhibited system [60]. This significant decrease in the cathodic Tafel slope indicates the action of the extract as a cathodic corrosion inhibitor, in a way that affects the mechanism of the cathodic reaction. Such behavior may be the result of the formation of a protective film on the surface of the steel, which reduces the diffusion of electroactive species of the cathodic reaction (H^+ ions) to the surface of the steel, increases the resistance on the surface of the metal, and reduces the rate of corrosion [58,61].

All of the above indicate the effect of the LMFE as a cathodic corrosion inhibitor on carbon steel corrosion in brine solution saturated with CO_2 in flow conditions.

3.1.2. Electrochemical Impedance Spectroscopy

Phenomena at the electrolyte–steel surface interface were investigated using the electrochemical impedance spectroscopy method. Figure 4 shows impedance spectra (Nyquist and Bode plots) recorded on the surface of carbon steel with and without the presence of LMF extract in a brine solution saturated with CO_2 in static and flow conditions.

By fitting the recorded impedance spectra with the assumed theoretical models (equivalent circuits), which can be seen in Figure 5, the values of the electrochemical parameters shown in Table 3 were obtained. On the graphs shown (Figure 5), points represent measured values, and lines represent fitted results. The fitted and measured values show good agreement (more precisely, the error in disagreement for all spectra is less than 5%). The listed electrical equivalent circuits (and similar) are used commonly in the interpretation of EIS results for the conditions and systems described in this paper [62].

Table 3. EIS parameters for the carbon steel sample in the simulated brine solution saturated with CO_2 with different concentrations of LMFE as a corrosion inhibitor in static and flow conditions (adapted from [43]).

Static Conditions									
γ (g/L)	R_s (Ω cm^2)	Y_{dl} ($\mu\Omega^{-1}s^n$)	n_1	R_p (Ω cm^2)	Y_{inh} ($\mu\Omega^{-1}s^n$)	n_2	R_{po} (Ω cm^2)	IE (%)	σ
0	3.12	408	0.83	142.78	-	-	-	-	-
1	5.56	407	0.77	359.05	-	-	-	60.23	2.97
2	7.04	331	0.73	371.56	-	-	-	61.57	2.94
3	8.34	82.2	0.75	634.37	77.3	0.94	10.85	77.49	2.11
4	5.31	77.5	0.70	1629.61	67.5	0.94	12.97	91.24	1.60
5	5.18	65.2	0.73	516.37	68.1	0.95	16.04	72.35	1.16
Flow Conditions									
γ (g/L)	R_s (Ω cm^2)	Y_{dl} ($\mu\Omega^{-1}s^n$)	n_1	R_p (Ω cm^2)	Y_{inh} ($\mu\Omega^{-1}s^n$)	n_2	R_{po} (Ω cm^2)	IE (%)	σ
0	12.31	1840	0.75	113.68	-	-	-	-	-
3	14.03	1570	0.97	219.87	310	0.83	371.86	48.30	2.18
4	14.20	230	0.79	384.28	291	0.71	512.91	70.42	2.57
5	14.81	140	0.83	2884.45	186	1.00	832.84	96.06	1.54
6	14.26	18220	0.80	1288.80	170	0.82	454.16	91.18	3.72

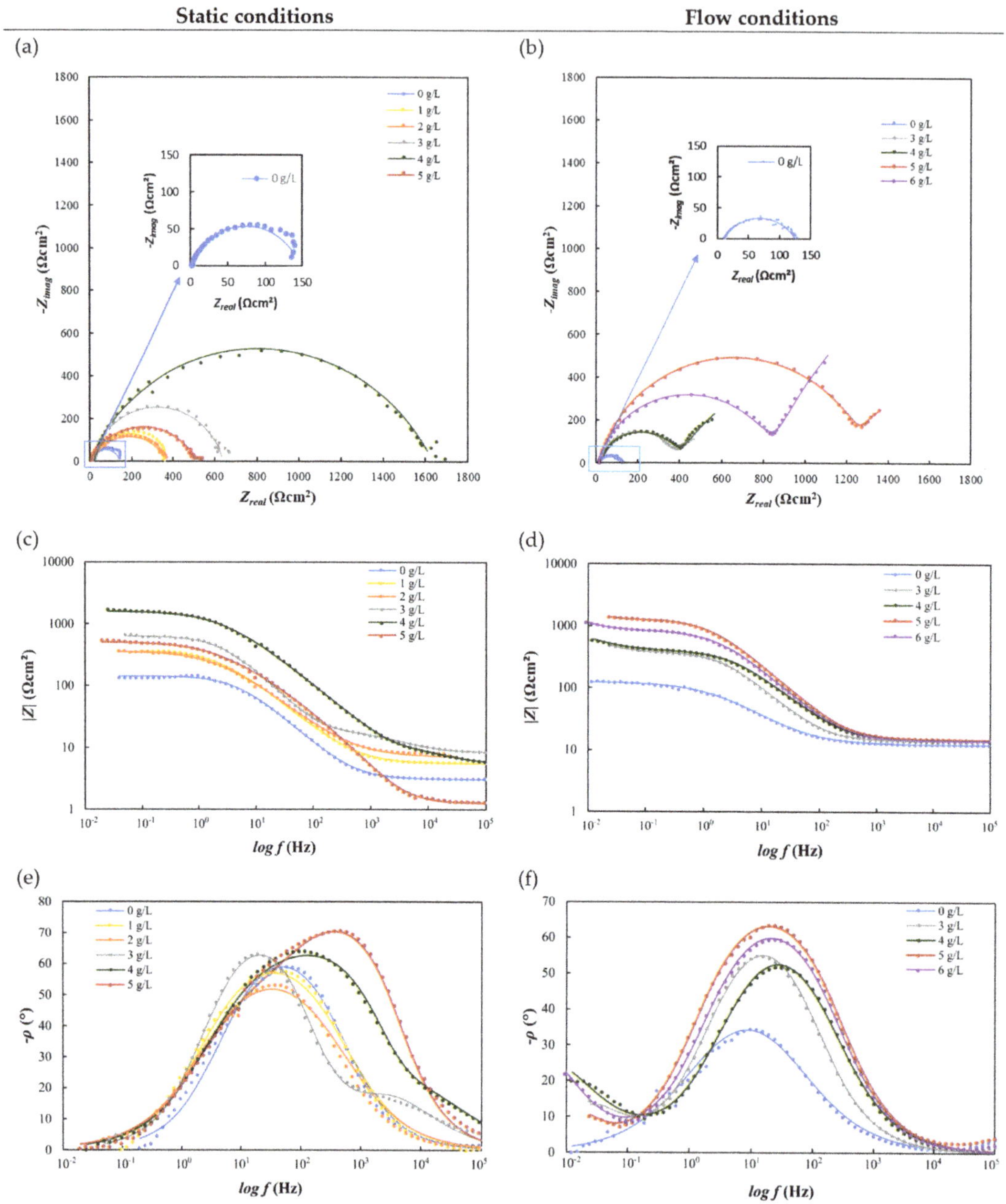

Figure 4. (a,b) Nyquist and (c–f) Bode plots for the carbon steel samples in the simulated brine solution saturated with CO_2 with different concentrations of LMFE as a corrosion inhibitor in static and flow conditions (adapted from [43]).

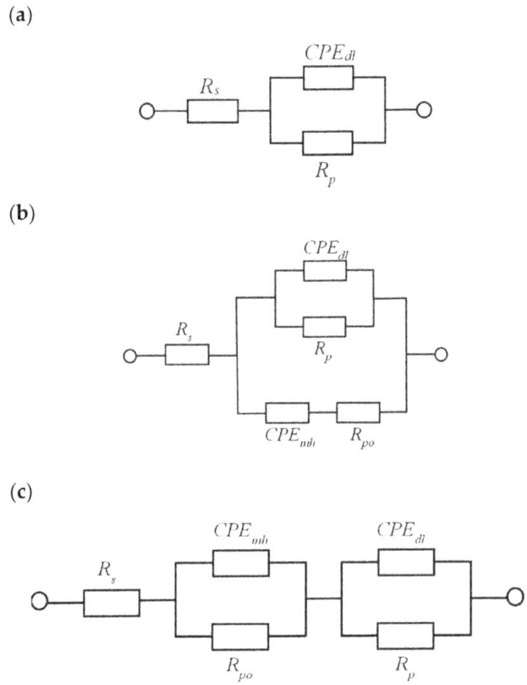

Figure 5. Equivalent circuit models used to analyze the impedance data; (**a**) uninhibited solution, (**b**) inhibited solution in static conditions, and (**c**) inhibited solution in flow conditions.

Three equivalent circuits were used to analyze the experimental data. For the measurements performed in static and flow conditions for the uninhibited solutions, the circuit system shown in Figure 5a was used. This theoretical model (Figure 5a) was also used for the solution that contained 1 and 2 g/L of LMFE in static conditions. For the solution with all the other LMFE concentrations (3–5 g/L) in static conditions, the equivalent circuit shown in Figure 5b was used. In the interpretation of test results of inhibited systems in flow conditions, the equivalent circuit shown in Figure 5c was used.

In the shown equivalent circuits, there were the following parameters: R_s, electrolyte resistance; R_p, charge transfer resistance; R_{po}, resistance of inhibitor film pores; CPE_{dl}, constant phase element representing the capacity of the electrochemical double layer; and CPE_{inh}, constant phase element representing the capacity of the inhibitor film. A constant phase element, *CPE*, replaces capacitors in equivalent circuits in order to compensate for surface inhomogeneity, roughness and geometric irregularity, porosity of electrodes, etc. [33,63].

Electrical equivalent circuits, EECs, for systems in the presence of LMFE represent a parallel/series connection of two *CPE*s, which refer to the metal surface and the coating film. The impedance of *CPE*s is defined by Equation (4), as follows:

$$Z_{CPE} = \frac{1}{Y}(j\omega)^{-n} \tag{4}$$

where Y is the proportionality factor, $j = \sqrt{-1}$, $\omega = 2\pi f$, and n is the dispersion coefficient associated with surface inhomogeneity. *CPE* can represent different electrical elements, depending on the value of n: for the value of $n = 0$, Y represents a resistor, $R = Y^{-1}$, and for $n = 1$, a capacitor, $C = Y$ [64–68].

The inhibitory effectiveness of the extract was calculated from the charge transfer resistance according to the expression (Equation (5)), as follows:

$$E(\%) = \frac{R_p^{inh} - R_p^0}{R_p^{inh}} \times 100 \quad (5)$$

where R_p^{inh} and R_p^0 are charge transfer resistances for carbon steel in a solution without and in the presence of the extract, respectively.

Nyquist plots recorded for the uninhibited system under static and flow conditions show a single capacitive half-circuit, as seen in Figure 4. The irregular shape of the Nyquist diagram is an indicator of the impedance frequency dispersion at the phase boundary and is characteristic of an inhomogeneous metal surface [69–74].

Since Y is proportional to C, changes in the value of Y correspond to changes in C. The value of Y_{dl} of an uninhibited system under flow conditions, Y_{dl} = 1840 $\mu\Omega^{-1}s^n$, is significantly higher than Y_{dl} = 408 $\mu\Omega^{-1}s^n$ under static conditions, which also means a higher value of C_{dl} under flow conditions compared to static conditions. Namely, the increase in the capacity of the double layer, C_{dl}, is related to the increase in the surface area available for the deposition of corrosion products. This is interpreted as the preferential dissolution of ferrite, while cementite (Fe_3C) stays behind and protrudes on the carbon steel surface [33,75].

The charge transfer resistance for uninhibited systems in static conditions, R_p = 142.78 Ω cm^2, and flow conditions, R_p = 113.68 Ω cm^2, are low, which indicates an active corrosion process and the formation of corrosion products that do not have protective properties. A lower resistance to charge transfer indicates a higher corrosion rate under flow conditions. The phase angle maximum, φ, of the Bode diagram under flow conditions for the uninhibited system takes on a lower value compared to the phase angle maximum under static conditions, which is attributed to the more intense corrosion of carbon steel under flow conditions. The existence of one phase angle peak (one time constant) on the Bode diagram indicates an active corrosion process [48].

The Nyquist diagrams show an increase in the real and imaginary components in the presence of the extract compared to the system without the extract. Also, this increase is consistent with the increase in extract concentration, which is attributed to the formation of a protective film on the steel surface in the presence of the LMFE [76]. The amount of impedance modules on the Bode diagram also increases with increasing extract concentration. The exception is the behavior of the extract at the highest concentrations (5 g/L in static conditions and 6 g/L in flow conditions) when, due to the CMC, there is a drop in resistance to charge transfer. For the same concentration values of LMF extract in static conditions, there are higher resistances to charge transfer compared to flow conditions (a comparison is only possible for the same concentrations in flow and static conditions, which are 3 g/L and 4 g/L, with the fact that 5 g/L was not taken into consideration since the CMC was reached in static conditions at that concentration).

In the presence of the extract in static conditions, the Bode diagram shows the formation of a time constant in the area of high frequencies, i.e., the presence of the extract changes the corrosion process with one time constant into a process with two time constants. The rate of corrosion is influenced by charge exchange at low frequencies and the formation of an inhibitory film at higher frequencies, which is evident from the shape of the Nyquist and Bode diagrams for static conditions, as shown in Figure 4 [18,77,78].

The existence of a time constant in the region of high frequencies means the existence of a protective inhibitory film on the surface of the electrode, while the shift of the phase angle maximum towards higher frequencies is associated with the growth of the inhibitory film [79]. In flow test conditions, although no clear formation of the second time constant is visible on the Bode diagram, an increase in the maximum of the phase angle is visible, as well as its shift towards higher frequencies, which is attributed to the formation of an inhibitory film on the carbon steel surface in flow conditions [48].

According to the values given in Table 3, for flow and static conditions, the values of Y_{dl} in the presence of the extract decrease (as does C_{dl}), which indicates the adsorption of the extract on the surface of the steel sample. The decrease is in accordance with the increase in the concentration of the extract. Also, the charge transfer resistances increase with an increase in the extract concentration, except for the highest concentrations, at which the drop in the charge transfer resistance, i.e., the decrease in the radius of the Nyquist diagram, is attributed to the achievement of the CMC.

Changes in the double layer capacitance and inhibitor film capacitance provide information about the inhibitor film. The capacity of the electrochemical double layer is calculated according to the Helmholtz model (Equation (6)), as follows:

$$C_{dl} = \frac{\varepsilon\varepsilon_0 A}{d} \tag{6}$$

where d is the thickness of the layer, ε is the dielectric constant of the medium, ε_0 is the permittivity of the vacuum, and A is the effective area of the electrode. In the presence of the extract, the decrease in the Y_{dl} value, or C_{dl}, can be interpreted as a decrease in the value of the dielectric constant due to the replacement of water molecules (with a high dielectric constant) with inhibitor molecules (with a low dielectric constant) on the surface of the steel sample. Also, this trend of C_{dl} value changes can be interpreted as an increase in layer thickness (according to Equation (6), the C_{dl} value will decrease), due to the adsorption of larger organic inhibitor molecules [80–84].

According to the literature, the values of the parameter n_1 range from 0.7 to 0.9, which was mostly achieved in this work, and they play a role in describing the experimental data [67]. The parameter n is a measure of the unevenness of the electrode surface, although its physical meaning has not been fully clarified [67].

The values of Y_{inh}, i.e., the capacities of the inhibitory film, C_{inh}, decrease with increasing LMFE concentration, while the pore resistance values, R_{po}, of the inhibitory film increase, which indicates a compact surface of the inhibitory film with protective properties against corrosion.

The electrolyte resistance values, R_s, are different for static and flow conditions. The reason for this is the different geometric performance of the electrolytic cell in static and flow conditions; that is, in the flow cell, the distance between the reference and working electrodes is greater than in the cell for static tests.

Charge transfer resistances increase with an increase in the concentration of the LMFE for static conditions up to 4 g/L, and up to 5 g/L for flow conditions, which means that the adsorbed extract increases the resistance to the dissolution of iron from the steel or blocks the reaction surface [33]. At the highest extract concentrations (5 g/L for static and 6 g/L for flow conditions), a drop in the resistance to charge transfer was observed due to the reaching of the CMC. The highest value of charge transfer resistance (R_p = 1629.61 Ω) and the highest inhibitory efficiency in static conditions (IE = 91.24%) was shown by the LMFE at a concentration of 4 g/L, and in flow conditions (R_p = 2884.45 Ω, IE = 96.06%) in a concentration of 5 g/L. The results obtained by the EIS method are comparable to the results of potentiodynamic polarization.

3.2. Scanning Electron Microscopy (SEM) and Fourier Transform Infrared Spectroscopy (FTIR)

Surface scans of the carbon steel sample, after its 4-h immersion in the uninhibited simulated brine solution saturated with CO_2 in static and flow conditions, in the presence of the most effective concentration of LMFE as a corrosion inhibitor (4 g/L in static conditions and 5 g/L in flow conditions), are shown in Figure 6a–d.

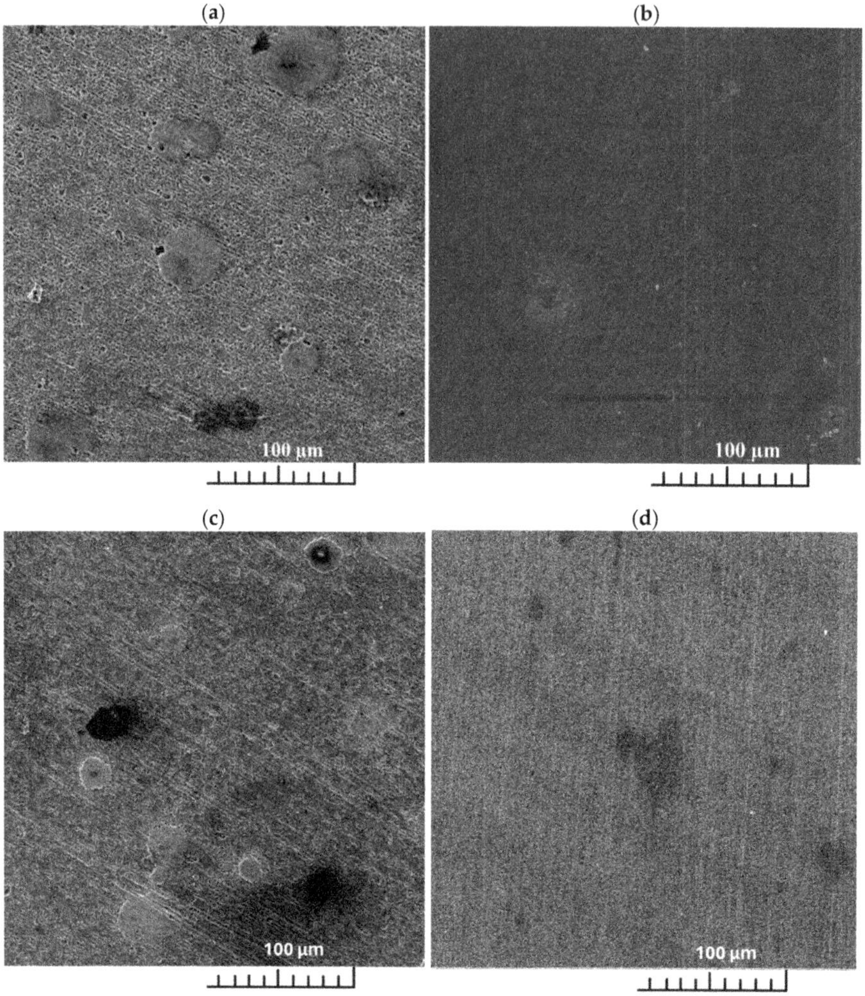

Figure 6. Carbon steel SEM scans after a 4-h immersion in the simulated brine solution saturated with CO_2 without the extract (**a**) in static and (**c**) in flow conditions and with LMFE (**b**) in static (4 g/L of LMFE) and (**d**) flow conditions (5 g/L of LMFE) (adapted from [43]).

Carbon steel surfaces after exposure to an uninhibited system in static and flow conditions (Figure 6a,c) are uneven and rough, which is a consequence of active corrosion, i.e., dissolution of iron from the steel. Under the test conditions, no corrosion products of protective properties are formed, so the rough surface can be interpreted as the dissolution of ferrite, while cementite rises from the surface of carbon steel [33,75].

In comparison with the exposure of steel to the inhibited solution, Figure 6b,d show that, after exposure to the action of LMFE, the carbon steel surface was smoother and more even, which can be related to the adsorption of LMFE on the carbon steel surface.

The mode of the adsorption of the LMFE compounds' functional groups on the carbon steel surface was examined using the FTIR method. The FTIR spectra of LMFE powder and the carbon steel surface, after its immersion for 4 h in the simulated brine solution saturated with CO_2 in the presence of 4 g/L of LMFE, are shown in Figure 7.

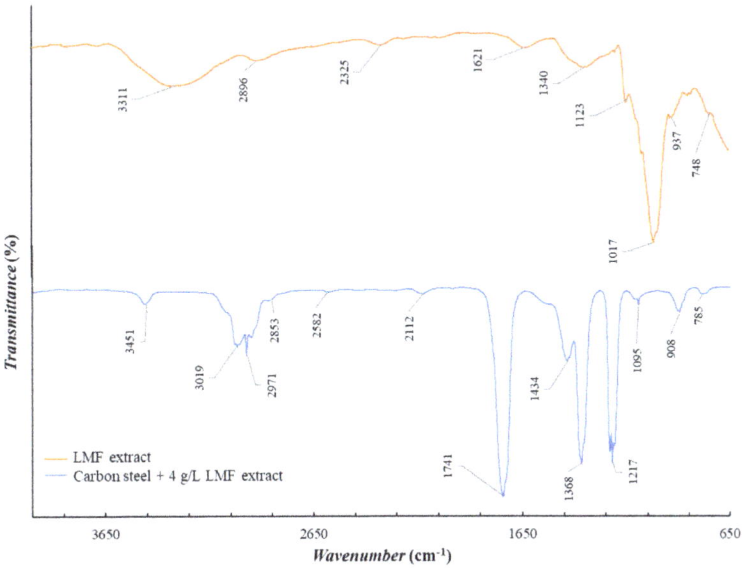

Figure 7. FTIR spectra of LMF extract powder and FTIR spectra of the carbon steel sample after 4 h of immersion in the simulated brine solution saturated with CO_2 and 4 g/L of the LMF extract (adapted from [43]).

According to the FTIR spectra of the LMFE, from the IR bands at 1017 cm^{-1}, 1340 cm^{-1}, and 3311 cm^{-1}, the dominant functional group was C-N, followed by the C-C and O-H functional groups, respectively. On the other hand, from the FTIR spectra of the inhibitors' film on the carbon steel sample surface, the vibration bands reveal the presence of some functional groups that were not seen on the LMFE spectra, but on the spectra of the inhibitors' film they were more noticeable. This means that, in the extract, some non-dominant functional groups exist, and through them, adsorption on the steel surface is achieved. From Figure 7, it can be seen that there was adsorption of the LMFE on the carbon steel sample surface through the C-N (1217 cm^{-1}), C-C (1368 cm^{-1}), C=O (1741 cm^{-1}), and C-H (2971–3019 cm^{-1}) functional groups [85]. The FTIR spectra results confirm the LMFE adsorption on the carbon steel sample surface, which was also proven by electrochemical methods.

3.3. Biodegradability and Toxicity Measurements

The chemical properties and toxicity of LMFE are given in Table 4, where *COD* is the chemical oxygen demand, *BOD* is the biochemical oxygen demand, EC_{50} is the effective concentration, and *TU* is the toxic unit.

Table 4. Chemical properties and toxicity of LMFE (adapted from [43]).

Parameter	BOD_5 (mgO_2/L)	COD (mgO_2/L)	BOD_5/COD	EC_{50} (%)	TU (%)
LMFE (γ = 4 g/L)	4209	4363	0.96	5.17	19.34

High values of the extracts' *COD*s indicated that the extract contains a high concentration of organic substance. The *BOD/COD* ratio indicated high biodegradability. Even though the 19.34% of the *TU* value puts the LMFE in the toxic category, those values need to be observed with respect to the applied concentration. A high *TU* value was the consequence of the applied high concentration of the LMFE (4 g/L). In the case of pipeline leakage and spillage of fluid into the environment, with LMFE as a corrosion inhibitor, no matter the high concentration and the *TU* value, the extract would be completely biodegraded.

The *COD* value indicated a high concentration of organic substances in the LMFE. The ratio of BOD_5 and *COD* was almost equal to one (0.96), which indicated a high biodegradability of LMFE [86]. In general, the compound is classified as toxic if the value is between 1% and 40% [86]. Although LMFE falls into the category of toxic compounds, the higher *TU* value (19.34%), in this case, depends mainly on the high LMFE concentration used and, thus, the increased presence of organic substances.

4. Conclusions

The extract from Lady's mantle flower has been studied as a GCI for carbon steel in simulated brine solution saturated with CO_2 in static and flow conditions, using potentiodynamic polarization and electrochemical impedance spectroscopy. The results showed that LMFE is an effective corrosion inhibitor in both static and flow conditions.

The tested LMFE achieved the best performance as a corrosion inhibitor at a concentration of 4 g/L in static test conditions and 5 g/L in flow conditions, with efficiencies greater than 90%. Both electrochemical test methods (EIS and potentiodynamic polarization) showed comparable results. Based on the results of potentiodynamic polarization, LMFE can be categorized as an anodic corrosion inhibitor in static conditions, while, in flow conditions, it can be categorized as a cathodic corrosion inhibitor.

The comparison of the SEM scans of the carbon steel surface after immersion in the uninhibited and the inhibited solutions show that the carbon steel surface, which was immersed in the inhibited solution, has a more uniform surface. The scans indicate that the inhibitor was absorbed on the tested carbon steel surface. Also, FTIR scans confirmed the adsorption of the LMFE by the functional groups C-N (1217 cm^{-1}), C-C (1368 cm^{-1}), C=O (1741 cm^{-1}), and C-H (2971–3019 cm^{-1}). The biodegradability and toxicity analysis show that LMF extract is almost completely biodegradable (BOD_5/COD = 0.96) and that it has a toxicity value of 19.34%. Even though this is a slightly higher value, due to the increased levels of organic substance present in the extract, it can be concluded that LMFE is environmentally acceptable as a GCI. Considering the obtained results, it can be concluded that LMFE proved to be a very effective corrosion inhibitor of carbon steel corrosion in a brine solution saturated with CO_2, in static and flow conditions.

Author Contributions: Conceptualization, K.Ž., L.H., G.B. and K.S.; methodology, G.B.; formal analysis, K.Ž., G.B., L.H. and K.S.; investigation, K.Ž.; resources, G.B. and L.H.; writing—original draft preparation, K.Ž. and G.B.; writing—review and editing, K.Ž., G.B., L.H. and K.S.; visualization, K.Ž.; supervision, G.B., L.H. and K.S.; project administration, G.B. and L.H.; funding acquisition, K.S. All authors have read and agreed to the published version of the manuscript.

Funding: This research was funded by the institutional projects The influence of nanoparticles on increasing the stability of the wellbore; Investigation of the influence of different lubricants on the properties of water-based drilling fluids-PODUIS; Ground water protection during well drilling-GWProtect, and Friction reduction during extended reach well drilling-FRERD, Faculty of Mining, Geology and Petroleum Engineering, University of Zagreb, Croatia.

Institutional Review Board Statement: Not applicable.

Informed Consent Statement: Not applicable.

Data Availability Statement: The data presented in this study are openly available in https://repozitorij.rgn.unizg.hr/islandora/object/rgn:1882 (accessed on 12 March 2024).

Conflicts of Interest: The authors declare no conflicts of interest. The funders had no role in the design of the study; in the collection, analyses, or interpretation of data; in the writing of the manuscript, or in the decision to publish the results.

References

1. Nesic, S.; Wang, S.; Cai, J.; Xiao, Y. Integrated CO_2 Corrosion-Multiphase Flow Model. In Proceedings of the SPE International Symposium on Oilfield Corrosion, Aberdeen, UK, 28 May 2004; OnePetro: Aberdeen, UK, 2004.
2. Koch, G.; Varney, J.; Thompson, N.; Moghissi, O.; Gould, M.; Payer, J. *International Measures and Prevention, Application and Economics of Corrosion Technologies Study*; NACE International: Houston, TX, USA, 2016.
3. CONCAWE. Performance of European Cross-Country Oil Pipelines—Details of Spillage Incidents from 2005. Available online: https://www.concawe.eu/publication/performance-of-european-cross-country-oil-pipelines-details-of-spillage-incidents-from-2005/ (accessed on 8 March 2024).
4. Haldhar, R.; Vanaraj, R.; Dagdag, O.; Berisha, A.; Kim, S.-C. *Convolvulus microphyllus* Extract as a Green, Effective, and Affordable Corrosion Inhibitor: Theoretical Calculations and Experimental Studies. *Coatings* **2023**, *13*, 860. [CrossRef]
5. Prasad, D.; Maithani, R.; El Ibrahimi, B. *Mimosa pudica* extract corrosion inhibitive nature for stainless steel in 0.5 mol L^{-1} sulfuric acid media. *J. Mol. Liq.* **2023**, *389*, 122940. [CrossRef]
6. Simović, A.R.; Grgur, B.N.; Novaković, J.; Janaćković, P.; Bajat, J. Black Pine (*Pinus nigra*) essential oil as a green corrosion inhibitor for carbon steel. *Metals* **2023**, *13*, 508. [CrossRef]
7. Gayakwad, N.; Patil, V.; Muralidhara Rao, B. The effect of Rhoeo discolor plant leaves extract on mild steel of corrosion inhibition in acid media. *Mater. Today Proc.* **2022**, *49*, 536–541. [CrossRef]
8. Manh, T.D.; Huynh, T.L.; Thi, B.V.; Lee, S.; Yi, J.; Nguyen, D.N. Corrosion Inhibition of Mild Steel in Hydrochloric Acid Environments Containing *Sonneratia caseolaris* Leaf Extract. *ACS Omega* **2022**, *7*, 8874–8886. [CrossRef]
9. Zehra, B.F.; Said, A.; Eddine, H.M.; Hamid, E.; Najat, H.; Rachid, N.; Toumert, L.I. *Crataegus oxyacantha* leaves extract for carbon steel protection against corrosion in 1M HCl: Characterization, electrochemical, theoretical research, and surface analysis. *J. Mol. Struct.* **2022**, *1259*, 132737. [CrossRef]
10. Gapsari, F.; Darmadi, D.B.; Setyarini, P.H.; Izzuddin, H.; Madurani, K.A.; Tanji, A.; Hermawan, H. *Nephelium lappaceum* Extract as an Organic Inhibitor to Control the Corrosion of Carbon Steel Weldment in the Acidic Environment. *Sustainability* **2021**, *13*, 12135. [CrossRef]
11. Mauro, A.C.; Ribeiro, B.D.; Garrett, R.; Borges, R.M.; da Silva, T.U.; de Paula Machado, S.; de Araujo, J.R.; de Oliveira Massafra, S.; de Oliveira Junior, F.O.R.; D'Elia, E. *Ziziphus joazeiro* Stem Bark Extract as a Green Corrosion Inhibitor for Mild Steel in Acid Medium. *Processes* **2021**, *9*, 1323. [CrossRef]
12. Nwosu, F.O.; Amusat, S.O. Corrosion Inhibition of Mild Steel Using *Parinari polyandra* Leave Extracts in Diluted Hydrochloric Acids. *Port. Electrochim. Acta* **2021**, *39*, 431–449. [CrossRef]
13. Saxena, A.; Prasad, D.; Thakur, K.K.; Kaur, J. PDP, EIS, and surface studies of the low-carbon steel by the extract of *Tinospora cordifolia*: A green approach to the corrosion inhibition. *Arab. J. Sci. Eng.* **2021**, *46*, 425–436. [CrossRef]
14. Dehghani, A.; Bahlakeh, G.; Ramezanzadeh, B.; Ramezanzadeh, M. Potential role of a novel green eco-friendly inhibitor in corrosion inhibition of mild steel in HCl solution: Detailed macro/micro-scale experimental and computational explorations. *Constr. Build. Mater.* **2020**, *245*, 118464. [CrossRef]
15. Saeed, M.T.; Saleem, M.; Niyazi, A.H.; Al-Shamrani, F.A.; Jazzar, N.A.; Ali, M. Carrot (*Daucus carota* L.) Peels Extract as an Herbal Corrosion Inhibitor for Mild Steel in 1M HCl Solution. *Mod. Appl. Sci.* **2020**, *14*, 97–112. [CrossRef]
16. Sedik, A.; Lerari, D.; Salci, A.; Athmani, S.; Bachari, K.; Gecibesler, I.H.; Solmaz, R. Dardagan Fruit extract as eco-friendly corrosion inhibitor for mild steel in 1 M HCl: Electrochemical and surface morphological studies. *J. Taiwan Inst. Chem. Eng.* **2020**, *107*, 189–200. [CrossRef]
17. Dehghani, A.; Blahlakeh, G.; Ramezanzadeh, B.; Ramezanzadeh, M. A combined experimental and theoretical study of green corrosion inhibition of mild steel in HCl solution by aqueous *Citrullus lanatus* fruit (CLF) extract. *J. Mol. Liq.* **2019**, *279*, 603–624. [CrossRef]
18. Khan, M.; Abdullah, M.M.S.; Mahmood, A.; Al-Mayouf, A.M.; Alkhathlan, H.Z. Evaluation of *Matricaria aurea* Extracts as Effective Anti-Corrosive Agent for Mild Steel in 1.0 M HCl and Isolation of Their Active Ingredients. *Sustainability* **2019**, *11*, 7174. [CrossRef]
19. Wang, Q.; Tan, B.; Bao, H.; Xie, Y.; Mou, Y.; Li, P.; Chen, D.; Shi, Y.; Li, X.; Yang, W. Evaluation of *Ficus tikoua* leaves extract as an eco-friendly corrosion inhibitor for carbon steel in HCl media. *Bioelectrochemistry* **2019**, *128*, 49–55. [CrossRef]
20. Saxena, A.; Prasad, D.; Haldhar, R.; Singh, G.; Kumar, A. Use of *Saraca ashoka* extract as green corrosion inhibitor for mild steel in 0.5 M H_2SO_4. *J. Mol. Liq.* **2018**, *258*, 89–97. [CrossRef]
21. Qiang, Y.; Zhang, S.; Tan, B.; Chen, S. Evaluation of *Ginkgo* leaf extract as an eco-friendly corrosion inhibitor of X70 steel in HCl solution. *Corros. Sci.* **2018**, *133*, 6–16. [CrossRef]
22. Saxena, A.; Prasad, D.; Haldhar, R.; Singh, G.; Kumar, A. Use of *Sida cordifolia* extract as green corrosion inhibitor for mild steel in 0.5 M H_2SO_4. *J. Environ. Chem. Eng.* **2018**, *6*, 694–700. [CrossRef]
23. Ugi, B.U.; Obeten, M.E.; Magu, T.O. Phytochemical constituents of *Taraxacum officinale* leaves as eco-friendly and nontoxic organic inhibitors for stainless steel corrosion in 0.2 M HCl acid Medium. *Int. J. Chem. Sci.* **2018**, *2*, 35–43.
24. Martinović, I.; Pilić, Z.; Zlatić, G.; Barišić, M.; Čelan, S. Corrosion Inhibition of Aluminium by *Alchemilla vulgaris* L. Extract in 3% NaCl Solution. *Croat. Chem. Acta* **2021**, *94*, 103–109. [CrossRef]
25. Barbouchi, M.; Benzidia, B.; Aouidate, A.; Ghaleb, A.; El Idrissi, M.; Choukrad, M. Theoretical modeling and experimental studies of Terebinth extracts as green corrosion inhibitor for iron in 3% NaCl medium. *J. King Saud Univ.—Sci.* **2020**, *32*, 2995–3004. [CrossRef]

26. Devikala, S.; Kamaraj, P.; Arthanareeswari, M.; Patel, M.B. Green corrosion inhibition of mild steel by aqueous Allium sativum extract in 3.5% NaCl. *Mater. Today Proc.* **2019**, *14*, 580–589. [CrossRef]
27. Devikala, S.; Kamaraj, P.; Arthanareeswari, M.; Pavithra, S. Green corrosion inhibition of mild steel by Asafoetida extract in 3.5% NaCl. *Mater. Today Proc.* **2019**, *14*, 590–601. [CrossRef]
28. Alibakhshi, E.; Ramezanzadeh, M.; Haddadi, S.A.; Bahlakeh, G.; Ramezanzadeh, B.; Mahdavian, M. Persian Liquorice extract as a highly efficient sustainable corrosion inhibitor for mild steel in sodium chloride solution. *J. Clean. Prod.* **2018**, *210*, 660–670. [CrossRef]
29. Prifiharni, S.; Nikitasari, A.; Priyotomo, G.; Musabikha, S.; Kusumastuti, R. Seaweed (*Sargassum polycystum*) extract as eco-friendly corrosion inhibitor for mild steel in CO_2-saturated 3.5% NaCl solution. *AIP Conf. Proc.* **2024**, *3003*, 020053. [CrossRef]
30. Žbulj, K.; Hrnčević, L.; Bilić, G.; Simon, K. Dandelion-Root Extract as Green Corrosion Inhibitor for Carbon Steel in CO_2-Saturated Brine Solution. *Energies* **2022**, *15*, 3074. [CrossRef]
31. Žbulj, K.; Bilić, G.; Hrnčević, L.; Simon, K. Potential of using plant extracts as green corrosion inhibitors in the petroleum industry. *Rud.-Geološko-Naft. Zb.* **2021**, *36*, 131–139. [CrossRef]
32. Aribo, S.; Olusegun, S.J.; Ibhadiyi, L.J.; Oyetunji, A.; Folorunso, D.O. Green inhibitors for corrosion protection in acidizing oilfield environment. *J. Assoc. Arab. Univ. Basic Appl. Sci.* **2017**, *24*, 34–38. [CrossRef]
33. Pustaj, G.; Kapor, F.; Jakovljević, S. Carbon dioxide corrosion of carbon steel and corrosion inhibition by natural olive leaf extract: Die Kohlendioxid-Korrosion von Kohlenstoffstahl und die Korrosionshemmung durch natürlichen Olivenblattextrakt. *Mater. Werkst.* **2017**, *48*, 122–138. [CrossRef]
34. Fazal, B.R.; Becker, T.; Kinsella, B.; Lepkova, K. A review of plant extracts as green corrosion inhibitors for CO_2 corrosion of carbon steel. *npj Mater. Degrad.* **2022**, *6*, 5. [CrossRef]
35. Shang, Z.; Zhu, J. Overview on plant extracts as green corrosion inhibitors in the oil and gas fields. *J. Mater. Res. Technol.* **2021**, *15*, 5078–5094. [CrossRef]
36. Chima, M.O.; Popoola, L.T.; Archibong, F.N.; Nwokeocha, T.O.; Nlemedim, P.U.; Allen, M.A. A novel application of CO_2 corrosion inhibitory potentials of banana peel extract on mild steel: Experimental studies. *S. Afr. J. Chem. Eng.* **2024**, *47*, 246–253. [CrossRef]
37. Tadic, V.; Krgović, N.; Žugić, A. Lady's mantle (*Alchemilla vulgaris* L., Rosaceae): A review of traditional uses, phytochemical profile, and biological properties. *Lek. Sirovine* **2020**, *40*, 66–74. [CrossRef]
38. Grlić, L. *Enciklopedija Samoniklog Jestivog Bilja*, 2nd ed.; August Cesarec: Zagreb, Croatia, 1990; p. 392.
39. Sastri, V.S. *Green Corrosion Inhibitors: Theory and Practice*; John Wiley & Sons, Inc.: Hoboken, NJ, USA, 2011.
40. Tran, T.; Brown, B.; Nesic, S. Corrosion of Mild Steel in an Aqueous CO_2 Environment—Basic Electrochemical Mechanisms Revisited. In Proceedings of the CORROSION 2015 Conference & Expo, Dallas, TX, USA, 15–19 March 2015; Paper No. 5671.
41. Linter, B.R.; Burstein, G.T. Reactions of pipeline steels in carbon dioxide solutions. *Corros. Sci.* **1999**, *41*, 117–139. [CrossRef]
42. Remita, E.; Tribollet, B.; Sutter, E.; Vivier, V.; Ropital, F.; Kittel, J. Hydrogen evolution in aqueous solutions containing dissolved CO_2: Quantitative contribution of the buffering effect. *Corros. Sci.* **2008**, *50*, 1433–1440. [CrossRef]
43. Žbulj, K. Extracts of Selected Plants as Steel Corrosion Inhibitors in Hydrocarbon Production and Transportation Systems. Ph.D. Thesis, Faculty of Mining, Geology and Petroleum Engineering, University of Zagreb, Zagreb, Croatia, 7 December 2021.
44. Jiang, X.; Zheng, Y.; Ke, W. Effect of flow velocity and entrained sand on inhibition performances of two inhibitors for CO_2 corrosion of N80 steel in 3% NaCl solution. *Corros. Sci.* **2005**, *47*, 2636–2658. [CrossRef]
45. Esih, I.; Sorić, T.; Pavlinić, Z. Time dependence of inhibiting effect of amidoamines in CO_2 saturated chloride solutions. *Br. Corros. J.* **1998**, *33*, 309–314. [CrossRef]
46. Pospelov, M.V.; Ignatov, A.V.; Yaminskiy, V.V.; Fokin, A.V. Hydrophobic interaction in inhibiting acid corrosion of steel by aliphatic amines. In Proceedings of the 6th European Symposium on Corrosion Inhibitors, Ferrara, Italy, 16–20 September 1985; Annali dell'Universita di Ferrara: Ferrara, Italy, 1985; pp. 455–464.
47. Ekawati, D.; Berntsen, T.; Seiersten, M.; Hemmingsen, T. Effect of Temperature, Bicarbonate, and MEG Concentrations on CO_2 Corrosion of Carbon Steels. *Corrosion* **2017**, *73*, 1157–1167. [CrossRef]
48. Desimone, M.P.; Grundmeier, G.; Gordillo, G.; Simison, S.N. Amphiphilic amido-amine as an effective corrosion inhibitor for mild steel exposed to CO_2 saturated solution: Polarization, EIS and PM-IRRAS studies. *Electrochim. Acta* **2011**, *56*, 2990–2998. [CrossRef]
49. Ahamad, I.; Prasad, R.; Quraishi, M.A. Inhibition of mild steel corrosion in acid solution by Pheniramine drug: Experimental and theoretical study. *Corros. Sci.* **2010**, *52*, 3033–3041. [CrossRef]
50. Bobina, M.; Kellenberger, A.; Millet, J.P.; Muntean, C.; Vaszilcsin, N. Corrosion resistance of carbon steel in weak acid solutions in the presence of l-histidine as corrosion inhibitor. *Corros. Sci.* **2013**, *69*, 389–395. [CrossRef]
51. Kahyarian, A.; Brown, B.; Nesic, S. Electrochemistry of CO_2 corrosion of mild steel: Effect of CO_2 on iron dissolution reaction. *Corros. Sci.* **2017**, *129*, 146–151. [CrossRef]
52. Tao, Z.; Zhang, S.; Li, W.; Hou, B. Corrosion inhibition of mild steel in acidic solution by some oxo-triazole derivatives. *Corros. Sci.* **2009**, *51*, 2588–2595. [CrossRef]
53. Satapathy, A.K.; Gunasekaran, G.; Sahoo, S.C.; Amit, K.; Rodrigues, P.V. Corrosion inhibition by *Justicia gendarussa* plant extract in hydrochloric acid solution. *Corros. Sci.* **2009**, *51*, 2848–2856. [CrossRef]
54. Ferreira, E.S.; Giacomelli, C.; Giacomelli, F.C.; Spinelli, A. Evaluation of the inhibitor effect of l-ascorbic acid on the corrosion of mild steel. *Corros. Sci.* **2004**, *83*, 129–134. [CrossRef]
55. Yan, Y.; Li, W.; Cai, L.; Hou, B. Electrochemical and quantum chemical study of purines as corrosion inhibitors for mild steel in 1 M HCl solution. *Electrochim. Acta* **2008**, *53*, 5953–5960. [CrossRef]

56. Li, W.H.; He, Q.; Zhang, S.T.; Pei, C.L.; Hou, B.R. Some new triazole derivatives as inhibitors for mild steel corrosion in acidic medium. *J. Appl. Electrochem.* **2008**, *38*, 289–295. [CrossRef]
57. Li, Z.; Zhang, J.; Cheng, J. The Influence of Critical Flow Velocity on Corrosion of Stainless Steel. *J. Fail. Anal. Prev.* **2017**, *17*, 1234–1240. [CrossRef]
58. Nam, N.D.; Hien, P.V.; Hoai, N.T.; Thu, V.T.H. A study on the mixed corrosion inhibitor with a dominant cathodic inhibitor for mild steel in aqueous chloride solution. *J. Taiwan Inst. Chem. Eng.* **2018**, *91*, 556–569. [CrossRef]
59. Pustaj, G.; Kapor, F.; Kvasnička, P. Inhibition properties and adsorption behavior of olive leaf extract on N80 carbon steel in CO_2-saturated brine solution. *Int. J. Mater. Res.* **2014**, *105*, 992–998. [CrossRef]
60. Khalil, M.M.H.; Ismail, E.H.; El-Magdoub, F. Biosynthesis of Au nanoparticles using olive leaf extract: 1st Nano Updates. *Arab. J. Chem.* **2012**, *5*, 431–437. [CrossRef]
61. Aballe, A.; Bethencourt, M.; Botana, F.J.; Marcos, M. $CeCl_3$ and $LaCl_3$ binary solutions as environment-friendly corrosion inhibitors of AA5083 Al–Mg alloy in NaCl solutions. *J. Alloys Compd.* **2001**, *323–324*, 855–858. [CrossRef]
62. El-Azazy, M.; Min, M.; Annus, P. (Eds.) *Electrochemical Impedance Spectroscopy*; IntechOpen: London, UK, 2020. [CrossRef]
63. Pustaj, G.; Kapor, F.; Veinović, Ž. Olive Leaf Extract as a Corrosion Inhibitor of Carbon Steel in CO_2-Saturated Chloride–Carbonate Solution. *Int. J. Electrochem. Sci.* **2016**, *11*, 7811–7829. [CrossRef]
64. Herrera Hernández, H.; González Díaz, F.; Fajardo SanMiguel, G.; Velázquez Altamirano, J.C.; González Morán, C.O.; Morales Hernández, J. Electrochemical impedance spectroscopy as a practical tool for monitoring the carbonation process on reinforced concrete structures. *Arab. J. Sci. Eng.* **2019**, *44*, 10087–10103. [CrossRef]
65. Perez, N. *Electrochemistry and Corrosion Science*, 1st ed.; Kluwer Academic Publishers: New York, NY, USA, 2004; 362p.
66. Mendoza-Flores, J.; Durán-Romero, R.; Genescá-Llongueras, J. Espectroscopía de Impedancia Electroquímica en Corrosión. In *Notas Técnicas Electroquímicas para el Control y Estudios de la Corrosión*; Llongueras, J.G., Ed.; Universidad Nacional Autónoma de México (UNAM): Mexico City, Mexico, 2002; pp. 55–91.
67. Herrera Hernández, H.; Ruiz Reynoso, A.M.; Trinidad González, J.C.; Gonzales Moran, C.O.; Miranda Hernandez, J.G.; Mandujano Ruiz, A.; Morales Hernandez, J.; Orozco Cruz, R. Electrochemical Impedance Spectroscopy (EIS): A Review Study of Basic Aspects of the Corrosion Mechanism Applied to Steels. In *Electrochemical Impedance Spectroscopy*; IntechOpen: London, UK, 2020. [CrossRef]
68. Mansfeld, F.; Shih, H.; Greene, H.; Tsai, C.H. Analysis of EIS data for common corrosion processes. In *Electrochemical Impedance: Analysis and Interpretation*, 1st ed.; Scully, J.R., Silverman, D.C., Kendig, M.W., Eds.; ASTM Publication: Fredericksburg, TX, USA, 1993; pp. 37–53.
69. Li, L.J.; Zhang, X.P.; Lei, J.L.; Pan, F.S. Osmanthus Fragran Leaves Extract as Corrosion Inhibitor for Carbon Steel in Hydrochloric Acid Solution. *Asian J. Chem.* **2012**, *24*, 1649–1653.
70. Larabi, L.; Harek, Y.; Traisnel, M.; Mansri, A. Synergistic Influence of Poly(4-Vinylpyridine) and Potassium Iodide on Inhibition of Corrosion of Mild Steel in 1M HCl. *J. Appl. Electrochem.* **2004**, *34*, 833–839. [CrossRef]
71. Mansfeld, F.; Kendig, M.W.; Tsai, S. Recording and Analysis of AC Impedance Data for Corrosion Studies. *Corrosion* **1982**, *38*, 570–580. [CrossRef]
72. Shih, H.; Mansfeld, F. A fitting procedure for impedance data of systems with very low corrosion rates. *Corros. Sci.* **1989**, *29*, 1235–1240. [CrossRef]
73. Martinez, S.; Metikoš Huković, M. A nonlinear kinetic model introduced for the corrosion inhibitive properties of some organic inhibitors. *J. Appl. Electrochem.* **2003**, *33*, 1137–1142. [CrossRef]
74. Bentiss, F.; Lebrini, M.; Vezin, H.; Chai, F.; Traisnel, M.; Lagrené, M. Enhanced corrosion resistance of carbon steel in normal sulfuric acid medium by some macrocyclic polyether compounds containing a 1,3,4-thiadiazole moiety: AC impedance and computational studies. *Corros. Sci.* **2009**, *51*, 2165–5173. [CrossRef]
75. Sun, J.B.; Zhang, G.A.; Liu, W.; Lu, M.X. The formation mechanism of corrosion scale and electrochemical characteristic of low alloy steel in carbon dioxide-saturated solution. *Corros. Sci.* **2012**, *57*, 131–138. [CrossRef]
76. Nam, N.D.; Bui, Q.V.; Mathesh, M.; Tan, M.Y.J.; Forsyth, M. A study of 4-carboxyphenylboronic acid as a corrosion inhibitor for steel in carbon dioxide containing environments. *Corros. Sci.* **2013**, *76*, 257–266. [CrossRef]
77. Ramachandran, S.; Jovancicevic, V. Molecular Modeling of the Inhibition of Mild Steel Carbon Dioxide Corrosion by Imidazolines. *Corrosion* **1999**, *55*, 259–267. [CrossRef]
78. Labjar, N.; Lebrini, M.; Bentiss, F.; Chihib, N.E.; El Hajjaji, S.; Jama, C. Corrosion inhibition of carbon steel and antibacterial properties of aminotris-(methylenephosphonic) acid. *Mater. Chem. Phys.* **2010**, *119*, 330–336. [CrossRef]
79. Heydari, M.; Javidi, M. Corrosion inhibition and adsorption behaviour of an amido-imidazoline derivative on API 5L X52 steel in CO_2-saturated solution and synergistic effect of iodide ions. *Corros. Sci.* **2012**, *61*, 148–155. [CrossRef]
80. Okafor, P.C.; Liu, X.; Zheng, Y. Corrosion inhibition of mild steel by ethylamino imidazoline derivative in CO_2-saturated solution. *Corros. Sci.* **2009**, *51*, 761–768. [CrossRef]
81. Webster, S.; Harrop, D.; McMahon, A.; Partridge, G.J. Corrosion inhibitor selection for oilfield pipelines. In Proceedings of the NACE Corrosion/1993, New Orleans, LA, USA, 8–12 March 1993. Paper No. 109.
82. Edwards, A.; Osborne, C.; Webster, S.; Klenerman, D.; Joseph, M.; Ostovar, P.; Doyle, M. Mechanistic studies of the corrosion inhibitor oleic imidazoline. *Corros. Sci.* **1994**, *36*, 315–325. [CrossRef]
83. Videm, K. The Anodic Behavior of Iron and Steel in Aqueous Solutions with CO_2, HCO_3^-, CO_2^- and Cl^-. In Proceedings of the NACE Corrosion/2000, Orlando, FL, USA, 26–31 March 2000; Paper No. 39.

84. Cao, C. On electrochemical techniques for interface inhibitor research. *Corros. Sci.* **1996**, *38*, 2073–2082. [CrossRef]
85. Pavia, D.L.; Lampman, G.M.; Kriz, G.S. *Introduction to Spectroscopy*, 5th ed.; Cengage Learning: Stamford, CT, USA, 2015.
86. Samudro, G.; Mangkoedihardjo, S. Review on BOD, COD and BOD/COD ratio: A triangle zone for toxic, biodegradable and stable levels. *Int. J. Acad. Res.* **2010**, *2*, 235–239.

MDPI AG
Grosspeteranlage 5
4052 Basel
Switzerland
Tel.: +41 61 683 77 34

Coatings Editorial Office
E-mail: coatings@mdpi.com
www.mdpi.com/journal/coatings

www.ingramcontent.com/pod-product-compliance
Lightning Source LLC
Chambersburg PA
CBHW061456210926
51850CB00008B/1739
9783725877232